Nucleoside Mimetics

ADVANCED CHEMISTRY TEXTS

A series edited by DAVID PHILLIPS, *Imperial College, London, UK,* PAUL O'BRIEN, *University of Manchester, UK* and STAN ROBERTS, *University of Liverpool, UK*

Volume 1
Chemical Aspects of Photodynamic Therapy
Raymond Bonnett

Volume 2
Transition Metal Carbonyl Cluster Chemistry
Paul J. Dyson and J. Scott McIndoe

Volume 3
Nucleoside Mimetics: Their Chemistry and Biological Properties
Claire Simons

This book is part of a series. The publisher will accept continuation orders which may be cancelled at any time and which provide for automatic billing and shipping of each title in the series upon publication. Please write for details.

Nucleoside Mimetics
Their Chemistry and Biological Properties

Claire Simons

Cardiff University, UK

Gordon and Breach Science Publishers

Australia • Canada • France • Germany • India • Japan
Luxembourg • Malaysia • The Netherlands • Russia
Singapore • Switzerland

Copyright © 2001 OPA (Overseas Publishers Association) N.V. Published by license under the Gordon and Breach Science Publishers imprint.

All rights reserved.

No part of this book may be reproduced or utilized in any form or by any means, electronic or mechanical, including photocopying and recording, or by any information storage or retrieval system, without permission in writing from the publisher. Printed in Singapore.

Amsteldijk 166
1st Floor
1079 LH Amsterdam
The Netherlands

British Library Cataloguing in Publication Data

ISBN: 90-5699-324-0
ISSN: 1027-3654

Contents

Preface		ix

Chapter 1 Introduction — 1

1.1	Background	1
1.2	Structure of Nucleosides	2
	1.2.1 Structure of the Sugar Moiety	3
	1.2.2 Structure of the Heterocyclic Base Moiety	4
1.3	Nucleoside Conformation	5
	1.3.1 Glycosyl Link Geometry	5
	1.3.2 C4′–C5′ Bond Geometry	6
	1.3.3 Ring Puckering	6
1.4	Nomenclature	7
1.5	Spectroscopy	8
	1.5.1 Mass Spectroscopy	8
	1.5.2 NMR Spectroscopy	9
1.6	Biological Properties	13
	1.6.1 RNA and DNA	14
	1.6.2 Purine and Pyrimidine Nucleotide Biosynthesis	15
1.7	Mechanisms of Action of Nucleoside Mimetics	19
	1.7.1 Inhibitors Targeted at Reverse Transcriptase	19
	1.7.2 Inhibitors Targeted at DNA Viruses	22
	1.7.3 Inhibitors Targeted at Nucleotide Metabolising Enzymes	23
1.8	References	27

Chapter 2 Synthesis of Conventional D-Nucleosides — 29

2.1	Introduction	29
	2.1.1 First Syntheses of the Natural Ribonucleosides	29
	2.1.2 Stereospecificity in Nucleoside Synthesis	30
2.2	Coupling Methods	33
	2.2.1 Coupling of Heterocyclic Salts with Halo-Sugars	33
	2.2.2 Vorbrüggen Procedure	35
	2.2.3 Nucleoside Synthesis from Glycals	38
	2.2.4 Glycosylamine Procedure	40
	2.2.5 Enzymatic Synthesis	43
2.3	Sugar Modifications	44
	2.3.1 Sugar Modifications via Anhydronucleosides	44

	2.3.2	Sugar Modifications via Epoxides	46
	2.3.3	Deoxygenation	46
	2.3.4	Azido and Amino Sugars	49
	2.3.5	Halogenation	51
	2.3.6	2'- and 3'-C-Nucleosides	53
2.4	Heterocyclic Base Modifications		54
	2.4.1	Pyrimidine Modifications	55
	2.4.2	Purine Modifications	60
2.5	References		60

Chapter 3 Sugar Modified Nucleosides 63

3.1	Introduction		63
3.2	Isonucleosides		64
	3.2.1	2'-Isonucleosides	64
	3.2.2	3'-Isonucleosides	65
3.3	4'-Substituted Nucleosides		66
	3.3.1	4'-Thionucleosides	66
	3.3.2	4'-Aza-nucleosides	68
3.4	Dioxo-, Oxathio- and Dithio-Nucleosides		70
3.5	Oxaza-, Isoxa- and Thiaza-Nucleosides		72
3.6	Ring Size		75
	3.6.1	Oxetanocin-A and Derivatives	75
	3.6.2	Pyranose Nucleosides	77
3.7	Spiro-Nucleosides		78
3.8	References		79

Chapter 4 Heterocyclic Base Modified Nucleosides 83

4.1	Introduction		83
4.2	5-Ring Heterocyclic Bases		83
	4.2.1	Imidazole Nucleosides	83
	4.2.2	Triazole Nucleosides	85
4.3	Purine-Modified Nucleosides		87
	4.3.1	Benzimidazole Nucleosides	87
	4.3.2	7-Deazapurine Nucleosides	88
	4.3.3	1-Deazapurine Nucleosides	89
	4.3.4	3-Deazapurine Nucleosides	91
	4.3.5	8-Azapurine Nucleosides	93
	4.3.6	Other Purine Modified Nucleosides	94
4.4	Pyrimidine-Modified Nucleosides		94
	4.4.1	Azapyrimidine and 3-Deazapyrimidine Nucleosides	96
	4.4.2	Fused Pyrimidine Nucleosides	97
4.5	References		99

Chapter 5 L-Nucleosides 103

5.1	Introduction	103
5.2	L-Oxathiolane and L-Dioxolane Nucleosides	104
	5.2.1 Mechanism of Action	106
5.3	2′,3′-Dideoxy-L-Nucleosides	106
5.4	D_4-L-Nucleosides	108
5.5	Other L-Nucleosides	109
	5.5.1 L-*Ribo*-Furanosyl-Nucleosides	110
	5.5.2 L-*Arabino*-Furanosyl-Nucleosides	111
	5.5.3 L-*Xylo*-Furanosyl-Nucleosides	112
	5.5.4 L-Hexopyranosyl-Nucleosides	114
5.6	References	114

Chapter 6 *C*-Nucleosides 117

6.1	Introduction	117
6.2	*C*-Nucleosides with a 5-Ring Heterocyclic Base Moiety	118
	6.2.1 Showdomycin	118
	6.2.2 Pyrazofurin	120
	6.2.3 Tiazofurin and Selenazofurin	122
6.3	*C*-Nucleosides with a 6-Ring Heterocyclic Base Moiety	122
	6.3.1 Pyrimidine *C*-Nucleosides	123
	6.3.2 Pyrazine *C*-Nucleosides	124
	6.3.3 Oxazine *C*-Nucleosides	126
6.4	*C*-Nucleosides with a Bicyclic Heterocyclic Base Moiety	127
	6.4.1 Formycin and Formycin B	127
	6.4.2 9-Deazapurine *C*-Nucleosides	129
	6.4.3 Pyrrolosine	129
	6.4.4 Thieno[3,4-*d*]pyrimidine *C*-Nucleosides	129
6.5	*C*-Arylglycosides as Nucleoside Mimetics	132
6.6	References	134

Chapter 7 Carbocyclic Nucleosides 137

7.1	Introduction	137
7.2	(–)-Aristeromycin and (–)-Neplanocin A	137
7.3	2′-Deoxycarbocyclic Nucleosides	142
	7.3.1 (+)-C-BVDU	143
	7.3.2 BMS-200475	144
	7.3.3 Conformationally Constrained 2′-Deoxycarbocyclic Nucleosides	145
7.4	Carbocyclic D_4-Nucleosides	145
	7.4.1 Carbovir and Abacavir	145
	7.4.2 Carbocyclic D_4-L-Nucleosides	148

7.5	Cyclopropyl Carbocyclic Nucleosides	149
7.6	Cyclobutyl Carbocyclic Nucleosides (Lobucavir)	150
7.7	References	151

Chapter 8   Acyclic Nucleosides — 155

8.1	Introduction	155
8.2	Acyclovir	156
	8.2.1 Synthesis of Acyclovir	156
	8.2.2 Acyclovir Prodrugs	157
8.3	Ganciclovir	158
	8.3.1 Mechanism of Action	158
	8.3.2 Synthesis of Ganciclovir	159
8.4	Penciclovir and Famciclovir	160
	8.4.1 Synthesis of Penciclovir and Famciclovir	160
8.5	Conformationally Restricted Acyclic Nucleosides	161
	8.5.1 Adenallene and Cytallene	162
	8.5.2 A-5021	164
8.6	Acyclic Nucleoside Phosphonates	164
	8.6.1 Mechanism of Action	164
	8.6.2 Synthesis of PMEA	166
	8.6.3 Synthesis of HPMPC and PMPA	167
	8.6.4 ANP Prodrugs	169
8.7	Base-Modified Acyclic Nucleosides	169
	8.7.1 HEPT Acyclic Nucleosides	170
	8.7.2 Benzothiadiazine Dioxide Acyclic Nucleosides	171
	8.7.3 Imidazo[1,2-a]pyridine Acyclic C-Nucleosides	172
8.8	References	172

Index   177

Preface

This book aims to provide a concise introduction to the chemistry of nucleoside mimetics, an area of research which has been extremely profitable in the development of new pharmaceuticals, for lecturers and students of advanced undergraduate, MSc and taught PhD courses. In addition the book will provide a convenient reference source for those beginning further research in this subject.

Nucleosides exhibit a broad spectrum of biological activity including antiviral, anticancer, antibacterial and antiparasitic activity, which generally results from their ability to inhibit specific enzymes. Nucleoside mimetics either interact with cellular enzymes involved in the biosynthesis of the RNA (ribonucleic acid) and DNA (deoxyribonucleic acid) precursor nucleotides or with specific viral enzymes, with the resulting biological activity/therapeutic effect observed dependent on the enzyme inhibited.

Both naturally occurring and synthetic nucleosides have been of value in antiviral chemotherapy, with nucleoside mimetics contributing significantly to the arsenal of agents for the treatment of diseases ranging from chickenpox to acquired immunodeficiency syndrome (AIDS). The structures of these nucleoside mimetics vary considerably resulting in the establishment of specific nucleoside classes.

The chemistry of nucleosides combines carbohydrate, heterocyclic and asymmetric synthesis resulting in a challenging but always stimulating field. Chapters 1 and 2 provide a general introduction to nucleosides covering areas such as structure, biosynthesis, enzyme targets, synthesis of nucleosides and more established chemical modifications. Chapters 3–8 cover each of the various classes of nucleosides — sugar modified, base modified, L-nucleosides, *C*-nucleosides, carbocyclic nucleosides and acyclic nucleosides respectively — each of these chapters is worth a book in its own right therefore key areas of interest with regards to the chemistry have been highlighted and additional reviews/publications indicated.

It is hoped that this overview will provide the reader with a useful insight into the diversity of nucleosides, the range of elegant chemistry involved and the continuing importance of nucleosides as both therapeutic agents and as probes for studying biochemical processes.

CHAPTER 1

INTRODUCTION

1.1 BACKGROUND

In 1869 Miescher extracted a substance called "nuclein" from pus cells, obtained from surgical bandages, and salmon sperm.[1] Twenty years later the term "nucleic acid" was introduced by Altmann,[2] who developed methods for the isolation of nucleic acids (formerly known as nuclein) from yeast and animal tissue. The existence of the two major forms of nucleic acids, ribonucleic acid (RNA) and deoxyribonucleic acid (DNA), was recognised during the early work of Miescher and Kossel[3] however, the exact components (*i.e.* heterocyclic bases and furanose sugars) took longer to determine.[4]

In 1909 the term "nucleoside' was proposed by Levene and Jacobs to describe carbohydrate derivatives of purines and pyrimidines.[5] The nucleosides of RNA and DNA can be obtained either by enzymatic or chemical hydrolysis of these nucleic acids, while numerous other naturally occurring nucleosides, such as the nucleoside antibiotics, have been isolated from plants and microorganisms.[6]

During the 1950s and 1960s an explosion in the chemistry of nucleosides and nucleotides occurred producing the conventional synthetic methodology on which nucleoside chemistry is based.[7,8] This explosion resulted in the availability of both natural and synthetic nucleosides, many of which have proved valuable as tools for elucidating the nucleotide biosynthetic pathways and produced numerous compounds with anticancer and antiviral activity. The potential of nucleosides as therapeutic agents emerged in the 1950s and 1960s with the discovery of anticancer agents such as the antibiotics arabinosyladenosine (ara-A), nucleocidin and toyocamycin and the synthetic nucleosides 5-fluoro-2'-deoxyuridine (FUDR), arabinosylcytidine (ara-C) and 8-azainosine.[6,9]

With the emergence of Human Immunodeficiency Virus (HIV), the main causative agent of Acquired Immunodeficiency Syndrome (AIDS), in the early 1980s,[10] the value of nucleoside mimetics as chemotherapeutic agents was confirmed. The first drug licensed for the treatment of HIV infection was AZT (Zidovudine®), a compound first synthesised at the Detroit Institute of Cancer Research as a potential anticancer agent.[11] AZT inhibits HIV-1 reverse transcriptase (see section 1.7.1) as did the next four licensed nucleoside mimetics, namely DDC (Zalcitabine®), DDI (Didanosine®), D$_4$T (Stavudine®) and 3TC™. The renewed interest in nucleosides as antiviral agents led to the generation of nucleosides with greater diversity in

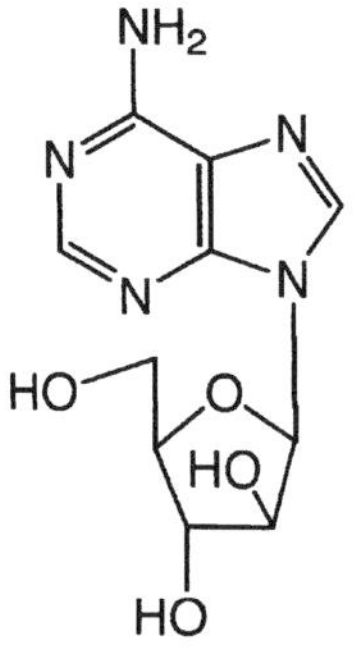

Ara-A
(Antiviral)

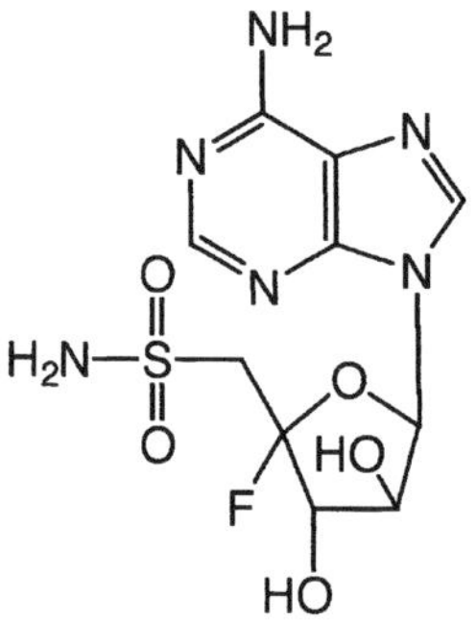

Nucleocidin
(Antitrypanosomal)

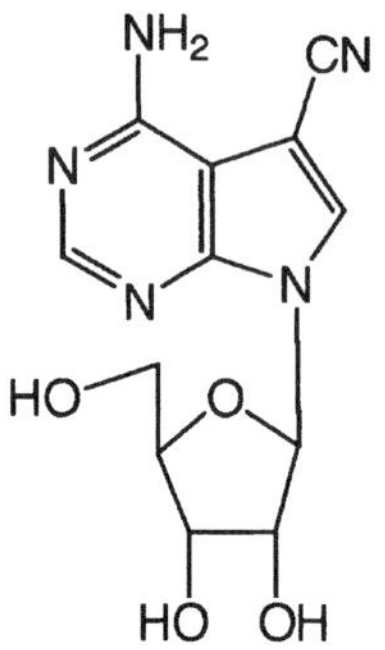

Toyocamycin
(Antitumour)

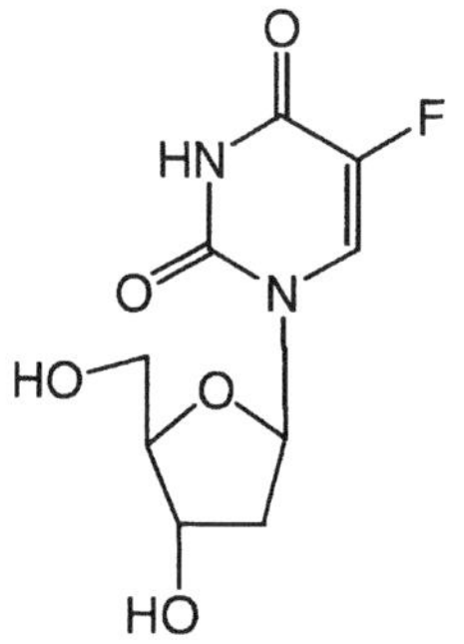

FUDR
(Anticancer)

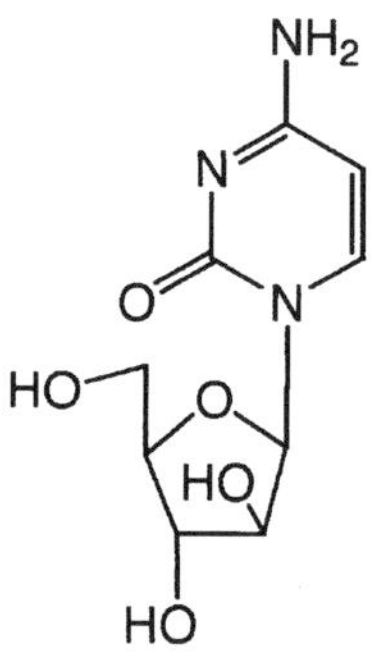

Ara-C
(Antileukemia)

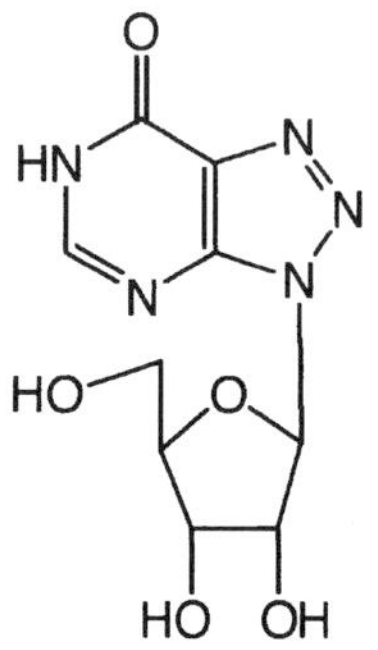

8-Azainosine
(Anticancer)

structure and mechanisms of antiviral/anticancer activity.[12–16] Of particular interest was the discovery of activity in L-nucleosides (such as 3TC™), the unnatural nucleoside enantiomers, and the non-nucleoside reverse transcriptase inhibitors (NNRTI's) TSAO-T and HEPT (see section 1.7.1.2)

Although there has been great interest in the reverse transcriptase inhibitors, there are many other known and potential enzymatic targets, including both cellular enzymes and virus-specified enzymes, with which nucleoside mimetics can interact as substrates/competitive inhibitors.[17] Antiviral specificity is often required for minimum toxicity, however the more specific inhibitors are often the most prone to the formation of virus-drug resistant strains. Therefore a balance is often required between drug specificity and toxicity in order to minimise the problem of drug resistance.

1.2 STRUCTURE OF NUCLEOSIDES

The nucleoside components of RNA and DNA are composed of a sugar moiety, ribose (**1.1**, β-D-ribofuranose) and 2-deoxyribose (**1.2**, 2-deoxy-β-D-ribofuranose)

Zidovudine®   **Zalcitabine®**   **Didanosine®**

Stavudine®   **3TC™**

respectively, linked to a purine or pyrimidine base through a β-*N*-glycosidic bond, through N^9 of the purine and N^1 of the pyrimidine heterocyclic base.

The purine bases adenine (**1.3**, 6-aminopurine) and guanine (**1.4**, 2-amino-6-oxypurine) are common to both RNA and DNA as is the pyrimidine base cytosine (**1.5**, 2-oxy-4-aminopyrimidine), however the pyrimidine base uracil (**1.6**, 2,4-dioxypyrimidine) only occurs in RNA and thymine (**1.7**, 2,4-dioxy-5-methyl pyrimidine) in DNA (Fig. 1.1). As well as the nucleosides of RNA and DNA, numerous naturally occurring nucleosides have been isolated and invariably the configuration at the anomeric centre (C-1') of the nucleoside is β.

1.2.1 Structure of the Sugar Moiety

The numbering system for the sugar moiety of nucleosides follows carbohydrate numbering with the slight difference that the numbers are primed (see structure **1.8**), with C-1' being the anomeric carbon attached to the aglycone (heterocyclic base of the nucleoside). The sugar moiety of the naturally occurring nucleosides is generally ribose (**1.1**), as in ribavirin[18] (**1.8**, 1-β-D-ribofuranosyl-1*H*-1,2,4-triazole-3-carboxamide) used in the treatment of respiratory syncytial virus (RSV), or 2-deoxyribose (**1.2**).

Other common naturally occurring sugar forms include β-D-arabinofuranose (**1.9**), the sugar component of the anti-herpes agent Vidarabine[19] (**1.10**, ara-A,

INTRODUCTION

1.1, R = OH, ribose
1.2, R = H, 2-deoxyribose

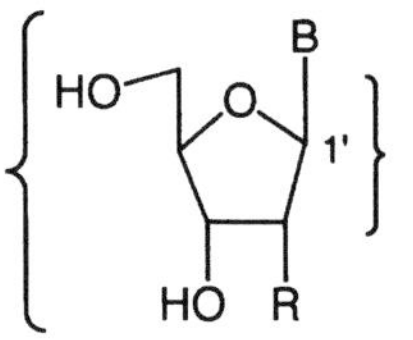

B = A, G, C, U or T
β-configuration

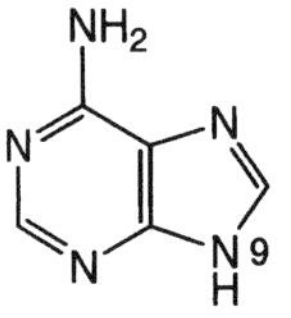

1.3 Adenine (A) 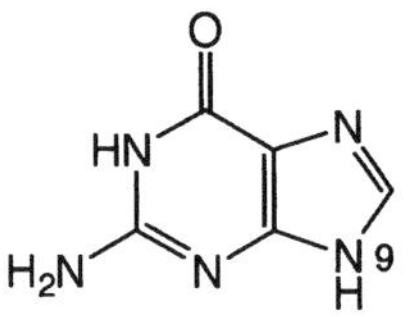   **1.4** Guanine (G) 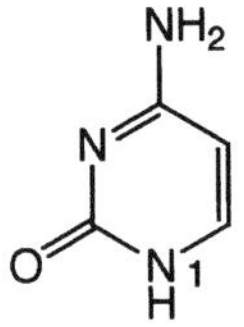   **1.5** Cytosine (C) 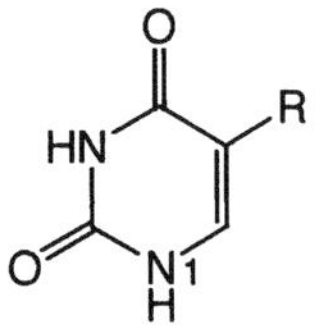   **1.6** R = H, Uracil (U)
1.7 R = CH$_3$, Thymine (T)

FIGURE 1.1 SUGAR AND BASE COMPONENTS OF RNA AND DNA NUCLEOSIDES.

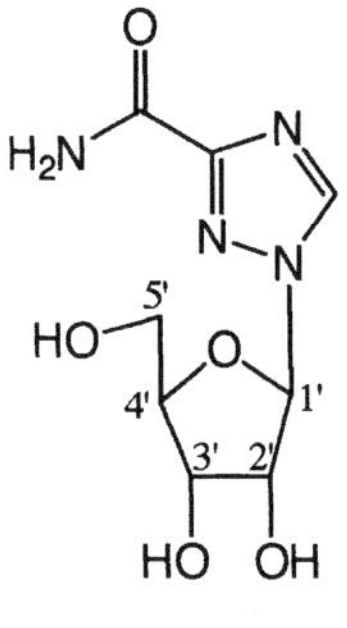

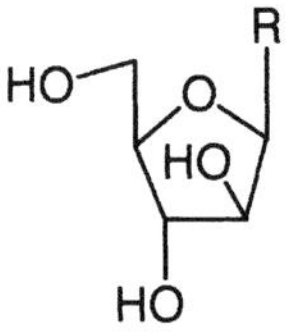

1.9 R = OH, 1-β-D-arabinofuranose
1.10 R = Adenine, Vidarabine

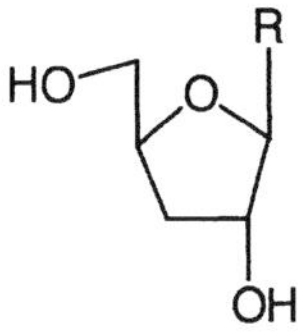

1.8 Ribavirin

1.11 R = OH, 3-deoxy-1-β-D-ribofuranose
1.12 R = Adenine, Cordycepin

FIGURE 1.2 SUGAR NUMBERING AND VARIATIONS.

9-β-D-arabinofuranosyladenine), and 3-deoxy-β-D-ribofuranose (**1.11**), the sugar component of the antibiotic Cordycepin[20] (**1.12**, 9-(3-deoxy-β-D-*erythro*-pentofuranosyl)adenine) (Fig. 1.2). There is considerably more variation with the synthetic nucleoside mimetics, which will be considered in Chapters 2 and 3.

1.2.2 Structure of the Heterocyclic Base Moiety

The purine and pyrimidine bases are numbered as shown in the general structures **1.13** and **1.14** respectively. In general trivial nomenclature is used although occasionally systematic nomenclature is necessary. Several other heterocyclic bases, other than those found in RNA and DNA, occur naturally. Hypoxanthine (6-oxypurine) is the heterocyclic base of inosine monophosphate (**1.15**), a key intermediate in the *de novo* synthesis of purine ribonucleotides (see section 1.6.2.1). An unusual naturally occurring purine base is the imidazooxazinone ring system found in the anticancer agent oxanosine[21] (**1.16**), which under alkaline conditions rearranges to xanthosine

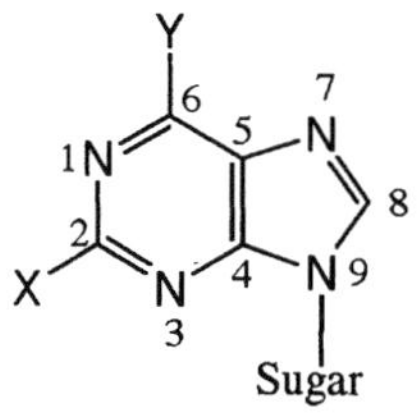

1.13

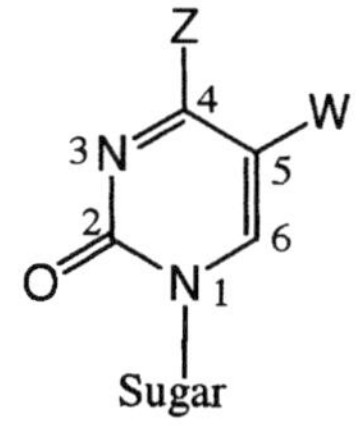

1.14

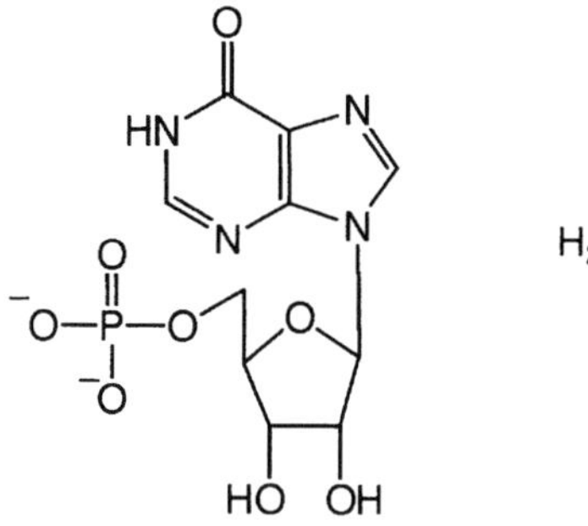

1.15 Inosine 5'-monophosphate

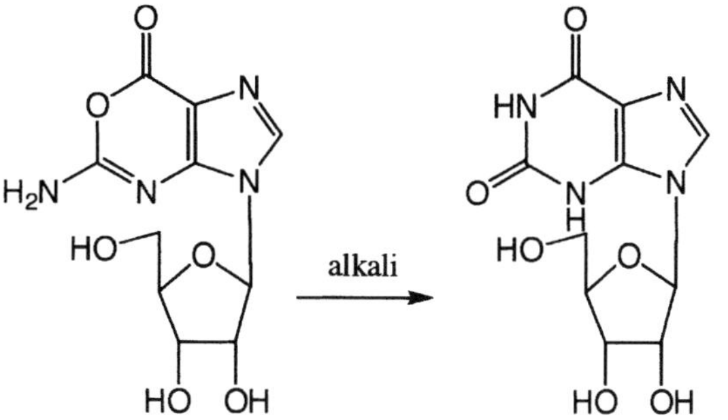

1.16 Oxanosine **1.17** Xanthosine

FIGURE 1.3 BASE NUMBERING AND VARIATIONS.

(**1.17**) which contains the 2,6-dioxypurine, xanthine (Fig. 1.3). As with the sugar moiety, there is considerable variation in the heterocyclic base component of the synthetic nucleoside mimetics, many of which will be considered in Chapters 2 and 4.

1.2.2.1 Tautomerism of Purines and Pyrimidines

Oxygenated purines and pyrimidines exist as tautomeric structures with the keto form being the major tautomer involved in hydrogen bonding between the bases in the nucleic acids RNA and DNA. The amine containing bases adenine and cytosine also exist as tautomeric structures with the amino form predominating as the tautomer involved in hydrogen bonding between the nucleic acid bases (Fig. 1.4).

1.3 NUCLEOSIDE CONFORMATION

The conformation of nucleosides and their analogues have been reviewed extensively,[22–25] with much renewed interest owing to the importance of nucleotide conformation on activity and function.[26–29] Three parameters are used to describe the conformation of nucleosides: the geometry of the glycosyl link between the base and ribose ring, the rotation about the exocyclic C4'–C5' bond, and the puckering of the sugar ring.

1.3.1 Glycosyl Link Geometry

The geometry of the glycosyl link between the base and the ribose ring is determined by the torsional angle χ(C2-N1-C1'-O4') and the value of χ describes whether the

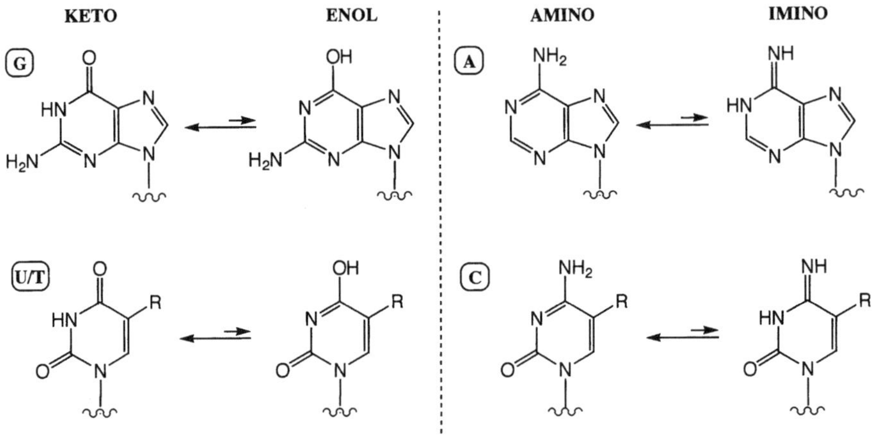

INTRODUCTION

FIGURE 1.4 TAUTOMERS OF PURINES AND PYRIMIDINES.

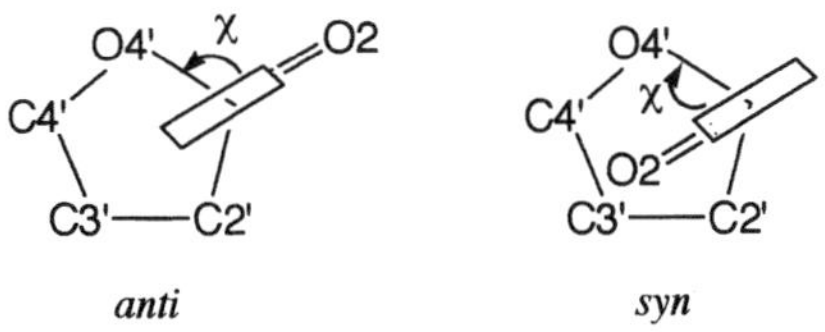

FIGURE 1.5 GLYCOSIDIC CONFORMATION.

conformation is *anti* ($0 \pm 90°$) or *syn* ($180 \pm 90°$) (Fig. 1.5). In pyrimidine nucleosides there is a preference for an *anti* conformation $-180 < \chi < -115°$ and in purine nucleosides high *anti* $\chi = -60°$ is allowed. These conformations are preferred over *syn* conformations, which would give rise to steric interactions between the base and the ribose ring.

1.3.2 C4'–C5' Bond Geometry

The geometry of the exocyclic bond relative to the ribose ring is described by the torsional angle γ(C3'-C4'-C5'-O5'). There are three possible orientations of the O5'-hydroxyl group, *ap* (trans, t), *+sc* (gauche, g+), and *-sc* (gauche, g-), with γ angles of approximately 180°, 60°, and –60° respectively (Fig.1.6). Pyrimidine nucleosides have a strong preference for *+sc*, while purine nucleosides have an equal preference for *ap* or *+sc*.

1.3.3 Ring Puckering

The puckering of the ribose ring gives rise to two main structural conformations, envelope (E) and twist (T). The puckering is determined by the pseudorotational phase angle P which describes which atoms are displaced out of the plane and the direction of displacement, and by the maximum torsional angle τ_{max} which describes the degree of displacement or pucker.

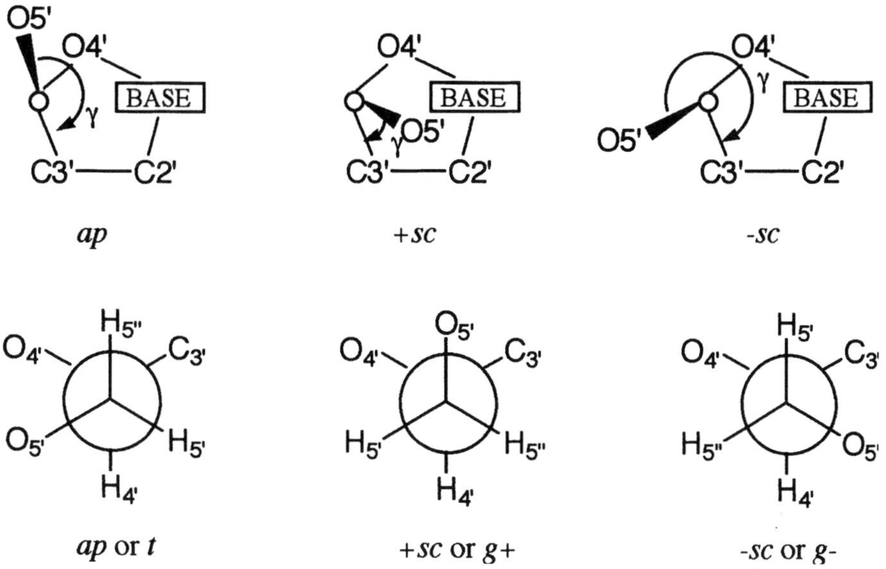

FIGURE 1.6 C4'–C5' CONFORMATION.

The conformational preference of the sugar ring is influenced by both steric effects and the *gauche* effect.[30] Approximate calculations of these structural conformations can be obtained by nuclear magnetic resonance coupling values,[31] or accurate values for the nucleoside in the solid phase can be determined from X-Ray crystallographic data.[32] The sugar moieties are classified as either North-type (N) (C2'-exo, C3'-endo) or South-type (S) (C2'-endo, C3'-exo) (Fig. 1.7). The two preferred conformations of ribose ring puckering are C3'-endo ($0 < P < 18°$) and C2'-endo ($162 < P < 180°$) with endo and exo describing the displacement of the atom above or below the plane of the other atoms in the sugar ring respectively. The preferred conformation of 2'-deoxyribose nucleosides is C2'-endo.

Combining all three conformational parameters gives the strongest preferences for C3'-endo with $-180 < \chi < -138°$ and C2'-endo with $-144 < \chi < -115°$.

1.4 NOMENCLATURE

Nucleosides are named with a mix of trivial and IUPAC nomenclature[33] and very often abbreviated names are used. Both the trivial (plain text) and systematic nomenclature (italics) for the nucleosides of RNA (**1.18–1.21**) and DNA (**1.22–1.25**) are shown in Fig. 1.8. In IUPAC nomenclature the sugar is described first with the position of attachment to the base indicated by the number in front of the sugar, *i.e.* 1- in pyrimidines and 9- in purines, followed by the name of the heterocyclic base. The nomenclature of other antiviral nucleosides will be described in the appropriate chapters.

INTRODUCTION

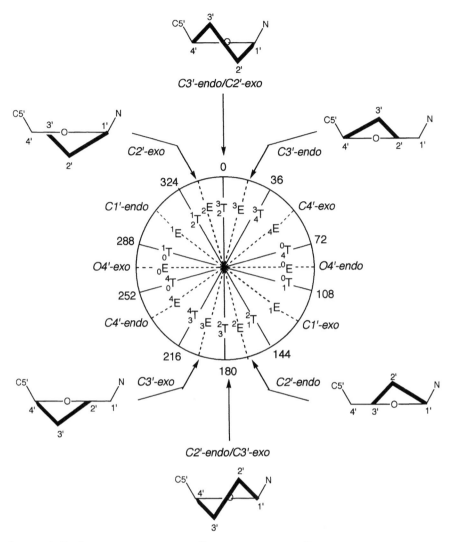

FIGURE 1.7 SUGAR CONFORMATIONS/PSEUDOROTATIONAL PARAMETERS.

1.5 SPECTROSCOPY

There are many analytical techniques available for the structure elucidation of nucleosides such as infra-red (IR) and UV spectroscopy, X-Ray crystallography and high performance liquid chromatography (HPLC), however the most useful techniques are Mass spectroscopy (MS) and Nuclear Magnetic Resonance spectroscopy (NMR).

1.5.1 Mass Spectroscopy

Mass spectroscopy has proved of value in the analysis of nucleosides.[34] Studies have shown that there are characteristic features of the chemistry of nucleosides in the gas phase regardless of the charge state of the molecular ions or the method of preparation.[35,36]

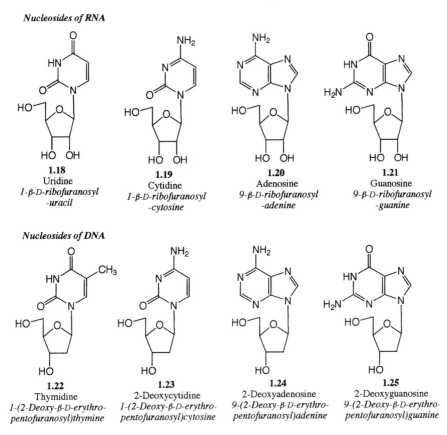

FIGURE 1.8 NOMENCLATURE OF THE NUCLEOSIDES OF RNA AND DNA.

Glycosidic cleavage commonly occurs with the heterocyclic base, in which the charge usually resides, the controlling element in most dissociation reactions. As a result of the stability of the heterocyclic base structural variations are difficult to study as there are very few product ions formed.

Ionisation by chemical ionisation (CI) or fast atom bombardment (FAB) results in intermolecular proton transfer to give the protonated molecular ion MH+, followed by spontaneous or collision induced dissociation to form ions.[37] The odd-electron molecular ion M•+ is produced in the gas phase by intramolecular abstraction of H• from the sugar hydroxyl groups (Fig. 1.9).[38]

1.5.2 NMR Spectroscopy

NMR methods including ^{1}H,[25,39,40] ^{13}C,[41,42] and N.O.E. (Nuclear Overhauser Effect)[43] have been used mainly to provide a qualitative determination of the glycosidic bond/sugar conformation and general structure characterisation.

1.5.2.1 ^{1}H NMR

Proton NMR spectra of the naturally occurring pyrimidine and purine nucleosides were determined in the 1960's and 1970's (Tables 1.1 and 1.2). Studies have provided

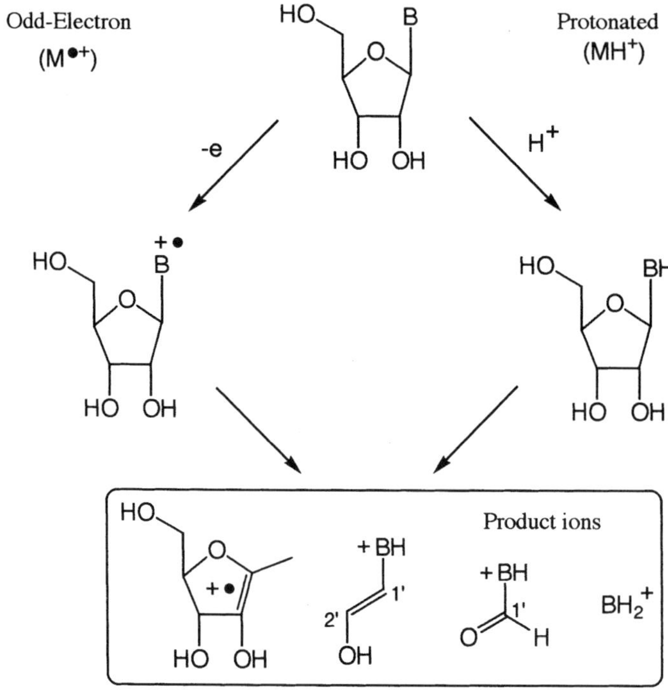

INTRODUCTION

FIGURE 1.9 PRODUCT IONS OF MS OF NUCLEOSIDES.

information on sugar conformation (*syn/anti*, S-type/N-type, α/β-glycosidic configuration) and base character (tautomerism, H-bonding, base stacking).[44]

Sugar moiety
The H-1' in ribose occurs at a higher field than deoxyribose owing to increased electrostatic shielding of the ribose H-1' by the 2'-hydroxy group. The ribose H-1' occurs as a doublet and the H-1' in deoxyribose occurs as a triplet as a result of spin-spin interaction with one and two C-2' protons respectively. H-4' and H-5' are reasonably constant in all the nucleosides with H-3' well separated downfield of H-4' and H-5'. Uridine and cytidine are exceptions with a smaller separation between H-3' and H-4'/H-5' owing to increased shielding.

Base moiety
In the pyrimidine nucleosides H-6, being less shielded, occurs downfield of H-5, of note is the observation that in DMSO-d_6 solution cytidine exists as the imino tautomer, with two NH signals observed (9.19 and 10.36 ppm).[39] The H-2 and H-8 protons of the purine nucleosides occur downfield (approx. 8.0–8.7 ppm) as a result of both C-2 and C-8 being bonded to two nitrogen atoms.

Anomeric determination
N.O.E. spectroscopy can be used to determine the anomeric configuration of nucleosides.[43] In β-anomers, with N- or S-type conformation, irradiation of H-1'

TABLE 1.1 ¹H NMR of Naturally Occurring Pyrimidine Nucleosides.

	H-1' $J_{1'2'/1'2''}$	H-2' $J_{2'3'/2''3'/2'2''}$	H-2''	H-3' $J_{3'4'}$	H-4' $J_{4'5'/4'5''}$	H-5' $J_{5'5''}$	H-5''	H-5 $J_{5,6}$	H-6	NH₂	NH	NH	Ref.
Ribose nucleosides													
U^a	5.79 5.1	4.03 4.8	—	3.97 4.3	3.85 2.8/3.0	3.63 12.3	3.56 8.2	5.64	7.88	—	11.66	—	40,45
C^b	5.91 4.0	4.32 5.2	—	4.22 6.0	4.14 2.8/4.3	3.94	3.83 12.7	6.05 7.6	7.86	7.11	—	—	40,46
2'-Deoxyribose nucleosides													
T^a	6.28 6.9/6.9	2.39 5.6/5.6/14.2	2.40	4.49 3.7	4.04 3.6/4.9	3.87 12.6	3.79	1.89_Me 1.1	7.68	—	11.58	—	40,47
dC^b	6.47 6.3/6.9	2.64 3.9/6.7/14.0	2.64	4.68 3.3	4.26 3.3/5.4	3.94–4.11 14.1	—	6.64 7.9	8.69	—	9.19	10.36	40,41

^a DMSO-d₆, ^b D₂O

TABLE 1.2 ¹H NMR OF NATURALLY OCCURRING PURINE NUCLEOSIDES IN D_2O.

	H-1' $J_{1'2'/2''}$	H-2' H-2" $J_{2'3'}$ /2"3' /2'2"	H-3' $J_{3'4'}$	H-4' $J_{4'5'/4'5''}$	H-5' H-5" $J_{5'5''}$		H-2	H-8	NH_2	NH	Ref.
Ribose nucleosides											
A	6.04	4.75 —	4.42	4.29	3.92	3.83	8.16	8.29	7.72	—	40,41,48
	6.1	5.2	3.7	2.6/3.0	8.6						
G	5.92	4.75 —	4.41	4.23	3.92	3.79	—	8.01	6.92	11.23	25,40,41
	4.0	4.8	5.1								
2'-Deoxyribose nucleosides											
dA	6.77	3.86–4.11	4.86	4.27	3.86–4.11		8.56	8.73	7.73	—	25,40,41
	6.3/7.5	3.3/6.3/14.0	2.8	3.1/4.3	11.7						
dG	6.55	3.78–4.11	4.74	4.20	3.78–4.11		—	8.36	6.91	11.20	25,40,41
				3.5/4.5	12.5						

TABLE 1.3 ^{13}C NMR of the Naturally Occurring Nucleosides in DMSO-d$_6$ (ref. 42).

Pyrimidine Nucleosides

Cmpd	C-2	C-4	C-5	C-6	CH$_3$	C-1'	C-2'	C-3'	C-4'	C-5'
U	151.88	164.20	102.54	141.68	—	88.58	70.58	74.34	85.42	61.70
T	151.06	164.37	109.98	136.80	12.72	84.53	39.90	71.13	87.90	61.83
C	156.42	166.17	95.15	142.29	—	89.59	70.04	74.60	84.84	61.35
dC	156.49	166.44	95.33	141.93	—	86.06	40.10	71.25	88.06	62.12

Purine Nucleosides

Cmpd	C-2	C-4	C-5	C-6	C-8	C-1'	C-2'	C-3'	C-4'	C-5'
A	152.92	149.54	119.85	156.59	140.62	88.62	71.22	74.12	86.44	62.11
dA	153.12	149.50	119.91	156.61	140.44	85.02	39.94	71.71	88.64	62.40
G	154.06	151.82	117.02	157.25	136.35	86.73	71.02	74.34	85.87	61.66
dG	154.18	151.48	117.15	157.63	136.13	83.33	39.90	71.31	88.17	62.35

often results in an N.O.E. of the H-4' signal as both H-1' and H-4' are situated on the same face (α) of the sugar molecule. In α-anomers irradiation of H-1' often results in an N.O.E. of the H-3' signal, with N-type conformers exhibiting a stronger enhancement than S-type conformers owing to the closer spacial proximity of H-1' and H-3' in N-type conformers.

1.5.2.2 ^{13}C NMR

The carbon-13 resonances of nucleosides (Table 1.3) can be divided into two distinct regions, those owing to the sugar moiety (≤ 90 ppm) and those of the heterocyclic base (≥ 95 ppm). The greatest difference in the ^{13}C spectra of ribose and 2'-deoxyribose nucleosides is the C-2' resonance signals. Exchanging the 2'-hydroxy group (~70 ppm) for a proton results in an upfield shift (~40 ppm), this also results in a β-substituent effect with the C-1' and C-3' resonances in deoxyribose which are shifted upfield 3–4 ppm relative to their positions in the ribosyl spectra.

The characteristic signals of the pyrimidine base are the C5–C6 resonances. In all the pyrimidine nucleosides C-6 occurs downfield of C-5 as a result of the β-effect in alkenes. A downfield shift in C-5 of thymidine, consistent with a CH$_3$-substituted carbon, distinguishes thymidine from uridine. C-8 in guanosine nucleosides is readily assigned being the only carbon bearing a proton, C-8 of adenosines are assigned by comparison, with the C-2 in adenosine nucleosides occurring downfield owing to its positioning between two nitrogen atoms. The C-4 and C-5 quaternary carbons of the purine nucleosides occur at similar shifts to those observed for C-4 and C-5 of the pyrimidines, and the quaternary carbons bearing substituents of high electron density (*e.g.* an amino group, C-6 in A, dA and C-2 in G, dG) are characteristically downfield.

1.6 BIOLOGICAL PROPERTIES

Nucleosides and their derivatives are involved in a diverse range of biological processes such as energy metabolism, with adenosine 5'-triphosphate (ATP) being the

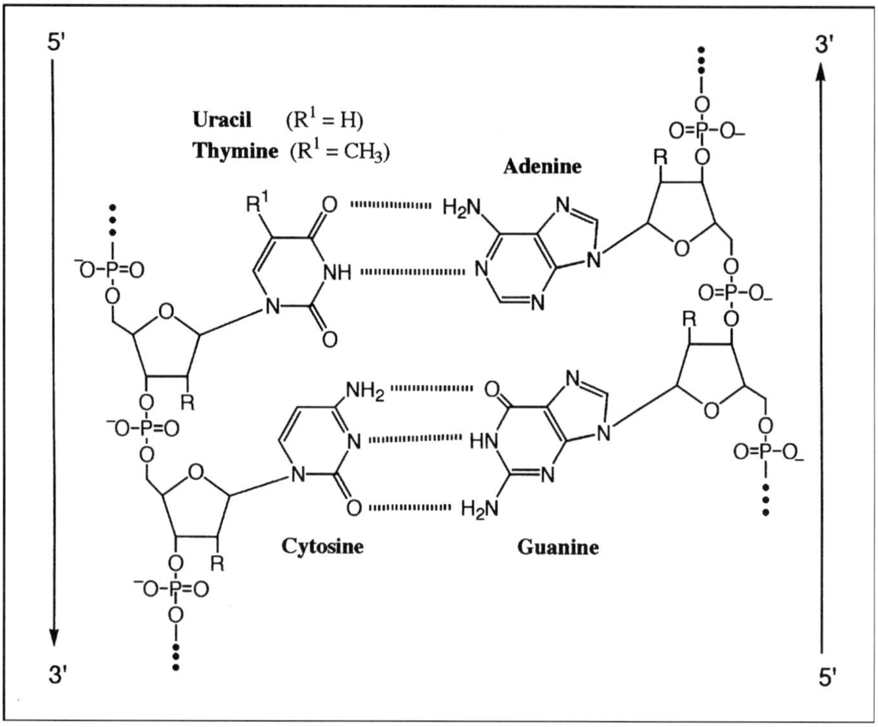

FIGURE 1.10 STRUCTURE OF RNA AND DNA DOUBLE HELIX.

principle form of chemical energy available to cells, as monomeric units of RNA and DNA and components of coenzymes *e.g.* nicotinamide adenine dinucleotide (NAD+), and as physiological mediators with adenosine 5'-diphosphate (ADP) being essential for platelet aggregation and guanosine 5'-triphosphate (GTP) being necessary for capping of messenger RNA (mRNA).

1.6.1 RNA and DNA

The 5'-monophosphates (nucleotides) of the nucleosides **1.18-1.25** are the monomeric units of RNA (Fig. 1.10, R = OH, R^1 = H) and DNA (Fig. 1.10, R = H, R^1 = CH_3). The double stranded complex is held together by hydrogen bonds between the pyrimidine and purine bases, with three hydrogen bonds between cytidine and guanine and two between adenine and either uracil or thymine, this is known as Watson-Crick base pairing. The nucleotides in the individual strands are linked by a phosphate backbone from a 3'-OH of one nucleotide to a 5'-OH of a second nucleotide, described as a 3'→ 5' phosphodiester bridge. Additional nucleotides are added on at the 3'-OH position, therefore a free hydroxyl at the 3'-position is essential for chain extension.

Nucleoside mimetics such 3'-azido-2',3'-dideoxythymidine (**1.26** AZT, Zidovudine) and 2',3'-dideoxycytidine (**1.27** DDC, Zalcitabine), which lack a 3'-hydroxy group, produce their antiviral effect against human immunodeficiency virus 1 (HIV-1) by incorporation of their respective triphosphates into the host DNA strand.[49]

1.26 AZT

1.27 DDC

This results in the inhibition of the viral enzyme reverse transcriptase (RT), which would normally initiate DNA synthesis from the 3'-OH end of a host primer eventually resulting in the complete transcription of viral RNA into DNA once inside the host cell (see section 1.7.1.1).

1.6.2 Purine and Pyrimidine Nucleotide Biosynthesis

The 5'-nucleotide derivatives are the main purine and pyrimidine compounds found in cells, with ATP occurring in the highest concentration. Both purine and pyrimidine nucleotide synthesis is highly regulated leading to fixed levels of the nucleotides, which is essential for the metabolic processes in which they are involved.[50,51]

1.6.2.1 De Novo Purine Nucleotide Synthesis

The *de novo* pathway starts from 5-phosphoribosyl-pyrophosphate (**1.28**, PRPP) and involves a series of reactions leading to the formation of inosine monophosphate (IMP, **1.15**) *via* the imidazole nucleotide AICAR (**1.29**). The synthesis of purine nucleotides requires amino acids as carbon and nitrogen donors and C_1 units are obtained from H_4 folate. IMP is the common precursor for both guanosine 5'-monophosphate (**1.31**, GMP) and adenosine 5'-monophosphate (**1.33**, AMP), with conversion to AMP and GMP requiring GTP and ATP as the energy source respectively (Scheme 1.1).

1.6.2.2 Pyrimidine Nucleotide Synthesis

Only uridine 5'-monophosphate (**1.36**, UMP) is formed by the *de novo* synthesis pathway. The pyrimidine nucleotides of cytidine and thymidine are both formed from uridine nucleotides, with cytidine nucleotides synthesised directly from uridine 5'-triphosphate (**1.37**, UTP) and thymidine 5'-monophosphate (**1.40**, TMP) from 2'-deoxyuridine 5'-monophosphate (**1.39**, dUMP).

De Novo Pyrimidine Nucleotide Synthesis

The synthesis of UMP by the *de novo* pathway involves the initial construction of the pyrimidine orotate (**1.34**) from the amino acids glutamine and aspartate. Reaction of orotate with PRPP (**1.28**) is catalysed by the enzyme orotate phosphoribosyl transferase leading to orotidine 5'-monophosphate (**1.35**, OMP), which on decarboxylation gives UMP (**1.36**) (Scheme 1.2).

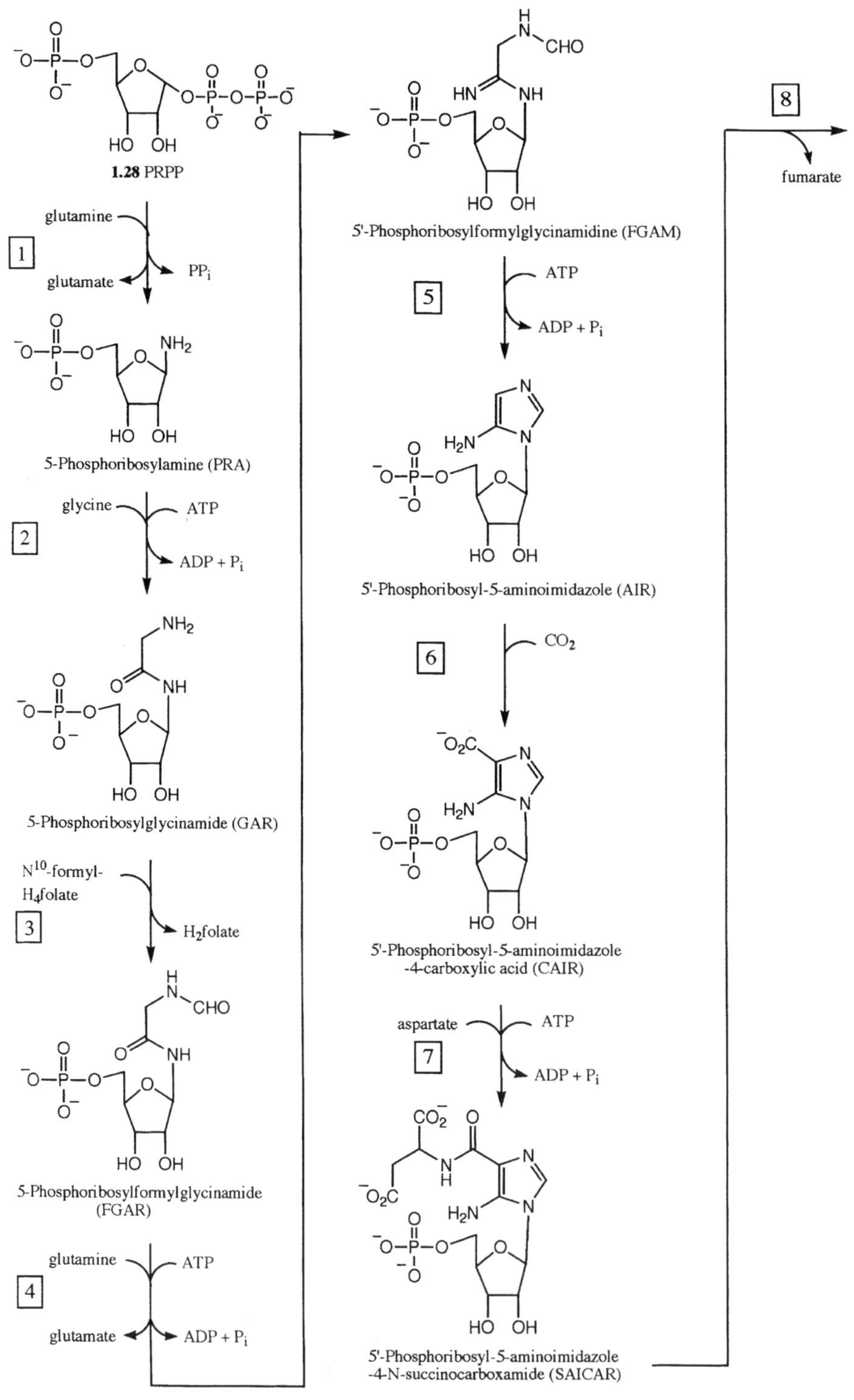

INTRODUCTION

1.28 PRPP

1. glutamine → glutamate, PP$_i$

5-Phosphoribosylamine (PRA)

2. glycine, ATP → ADP + P$_i$

5-Phosphoribosylglycinamide (GAR)

3. N^{10}-formyl-H$_4$folate → H$_2$folate

5-Phosphoribosylformylglycinamide (FGAR)

4. glutamine, ATP → glutamate, ADP + P$_i$

5'-Phosphoribosylformylglycinamidine (FGAM)

5. ATP → ADP + P$_i$

5'-Phosphoribosyl-5-aminoimidazole (AIR)

6. CO_2

5'-Phosphoribosyl-5-aminoimidazole-4-carboxylic acid (CAIR)

7. aspartate, ATP → ADP + P$_i$

5'-Phosphoribosyl-5-aminoimidazole-4-N-succinocarboxamide (SAICAR)

8. → fumarate

16 NUCLEOSIDE MIMETICS

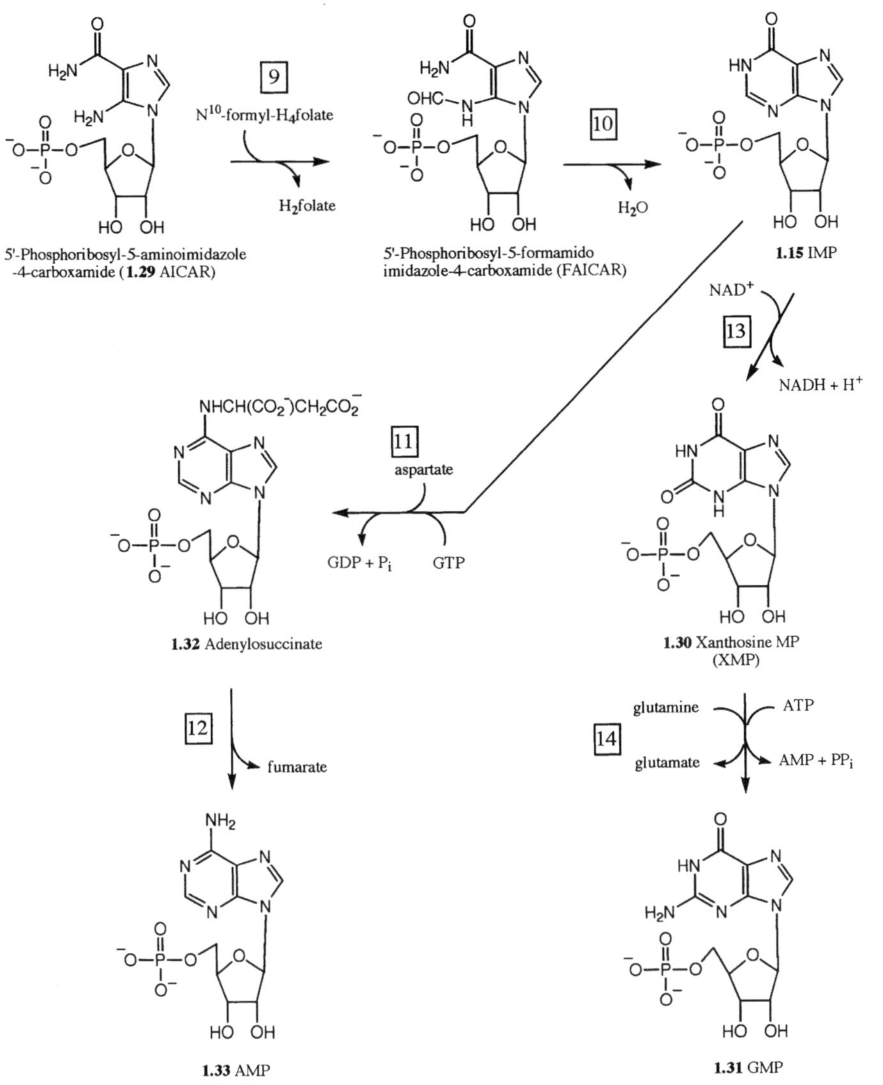

SCHEME 1.1 DE NOVO BIOSYNTHESIS OF GMP AND AMP.
CATALYSING ENZYMES: [1] GLUTAMINE PRPP AMIDOTRANSFERASE; [2] GAR SYNTHETASE; [3] GAR TRANSFORMYLASE; [4] FGAM SYNTHETASE; [5] AIR SYNTHETASE; [6] AIR CARBOXYLASE; [7] SAICAR SYNTHETASE; [8] ADENYLOSUCCINATE LYASE; [9] AICAR TRANSFORMYLASE; [10] IMP CYCLOHYDROLASE; [11] ADENYLSUCCINATE SYNTHETASE; [12] ADENYLOSUCCINASE; [13] IMP DEHYDROGENASE; [14] GMP SYNTHETASE.

Synthesis of Cytidine and Thymidine Nucleotides

The precursor of cytidine 5'-triphosphate (**1.38**, CTP) is uridine 5'-triphosphate (**1.37**, UTP). UMP (**1.36**) is first converted to UTP by the action of nucleotide diphosphokinase, with CTP synthesised directly from UTP by the action of CTP synthetase (Scheme 1.2). The formation of thymidine 5'-monophosphate (**1.40**, TMP) from 2'-deoxyuridine 5'-monophosphate (**1.39**, dUMP) is catalysed by thymidylate synthase which transfers a C_1 unit from N^5,N^{10}-methylene-H_4folate

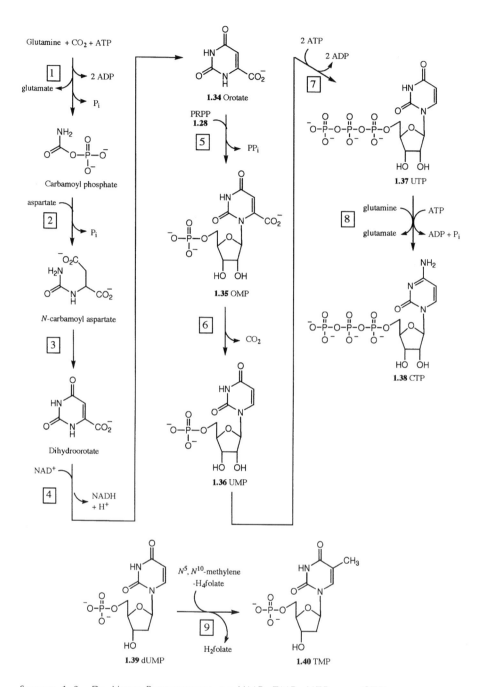

SCHEME 1.2 DE NOVO BIOSYNTHESIS OF UMP, TMP, UTP AND CTP.
CATALYSING ENZYMES: [1] CARBAMOYL PHOSPHATE SYNTHETASE II; [2] ASPARTATE CARBAMOYL TRANSFERASE; [3] DIHYDROOROTASE; [4] DIHYDROOROTATE DEHYDROGENASE; [5] OROTATE PHOSPHORIBOSYLTRANSFERASE; [6] OMP DECARBOXYLASE; [7] NUCLEOTIDE DIPHOSPHOKINASE; [8] CTP SYNTHETASE; [9] THYMIDYLATE SYNTHASE.

TABLE 1.4 BASE + PRPP → NUCLEOSIDE 5'-MONOPHOSPHATE (NMP) + PPi.

Base	PRTase	NMP
Hypoxanthine	HGPRTase	IMP
Guanine	HGPRTase	GMP
Adenine	APRTase	AMP
Orotate	PPRTase	OMP
Uracil	PPRTase	UMP
Thymine	PPRTase	TMP

to dUMP. The C_1 unit is then reduced to a methyl group by the folate reactant (Scheme 1.2).

1.6.2.3 Salvage Pathways for Nucleotide Synthesis
Distinct salvage pathways exist for the purine and pyrimidine nucleoside 5'-monophosphates, the general reaction of which is shown above (Table 1.4). Either a purine or pyrimidine base reacts with PRPP (**1.28**) to give the corresponding nucleoside 5'-monophosphate. Specific phosphoryl transferases (PRTase) are required to effect the salvage process.

Hypoxanthine and guanine are substrates for hypoxanthine-guanine PRTase (HGPRTase), adenine is a substrate for adenine PRTase (APRTase), and the pyrimidines orotate, uracil and thymine are substrates for pyrimidine PRTase (PPRTase), cytosine however is not a substrate for any of the phosphoryl transferases.

1.7 MECHANISMS OF ACTION OF NUCLEOSIDE MIMETICS

Nucleoside mimetics interact with either specific viral enzymes or with cellular enzymes involved in the biosynthesis of the RNA and DNA precursor nucleotides, with the resulting therapeutic effect observed dependent on the enzyme inhibited.[14,16] Modification of conventional nucleosides can produce nucleoside classes with very specific enzyme targets. With the increasing interest in nucleoside mimetics, this has led to a large variety of nucleoside structures and an ever increasing range of target enzymes and improved specificity.

1.7.1 Inhibitors Targeted at Reverse Transcriptase

The virally encoded enzyme reverse transcriptase (RT) plays a pivotal role in the replicative process of HIV. RT is a multifunctional enzyme which has both RNA-dependent DNA polymerase and DNA-dependent DNA polymerase activity as well as RNase H activity. RT is responsible for transcribing the genetic material contained within the two copies of (+)-strand RNA, located within the core of the virus, into DNA once inside the host cell. Many of the antiviral nucleoside mimetics interact at the substrate binding site of the HIV RT where they act as competitive inhibitors/alternate substrates (chain terminators).[49]

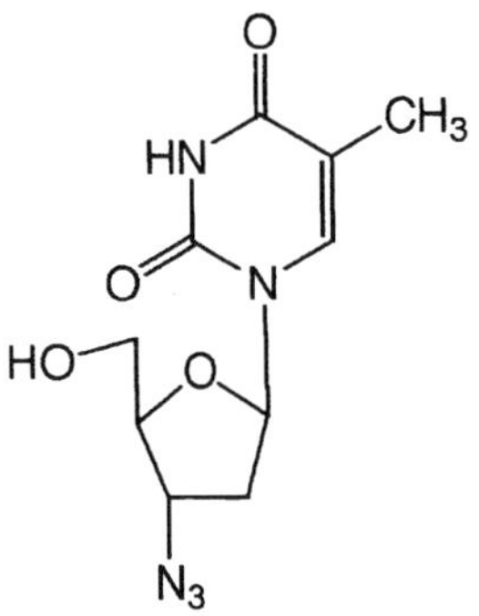

INTRODUCTION

1.26 AZT
(Zidovudine)

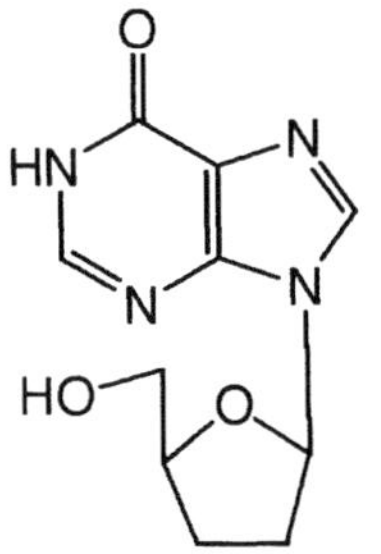

1.41 DDI
(Didanosine)

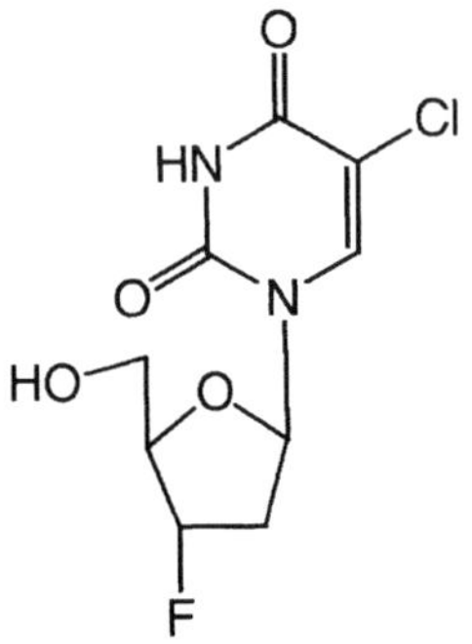

1.42 FddClU

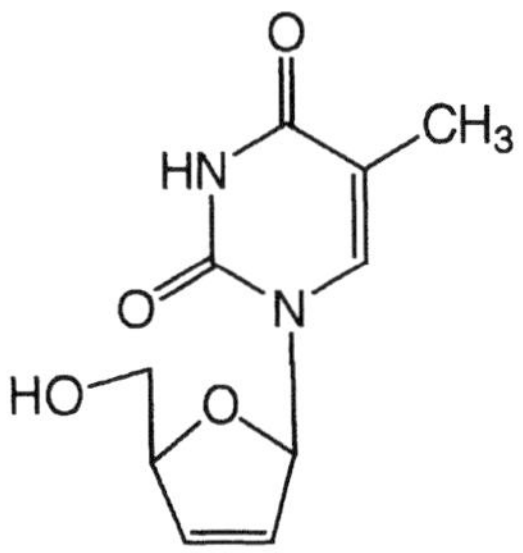

1.43 D4T
(Stavudine)

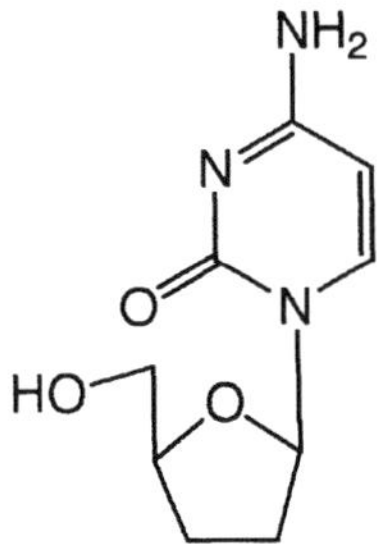

1.27 DDC
(Zalcitabine)

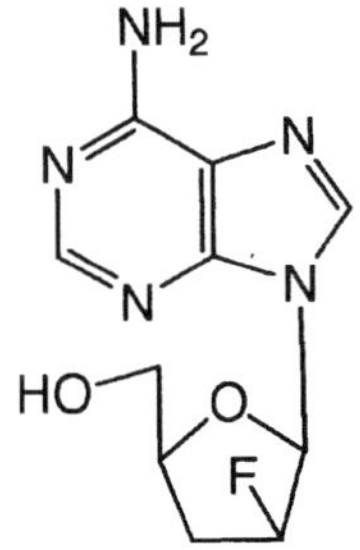

1.44 FddAraA

FIGURE 1.11 ALTERNATE SUBSTRATES/CHAIN TERMINATORS.

1.7.1.1 Chain Terminators

The first compounds approved for the treatment of AIDS were the nucleoside mimetics 3'-azido-2',3'-dideoxythymidine (**1.26**, AZT), 2',3'-dideoxyinosine (**1.41**, DDI) and 2',3'-dideoxycytidine (**1.27**, DDC).[52,53] These 2',3'-dideoxynucleosides (DDN's) all lack a 3'-hydroxy group and are active in their triphosphate forms (DDNTP). These DDNTP analogues interact at the substrate (dNTP) binding site of the HIV RT where they act as competitive inhibitors/alternate substrates. A 3'-hydroxy group is essential for the extension of the viral DNA chain, therefore once incorporated into the growing DNA chain, DDNTPs inhibit the action of RT by premature chain termination. Many of the antiviral nucleoside mimetics which exhibit this mechanism of action are shown in Figure 1.11.

The DDN's show differences in their ability to suppress HIV, and this has been shown to be directly related to their intracellular metabolism to the 5'-triphosphate form.[54] For D_4T (**1.43**) the rate-limiting step (r.l.s.) is the formation of D_4T-MP, however for AZT the r.l.s. is the formation of AZT-DP from AZT-MP, this is owing to the ability of AZT-MP to inhibit dTMP kinase, the enzyme which catalyses the conversion of AZT-MP to AZT-DP (Fig. 1.12).

INTRODUCTION

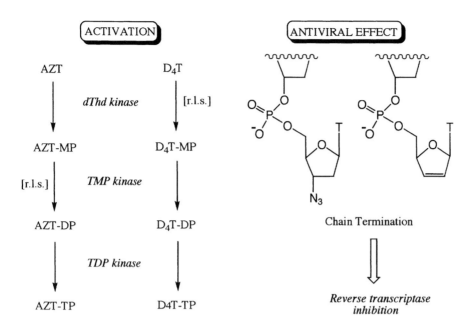

FIGURE 1.12 ACTIVATION AND ANTIVIRAL EFFECT OF AZT AND D₄T.

1.7.1.2 HIV-1 Specific Reverse Transcriptase Inhibitors
Two compounds of particular interest are 1-(2-hydroxyethoxymethyl)-6-phenylthiothymidine (**1.45**, HEPT)[55] and 2',5'-bis-O-(*tert*-butyldimethylsilyl)-3'-spiro-5"-(4"-amino-1",2"-oxathiole-2",2"-dioxide)thymidine (**1.46**, TSAO-T) (Fig. 1.13).[56]

These compounds are specific inhibitors of HIV-1 RT and do not affect the replication of other retroviruses. HEPT and TSAO-T are described as non-nucleoside RT inhibitors (NNRTI's) as they do not bind at the substrate (dNTP) binding site. The selectivity of the NNRTI's is owing to a specific interaction with a non-substrate binding site of the HIV-1 RT.

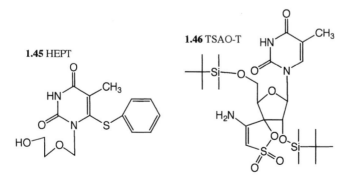

FIGURE 1.13 SPECIFIC HIV-1 RT INHIBITORS.

NUCLEOSIDE MIMETICS   21

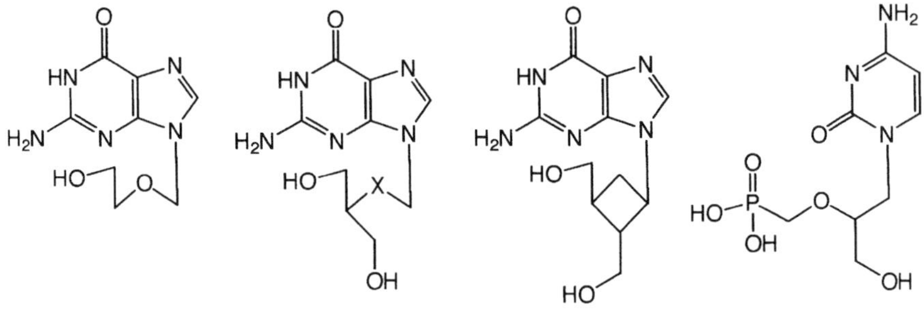

1.47 ACV (Zovirax®)
1.48 X = O, GCV (Cymevene®)
1.49 X = CH₂, PCV (Vectavir®)
1.50 BHCG (Lobucavir)
1.51 HPMPC (Cidofovir)

FIGURE 1.14 VIRAL DNA CHAIN TERMINATORS.

1.7.2 Inhibitors Targeted at DNA Viruses

The mechanism of action of nucleoside mimetics which act as inhibitors of DNA viruses such as Hepatitis B virus (HBV), Herpes Simplex virus (HSV), Epstein Barr virus (EBV), Varicella Zoster virus (VZV) and Cytomegalovirus (CMV), varies depending on the structure class to which the nucleoside mimetic belongs. These mimetics either act as inhibitors/substrates of the viral DNA polymerase, resulting in chain termination, or DNA strand breakage.

1.7.2.1 DNA Polymerase Inhibitors/Chain Terminators

The acyclic nucleosides 9-(2-hydroxyethoxymethyl)guanine (**1.47**, ACV) and 9-(1,3-dihydroxy-2-propoxymethyl)guanine (**1.48**, GCV) and the carboacyclic nucleosides 9-(4-hydroxy-3-hydroxymethylbutyl)guanine (**1.49**, PCV) and carbaoxetanocin G (**1.50**, BHCG) are recognised as substrates by the HSV- and VZV-encoded thymidine kinase (TK)[16] (Figure 1.14).

The viral TK converts these mimetics to their monophosphates, and then cellular enzymes convert to the di- and then tri-phosphates, which is the required form for activity. The triphosphates act as substrates of the viral DNA polymerase and after incorporation into the viral DNA, chain termination occurs (Fig. 1.15). The acyclic nucleoside phosphonates can also act in this manner, 1-[(S)-3-hydroxy-2-(phosphonomethoxy)propyl]cytosine (**1.51**, HPMPC) has good activity against CMV and is active as its diphosphate HPMPC-DP.[57] HPMPC-DP may act as a competitive inhibitor/substrate of viral DNA polymerase being incorporated either terminally, and so causing chain termination, or internally resulting in the formation of aberrant DNA, the overall effect being the impaired synthesis of the viral DNA (Fig. 1.15).

1.7.2.2 Viral DNA Strand Breakage

(E)-5-(2-Bromovinyl)-2'-deoxyuridine (**1.52**, BVDU) and its analogues are substrates for HSV-1 TK and VZV-TK (Fig.1.16), this specificity is owing to the (E)-5-(2-bromovinyl) substituent.[58] BVDU is converted to the diphosphate by either HSV-1 TK or VZV-TK and then to its active triphosphate form by cellular kinases. The triphosphate acts as a substrate of the viral DNA polymerase and is incorporated into the interior of the DNA chain which results in DNA strand breakage.[59]

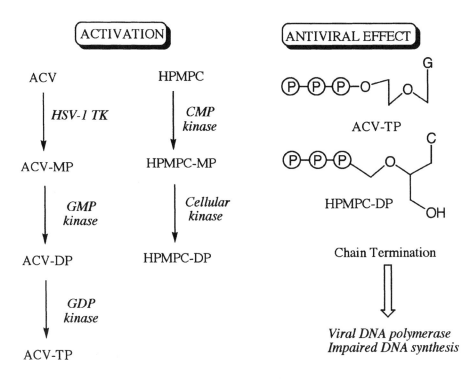

FIGURE 1.15 ACTIVATION AND ANTIVIRAL EFFECT OF ACV AND HPMPC.

FIGURE 1.16 VIRAL DNA STRAND BREAKERS.

1.7.3 Inhibitors Targeted at Nucleotide Metabolising Enzymes

Nucleotide metabolising agents play an important role in the cell cycle and the rate of cell division. During the cell cycle all the components of the cell double and eventually cell division occurs leading to the formation of two daughter cells. RNA synthesis and DNA replication occur at different stages of the cell cycle, with RNA synthesis occurring during G1, S and G2 phases of the cell cycle and DNA replication in the S phase. An increase in the levels of the enzymes involved in nucleotide metabolism is observed during these phases and in particular during S phase. The levels of these nucleotide metabolising enzymes are elevated further in rapidly

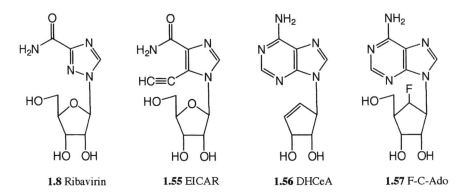

FIGURE 1.17 INHIBITORS OF IMP DEHYDROGENASE, PNP AND AdoHcy HYDROLASE.

growing cells such as tumour cells, therefore inhibition of these enzymes by nucleoside mimetics can be beneficial in cancer therapy as well as antiviral therapy.

1.7.3.1 Inhibitors of Purine Nucleotide Metabolising Enzymes
The main enzymatic targets involved in purine biosynthesis are inosine monophosphate dehydrogenase (IMP dehydrogenase, IMPDH), purine nucleoside phosphorylase (PNP) and S-adenosyl-L-homocysteine (AdoHcy, SAH) hydrolase. Some of the nucleoside mimetics which inhibit these key enzymes are shown in Figure 1.17.

IMPDH
IMP dehydrogenase (IMPDH, E.C. 1.1.1.205), which exists in two isoforms IMPDH-h1 and IMPDH-h2,[60,61] is linked with cell proliferation and malignancy[62] and is a key enzyme in the biosynthesis of GMP (**1.31**) from IMP (**1.15**) (*via* XMP (**1.30**)) (Fig. 1.18), therefore inhibition of this enzyme results in a depletion of the guanylate (GMP, GDP, GTP, dGTP) pools.

A depletion of GTP has a 'knock-on' effect on the formation of AMP (**1.33**) which is dependent on GTP as an energy source for its biosynthesis.[63] IMPDH-h2 expression is up-regulated in rapidly proliferating human leukaemia cells and solid tumour tissues,[64] IMPDH-h1 activity is increased in activated lymphocytes.[65] Overall inhibitors of IMP dehydrogenase such as Ribavirin (**1.8**) and EICAR (**1.55**), which are active as their monophosphates, cause a depletion of purine nucleotides, resulting in a broad spectrum of activity against RNA and DNA viruses,[66] and tumour cell proliferation.[67]

PNP
PNP (E.C. 2.4.2.1) catalyses the reversible phosphorolysis of inosine, guanosine and their 2'-deoxy derivatives to the free bases and (2-deoxy)ribose-1-phosphate (Fig. 1.19), and is an essential enzyme of the purine salvage pathway. Many anticancer and antiviral purine nucleosides, such as arabinosylguanine, 2'-deoxy-6-thioguanosine and 2', 3'-dideoxyinosine (**1.41**), are substrates for PNP and are cleaved and therefore inactivated *in vivo*, for this reason PNP inhibitors are often given in conjunction with anticancer and antiviral nucleosides.[68]

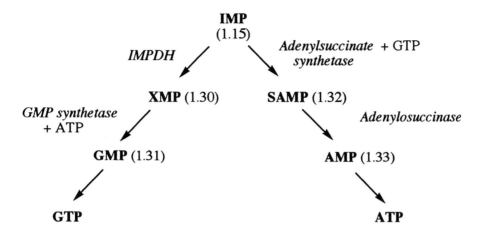

FIGURE 1.18 TWO COMPETING ROUTES FOR IMP.

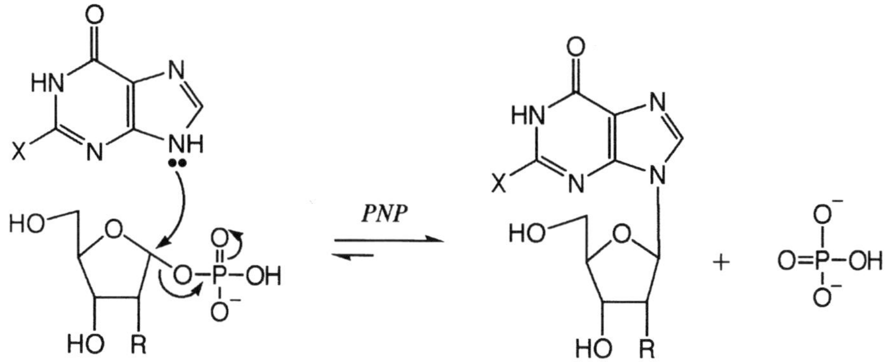

FIGURE 1.19 REVERSIBLE ACTION OF PNP (R = H, OH; X = H, NH$_2$).

A report on patients with a genetic deficiency in PNP activity which results in a lack of T-lymphocytic function and accumulation of cytotoxic levels of dGTP in certain lymphocytes,[69] led to an increasing interest in the potential of PNP inhibitors for the treatment of T-cell leukaemias, T-cell autoimmune diseases such as rheumatoid arthritis, and to prevent rejection after organ transplantation.[70] Inhibitors of PNP can potentiate the activity of purine anticancer and antiviral agents and are capable of providing specific suppression of the immune system.

AdoHcy Hydrolase

The cellular enzyme S-adenosyl-L-homocysteine (AdoHcy) hydrolase (EC 3.3.1.1) is responsible for the cleavage of AdoHcy into its two components, adenosine (Ado) and L-homocysteine (Hcy) (Fig. 1.20).[71] Homocysteine is required for the formation of the methyl donor S-adenosylmethionine (SAM), which acts as a methyl donor for viruses which require methylations (5'-capping) for the maturation of their viral

INTRODUCTION

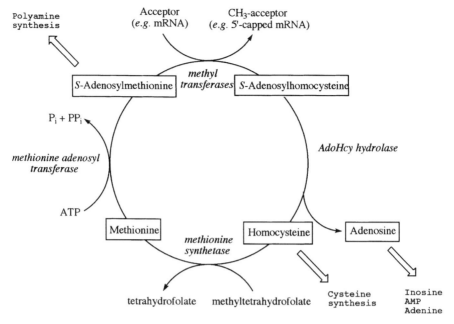

FIGURE 1.20 METABOLIC PATHWAY OF ADOHCY.

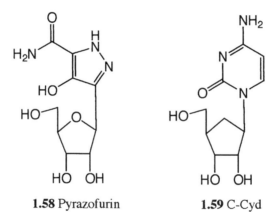

FIGURE 1.21 INHIBITORS OF OMP DECARBOXYLASE AND CTP SYNTHETASE.

mRNA. Inhibition of AdoHcy hydrolase results in a depletion of homocysteine and therefore a depletion of SAM. The SAH hydrolase inhibitors DHCeA (**1.56**)[72] and F-C-Ado (**1.57**)[73] are active against (–) RNA viruses which require methylation by SAM for maturation of viral mRNA.

1.7.3.2 Inhibitors of Pyrimidine Nucleotide Metabolising Enzymes

OMP decarboxylase which catalyses the final step in the biosynthesis of UMP (**1.36**), and CTP synthetase which converts UTP (**1.37**) to CTP (**1.38**), are two targets in the biosynthesis of pyrimidine nucleotides.

The OMP decarboxylase inhibitor Pyrazofurin (**1.58**, Figure 1.21) which is active against a broad spectrum of viruses including influenza A, B, C[74] and murine leukemia,[75] causes cessation of the synthesis of all pyrimidines resulting in impaired cell growth and metabolism. Cyclopentylcytosine (C-Cyd) (**1.59**, Figure 1.21), which is active in its triphosphate form, is selective in targeting CTP synthetase.[76] Inhibition of CTP synthetase causes a depletion of CTP, CDP, dCDP and dCTP pools.

1.8 REFERENCES

1. F. Miescher, *Hoppe-Seyler's Med. Chem. Unters.*, **1871**, 441.
2. R. Altmann, *Arch. Anat. u. Physiol., Physiol. Abt.*, **1889**, 524.
3. A. Kossel, *Arch. Anat. u. Physiol., Physiol. Abt.*, **1891**, 181.
4. A.M. Michelson, *"The Chemistry of Nucleosides and Nucleotides"*, Academic Press, London, **1963**.
5. P.A. Levene, W.A. Jacobs, *Ber.*, **1909**, *42*, 2474–2478.
6. J.G. Buchanan, R.H. Wightman in "*Topics in Antibiotic Chemistry*", P. Sammes (ed.), Ellis Horwood, Chichester, **1982**, *6*, pp 229–339.
7. J.J. Fox, I. Wempen, *Adv. Carbohydr. Chem.*, **1959**, *14*, 283–380 and references cited therein.
8. J.A. Montgomery, H.J. Thomas, *Adv. Carbohydr. Chem.*, **1962**, *17*, 301–369 and references cited therein.
9. J.A. Montgomery, *Med. Res. Rev.*, **1982**, *2*, 271–308.
10. F. Barresinoussi, J.C. Chermann, F. Rey, M.T. Nugeyre, S. Chamaret, J. Gruest, C. Dauguet, C. Axlerblin, F. Vezinetbrun, C. Rouzioux, W. Rozenbaum, L. Montagnier, *Science*, **1993**, *220*, 868–871.
11. J.P. Horwitz, J. Chua, M. Noel, *J. Org. Chem.*, **1964**, *29*, 2076–2078.
12. T.S. Mansour, R. Storer, *Current Pharmaceutical Design*, **1997**, *3*, 227–264.
13. J. Balzarini, R.F. Schinazi, D. Kinchington, *Int. Antiviral News*, **1997**, *5*, 74–81.
14. J. Saunders, J.M. Cameron, *Med. Res. Rev.*, **1995**, *15*, 497–531.
15. E. De Clercq, *J. Med. Chem.*, **1995**, *38*, 2491–2517.
16. E. De Clercq, *Nucleosides & Nucleotides*, **1994**, *13*, 1271–1295.
17. E. Helgstrand, B.Öberg, *Antibiot. Chemother.*, **1980**, *27*, 22–69.
18. J.T. Witkowski, R.K. Robins, R.W. Sidwell, L.N. Simon, *J. Med. Chem.*, **1972**, *15*, 1150–1154.
19. R.A. Buchanan, F. Hess in "*Viral Chemotherapy*", D. Hugar (ed.), Pergamon Press, Oxford, **1985**, *2*, pp 109–143.
20. K.G. Cunningham, S.A. Hutchinson, W. Manson, F.S. Spring, *J. Chem. Soc.*, **1951**, 2299–2300.
21. H. Nakamura, N. Yagisawa, H. Shimada, T. Takita, H. Umezawa, Y. Iitaka, *J. Antibiot.*, **1981**, *34*, 1219–1221.
22. P. Van Roey, E.W. Taylor, C.K. Chu, R.F. Schinazi, *Ann. N.Y. Acad. Sci.*, **1990**, *616*, 29–39.
23. D.A. Pearlman, S.-H. Kim, *J. Biomol. Struct. Dyn.*, **1985**, *3*, 99–125.
24. W. Saenger in "*Principles of Nucleic Acid Structure*", Springer-Verlag, New York, **1984**, pp 9–104.
25. D.B. Davies, *Progress in NMR Spectroscopy*, **1978**, *12*, 135–225.
26. C. Thibaudeau, J. Chattopadhyaya, *Nucleosides & Nucleotides*, **1997**, *16*, 523–529.
27. J. Plavec, C. Thibaudeau, J. Chattopadhyaya, *Pure Appl. Chem.*, **1996**, *68*, 2137–2144.
28. C. Thibaudeau, J. Plavec, J. Chattopadhyaya, *J. Org. Chem.*, **1996**, *61*, 266–286.
29. J. Plavec, W.M. Tong, J. Chattopadhyaya, *J. Am. Chem. Soc.*, **1993**, *115*, 9734–9746.
30. C. Altona, M. Sundaralingam, *J. Am. Chem. Soc.*, **1973**, *95*, 2333–2344.
31. C. Altona, M. Sundaralingam, *J. Am. Chem. Soc.*, **1972**, *94*, 8205–8212.
32. W.K. Olson, *J. Am. Chem. Soc.*, **1987**, *104*, 278–286.
33. A.D. McNaught, *J. Carbohydr. Chem.*, **1996**, *16*, 1191–1280.
34. E. Font, S. Lasanta, O. Rosario, J.F. Rodriguez, *Nucleosides & Nucleotides*, **1998**, *17*, 845–853.
35. P.F. Crain, *Mass Spectrum Rev.*, **1990**, *9*, 505–554.
36. J.A. McCloskey, *Acc. Chem. Res.*, **1991**, *24*, 81–88.
37. M.S. Wilson, J.A. McCloskey, *J. Am. Chem. Soc.*, **1975**, *97*, 3436–3444.
38. S.J. Shaw, D.M. Desiderio, K. Tsuboyama, J.A. McCloskey, *J. Am. Chem. Soc.*, **1970**, *92*, 2510–2522.
39. L. Gatlin, J.C. Davis, *J. Am. Chem. Soc.*, **1964**, *84*, 4464–4470.
40. D.B. Davies, A. Rabczenko, *J. Chem. Soc., Perkin Trans. II*, **1975**, 1703–1711.
41. A.J. Jones, M.W. Winkley, D.M. Grant, R.K. Robins, *Proc. Natl. Acad. Sci. U.S.A.*, **1970**, *65*, 27–30.
42. A.J. Jones, D.M. Grant, M.W. Winkley, R.K. Robins, *J. Am. Chem. Soc.*, **1970**, *92*, 4079–4087.
43. H. Rosemeyer, G. Tóth, F. Seela, *Nucleosides & Nucleotides*, **1989**, *8*, 587–597.
44. O. Jardetzky, G.C.K. Roberts, "*NMR in Molecular Biology*", Academic Press, New York, New York, **1981**, pp 187–217.

45. R. Deslauriers, I.C.P. Smith, *Can. J. Chem.*, **1973**, *51*, 833–838.
46. F.E. Hruska, A. Mak, H. Singh, D. Shugar, *Can. J. Chem.*, **1973**, *51*, 1099–1106.
47. D.J. Wood, F.E. Hruska, K.K. Ogilvie, *Can. J. Chem.*, **1974**, *52*, 3353–3366.
48. D.B. Davies, S.S. Danyluk, *Biochemistry*, **1975**, *14*, 543–553.
49. E. De Clercq, *Aids Research and Human Retroviruses*, **1992**, *8*, 119–134.
50. H. Zalkin, J.E. Dixon, *Prog. Nucleic Acid Res.*, **1992**, *42*, 259.
51. J.G. Cory in "Biochemistry", L. Moran, K.G. Scrimgeour (eds.), Prentice Hall, **1996**, pp 489–523.
52. H. Mitsuya, K.J. Weinhold, P.A. Furman, M. H. St. Clair, S. Nusinoff Lehrman, R.C. Gallo, D. Bolognesi, D.W. Barry, S. Broder, *Proc. Natl. Acad. Sci. USA*, **1985**, *82*, 7096–7100.
53. H. Mitsuya, S. Broder, *Proc. Natl. Acad. Sci. USA*, **1986**, *83*, 1911–1915.
54. J. Balzarini, P. Herdewijn, E. De Clercq, *J. Biol. Chem.*, **1989**, *264*, 6127–6133.
55. T. Miyasaka, H. Tanaka, M. Baba, H. Hayakawa, R.T. Walker, J. Balzarini, E. De Clercq, *J. Med. Chem.*, **1989**, *32*, 2507–2509.
56. J. Balzarini, M.-J. Pérez-Pérez, A. San-Félix, D. Schols, C.-F. Perno, A.-M. Vandamme, M.-J. Camarasa, E. De Clercq, *Proc. Natl. Acad. Sci. USA*, **1992**, *89*, 4392–4396.
57. H-T. Ho, K.L. Woods, J.J. Bronson, H. De Boeck, J.C. Martin, M.J.M. Hitchcock, *Molec. Pharmacol.*, **1992**, *41*, 197–202.
58. E. De Clercq, R.T. Walker, *Prog. Med. Chem.*, **1986**, *23*, 187–218.
59. J. Balzarini, R. Bernaerts, A. Verbruggen, E. De Clercq, *Mol. Pharmacol.* **1990**, *37*, 402–407.
60. F.R. Collart, E. Hubermann, *J. Biol. Chem.*, **1988**, *263*, 15769–15772.
61. Y. Natsumeda, S. Ohno, S. Kawasaki, Y. Konno, G. Weber, K. Suzuki, *J. Biol. Chem.*, **1990**, *265*, 5292–5295.
62. R.C. Jackson, G. Weber, *Nature*, **1975**, *256*, 331–333
63. R.W. Sidwell, J.H. Huffman, G.P. Khare, L.B. Allen, J.T. Witkowski, R.K. Robins, *Science*, **1972**, *177*, 705–706.
64. F.R. Collart, C.B. Chubb, B.L. Mirkin, E. Hubermann, *Cancer Res.*, **1992**, *52*, 5826–5828.
65. J.S. Dayton, T. Linsten, C.B. Thompson, S.B. Mitchell, *J. Immunol.*, **1994**, *152*, 984–991.
66. J. Balzarini, C.-K. Lee, P. Herdewijn, E. De Clercq, *J. Biol.Chem.*, **1991**, *266*, 21509–21514.
67. N. Minakawa, T. Takeda, T. Sasaki, A. Matsuda, T. Ueda, *J. Med. Chem.*, **1991**, *34*, 778–786.
68. J.C. Sircar, R.B. Gilbertsen, *Drugs of the Future*, **1988**, *13*, 653.
69. E.R. Giblett, A.J. Ammann, D.W. Wara, R. Sandman, L.K. Diamond, *Lancet*, **1975**, *i*, 1010–1013.
70. M.L. Markert, *Immunodeficiency Rev.*, **1991**, *3*, 45.
71. P.M. Ueland, *Pharmacol. Rev.*, **1982**, *34*, 223–253.
72. E. De Clercq, M. Cools, J. Balzarini, V.E. Marquez, D.R. Borcherding, R.T. Borchardt, J.C. Drach, S. Kitaoka, T. Konno, *Antimicrob. Ag. Chemother.*, **1989**, *33*, 1291–1297.
73. M. Cools, J. Balzarini, E. De Clercq, *Mol. Pharmacol.*, **1991**, *39*, 718–724.
74. W.M. Shannon, *Ann. N.Y. Acad. Sci.*, **1977**, *284*, 472–507.
75. S. Shigeta, K. Konno, T. Yokota, K. Nakamura, E. De Clercq, *Antimicrob. Ag. Chemother.*, **1988**, *32*, 906–911.
76. E. De Clercq, R. Bernaerts, Y.F. Shealy, J.A. Montgomery, *Biochem. Pharmacol.*, **1990**, *39*, 319–325.

CHAPTER 2

SYNTHESIS OF CONVENTIONAL D-NUCLEOSIDES

2.1 INTRODUCTION

The first syntheses of the natural ribonucleosides were published in the late 1940s. Since this time considerable effort has been expended towards improving synthetic methodology in order to optimize yields and stereospecificity in the glycosylation reaction. This effort has resulted in some elegant methodology, involving both chemical and enzymatic coupling procedures, which can be applied to the syntheses of a broad range of nucleosides. Also of considerable importance was the establishment of a series of rules which explained the mechanisms of action of the coupling procedures and allowed the prediction of the stereochemistry of the nucleoside produced.

2.1.1 First Syntheses of the Natural Ribonucleosides

The first syntheses of uridine (**1.18**) and cytidine (**1.19**) employed the methodology of Hilbert and Johnson,[1] using 2,6-diethoxypyrimidine[2] (**2.1**) which was readily prepared by the reaction of 2,6-dichloropyrimidine with sodium ethoxide. Coupling of **2.1** with the acylated ribosyl bromide (**2.2**) produces the intermediate (**2.3**), which is unstable at 50°C and rapidly forms the nucleoside (**2.4**) with elimination of ethyl bromide. Treatment of **2.4** with methanolic ammonia results in simultaneous de-ethylation and deacetylation to give cytidine[3] (**1.19**), which can be converted to uridine (**1.18**) on reaction with cytidine deaminase[3] (Scheme 2.1).

The Fischer-Helferich procedure[4] was used for the first syntheses of adenosine[5] (**1.20**) and guanosine[6] (**1.21**) and involved the condensation of the silver salt of 2,8-dichloroadenine (**2.5**) with the acylated ribosyl chloride (**2.6**). The resulting nucleoside (**2.7**) was then deprotected to give 9-β-D-ribofuranosyl-2,8-dichloroadenine (**2.8**) and converted to adenosine (**1.20**) by dehalogenation (Scheme 2.2). Careful monitoring of the catalytic hydrogenation of **2.8** results in selective removal of the chlorine at C-8 (**2.9**). Diazotization of **2.9** with nitrous acid produces the diazonium ion (**2.10**) which undergoes unimolecular decomposition. Aqueous quenching of the resulting aryl cation (**2.11**) replaces the amino group with a hydroxyl group (**2.12**). Finally reaction of **2.12** with ethanolic ammonia produces guanosine (**1.21**) (Scheme 2.2).

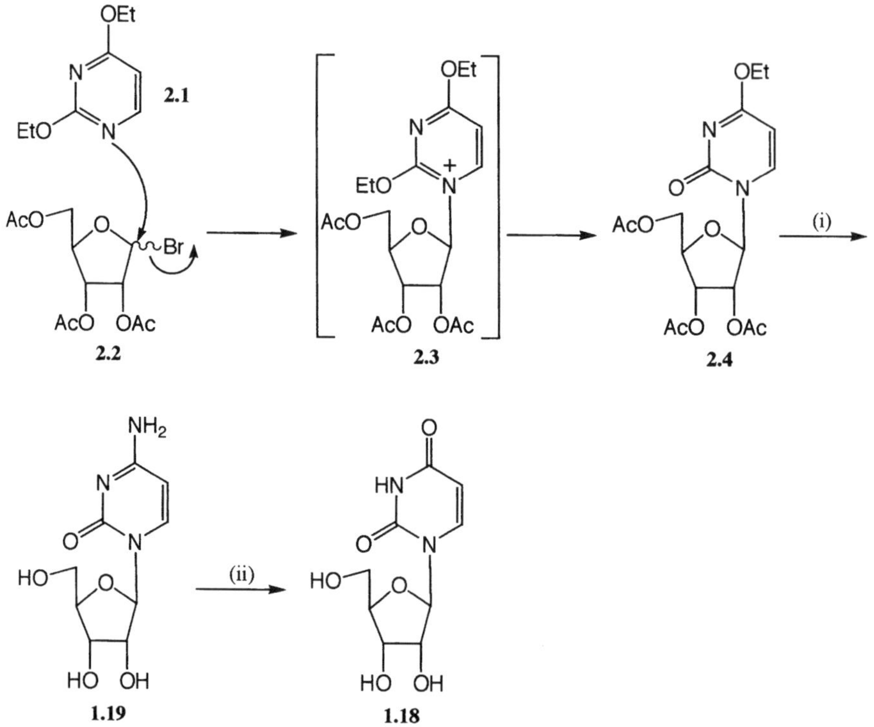

SCHEME 2.1 REAGENTS AND CONDITIONS: (I) MeOH, NH$_3$, 65°C (II) CYTIDINE DEAMINASE.

The stereochemistry of the glycosylation reaction follows the *trans rule*[7] which states that "condensation of a heavy metal salt of a purine or pyrimidine with an acylated glycosyl halide will form a C1–C2 *trans* configuration in the sugar moiety regardless of the original configuration at C1–C2".

2.1.2 Stereospecificity in Nucleoside Synthesis

The *trans rule* is used to explain the stereochemical control of the reaction between a poly-*O*-acyl glycosyl halide and heavy metal salts (Ag, Hg, Sn) of purines or pyrimidines. The nitrogen heterocycle will attach at C1 of the sugar so that it is *trans* to the C2 acyloxy group regardless of the anomeric configuration of the glycosyl halide.

If the poly-*O*-acyl-D-glycosyl halide (**2.13**) has a C1–C2 *cis* configuration, the nitrogen heterocycle (B) will displace the halogen atom with S_N2/Walden inversion producing the *trans* β-nucleoside (**2.14**). With the C1–C2 *trans* poly-*O*-acyl-D-glycosyl halide (**2.15**), the *trans* nucleoside (**2.14**) is still produced *via* the carbonium ion intermediate (**2.16**). With the α-face of the sugar blocked, attack by the purine or pyrimidine is from the top face resulting in the formation of the *trans* β-nucleoside (**2.14**) (Scheme 2.3).

This carbonium ion intermediate (**2.16**) is also observed in the reaction of poly-*O*-acyl-D-glycosides (**2.17**) with nitrogen heterocycles (Scheme 2.3). Coupling of

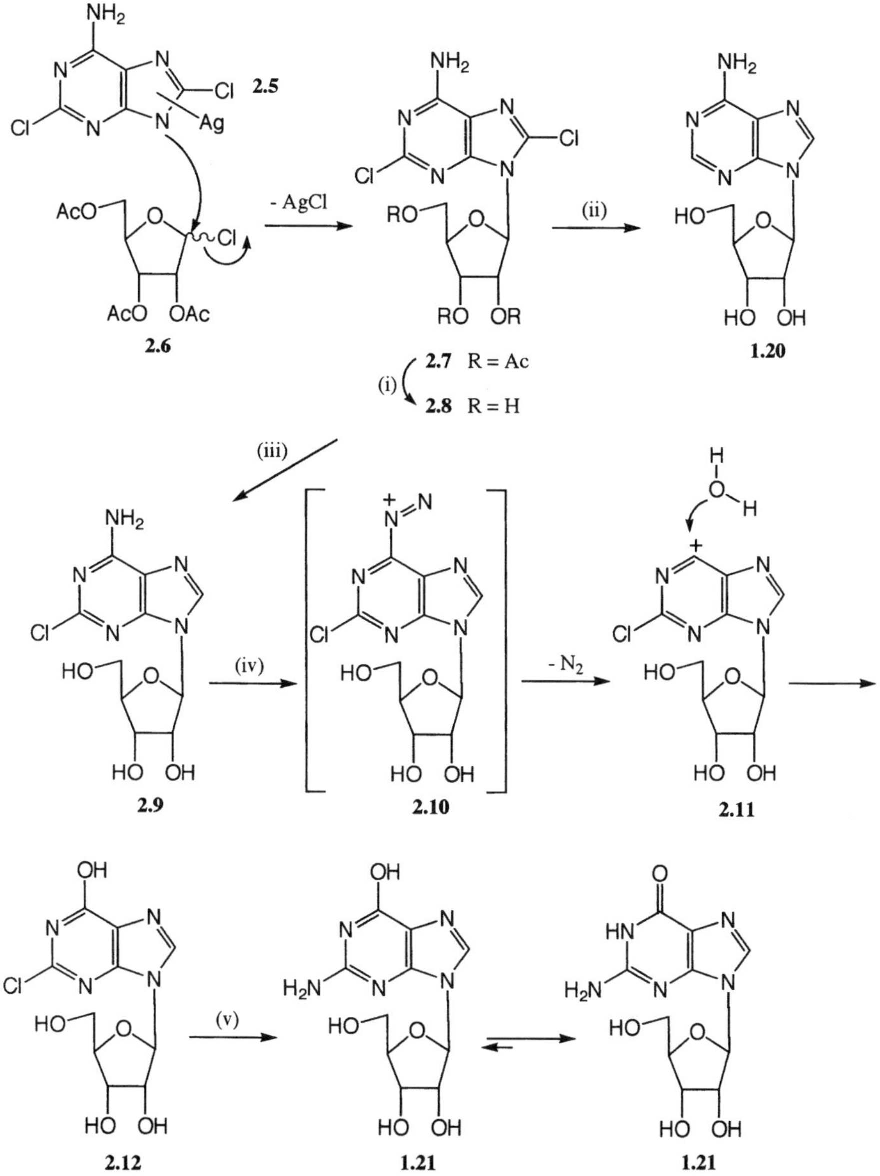

SCHEME 2.2 REAGENTS AND CONDITIONS: (I) MeOH, NH₃ (II) H₂, Pd/BaSO₄, NaOH, H₂O, 7 h (III) H₂, Pd/BaSO₄, NaOH, H₂O, 9 min. (IV) NaNO₂, AcOH, H₂O (V) EtOH, NH₃, 150°C.

D-ribose sugars with a pyrimidine or purine results almost exclusively in formation of the β-nucleoside. This is a result of the neighbouring group participation of the C2-acyloxy group which directs the position of attack of the incoming nitrogen heterocycle anion.

In accordance with the *trans rule*, coupling of an acylated D-arabinosyl halide (**2.18**) or acetate with a nitrogen heterocycle will produce the α-nucleoside (**2.20**)

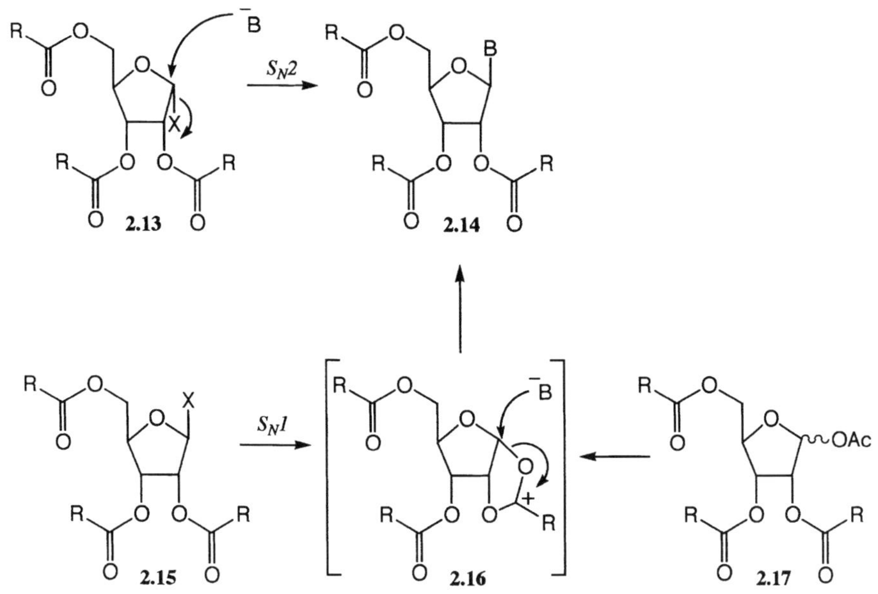

SCHEME 2.3 S_N1 AND S_N2 MECHANISMS IN RIBONUCLEOSIDE SYNTHESIS.

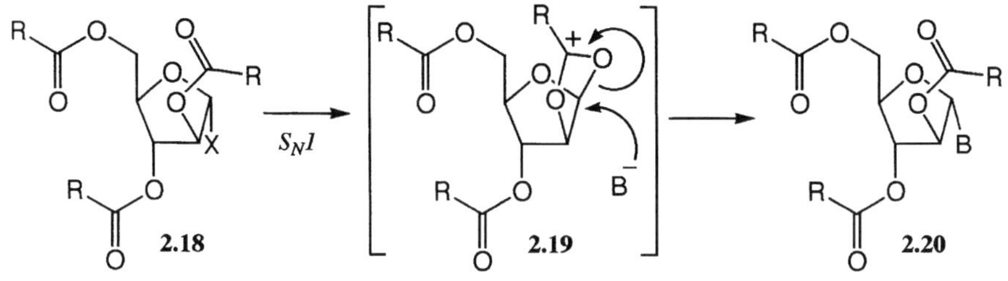

SCHEME 2.4

with C1–C2 *trans* configuration, *via* the intermediate carbonium ion (**2.19**) (Scheme 2.4). The nucleoside formed on reaction of a nitrogen heterocycle with a sugar lacking a C2 acyloxy or directing group is determined by the sugar used. If a 2'-deoxy-D-glycosyl halide (**2.21α** or **2.21β**) is used, the reaction proceeds by S_N2 displacement with the nucleoside formed with inversion of configuration, *i.e.* **2.21α→2.22β** and **2.21β→2.21α** (Scheme 2.5). However if a 1-*O*-acyl-2-deoxy-D-glycoside (**2.23**) is used, the reaction proceeds *via* a S_N1 mechanism with the formation of an oxonium intermediate (**2.24**). The nucleophilic nitrogen heterocycle can attack **2.18** from either face of the sugar ring resulting in the formation of the nucleoside as an anomeric mixture, *i.e.* **2.23→2.22α** and **2.22β** (Scheme 2.5).

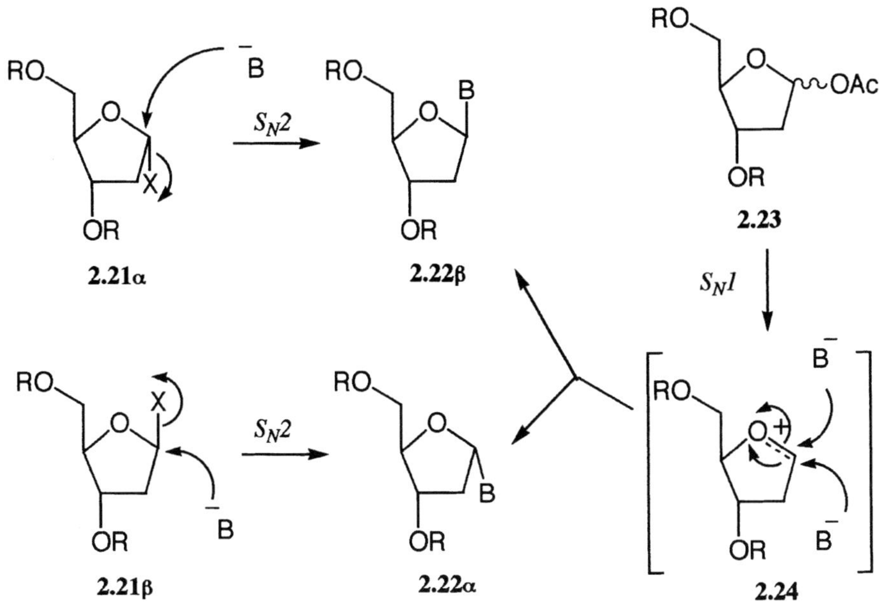

SCHEME 2.5

2.2 COUPLING METHODS

There are five main methods for the synthesis of conventional D-nucleosides (Fig. 2.1, appropriate section indicated). The choice of method is dependent on the sugar/heterocyclic base employed, the stability of the nucleoside towards the reagents employed, and the stereo- and regio-selectivity required.

2.2.1 Coupling of Heterocyclic Salts with Halo-Sugars

The use of purine/pyrimidine salts in coupling reactions with sugars is of most benefit when the sugar employed is a 1-halo-2-deoxy-furanoside, as this method allows stereospecific formation of β-nucleosides *via* direct Walden inversion at the anomeric carbon by the anionic heterocyclic nitrogen. The earlier methods of nucleoside synthesis primarily involved heavy metal salts, which are still of value but are generally no longer employed owing to the costs and toxicity associated with the heavy metals themselves.

2.2.1.1 Chloromercuri procedure

The silver salt methodology was superseded by the chloromercuri method as a result of studies involving the reactions of various metal derivatives of purines with acetylglycosyl halides in boiling xylene[8]. The chloromercuri derivatives were found to be the most satisfactory for glycosyl purine synthesis with improved yields (30–40%) of nucleosides. An example of the synthesis of a purine[9] nucleoside using this methodology is shown in Scheme 2.6.

SYNTHESIS OF CONVENTIONAL D-NUCLEOSIDES

SECTION

2.2.1

2.2.2

2.2.3

2.2.4

2.2.5

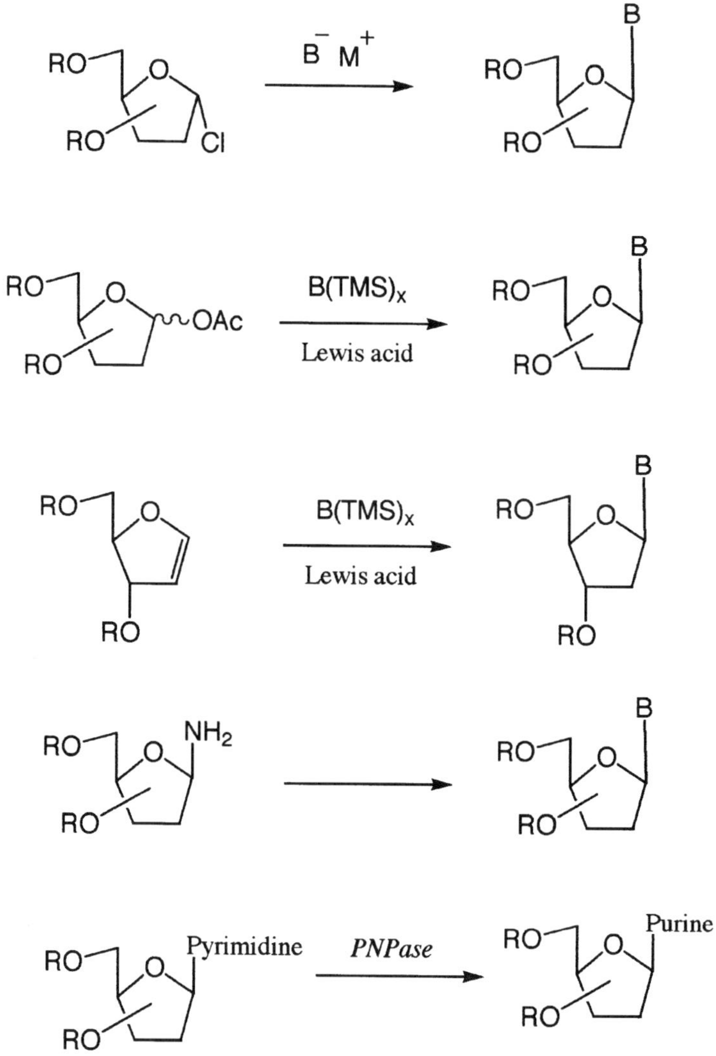

FIGURE 2.1   COUPLING METHODS OF CONVENTIONAL D-NUCLEOSIDES (B = A, G, U, C, T).

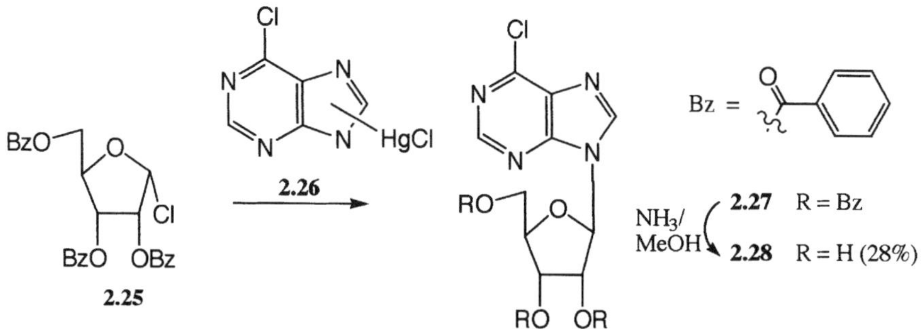

SCHEME 2.6   CHLOROMERCURI-PROCEDURE

2.2.1.2 Sodium-salt-glycosylation procedure

The use of sodium purine/pyrimidine salts was an improvement over the Fischer–Helferich and chloromercuri procedures, as it avoided the use of expensive silver reagents and toxic mercuric reagents, with considerably higher yields obtained. The chlorosugar generally employed for the synthesis of 2'-deoxynucleosides is 1-chloro-2-deoxy-3,5-di-*O*-*p*-toluoyl-α-D-*erythro*-pentofuranoside (**2.29**),[10] and the sodium salt is prepared *in situ* by reaction of the heterocyclic base with sodium hydride in acetonitrile. Reaction of the chlorosugar **2.29** with the sodium salt of 6-chloropurine (**2.30**) produced a mixture of N^9- and N^7-nucleosides (**2.31** and **2.32** respectively), with the N^9-isomer being the major product (59%).[11] Treatment of **2.31** with methanolic ammonia resulted in deprotection of the sugar moiety with concomitant nucleophilic displacement of the 6-chloro group to give 2'-deoxyadenosine (**1.24**) (Scheme 2.7).

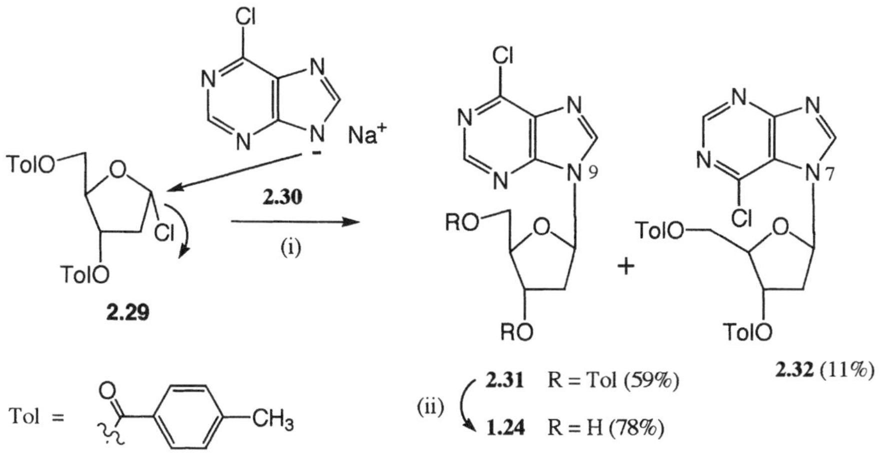

SCHEME 2.7 REAGENTS AND CONDITIONS: (I) CH_3CN, 50°C (II) MeOH, NH_3 100°C.

2.2.2 Vorbrüggen Procedure

The reaction of persilylated heterocyclic bases with peracylated sugars in the presence of a Lewis acid catalyst such as $SnCl_4$[12] or $(CH_3)_3SiSO_3CF_3$ (TMSOTf)[13] has become a routine synthetic method for the synthesis of purine and pyrimidine nucleosides. Silylation of the base can be achieved with either hexamethyldisilazane[12] (HMDS) or *N*,*O*-bis(trimethylsilyl) acetamide[14] (BSA), though the latter is preferred owing to the difficulties in removing the excess HMDS which, if still present in the reaction mixture, can inactivate the catalyst. TMSOTf is often the preferred catalyst, in practical terms it avoids the problems associated with tin residues during the work-up procedure.

2.2.2.1 Mechanism of Pyrimidine Synthesis

The mechanism[15] of pyrimidine nucleoside synthesis employing the Vorbrüggen procedure first involves the conversion of the peracylated sugar (**2.33**), on reaction

SYNTHESIS OF CONVENTIONAL D-NUCLEOSIDES

with the Lewis acid (TMSOTf), into the stable 1,2-acyloxonium salt (**2.34**), with generation of triflate anion ($CF_3SO_3^-$) and silylated acetic acid. The silylated pyrimidine (**2.35**) reacts with the triflate anion, to regenerate TMSOTf, with simultaneous attack of the electrophilic sugar cation to give the nucleoside **2.36**. The remaining trimethylsilyl (TMS) group is readily cleaved during aqueous work-up (Scheme 2.8).

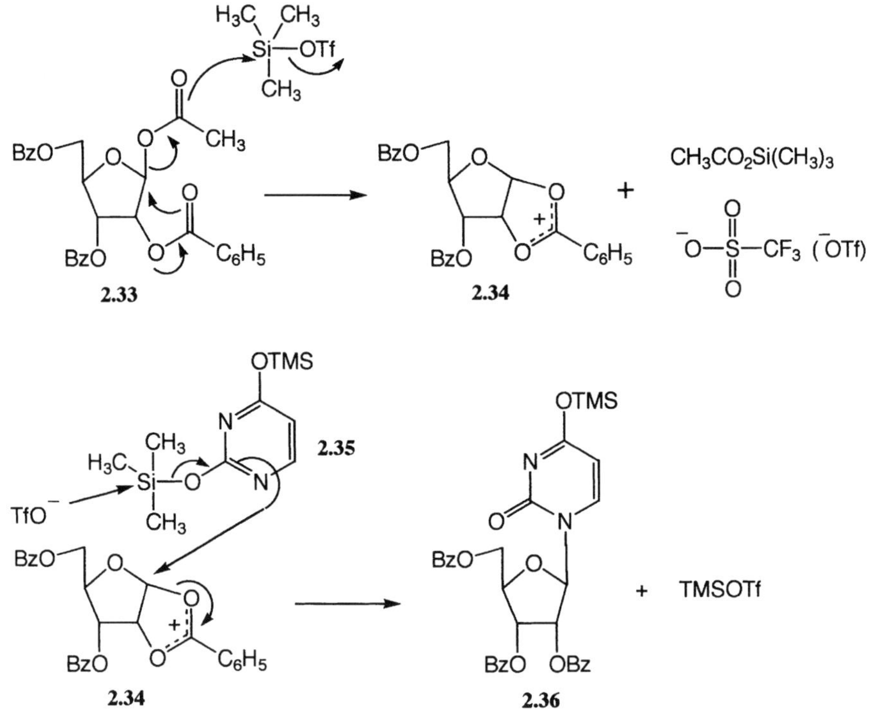

SCHEME 2.8   MECHANISM OF VORBRÜGGEN PYRIMIDINE NUCLEOSIDE SYNTHESIS.

2.2.2.2   Mechanism of Purine Synthesis

This method has been applied to the regioselective synthesis of N^9-guanosine,[16] and involves a mechanism[15] similar to that described for pyrimidine synthesis. In contrast to pyrimidine glycosylations, where the protocols for regiochemical control have been established, regiochemical control of purine glycosylations is more problematic, often yielding unfavourable mixtures of N^7- to N^9- (desired) isomers if $SnCl_4$ is used as the Lewis acid.[15] However the use of TMSOTf minimizes the production of the N^7-product and in conjunction with the influence of suitable protecting groups can eliminate N^7-product formation.

Silylated 2-*N*-acetyl-6-*O*-diphenylcarbamoylguanosine (**2.37**) was reacted with tetra-*O*-acetyl-D-ribose (**2.38**) in the presence of catalytic TMSOTf (Scheme 2.9) to give the N^9-nucleoside **2.40**, which after aqueous work-up and deprotection of the acetyl and diphenylcarbamoyl groups with methanolic ammonia gave N^9-

guanosine (**1.21**). This reaction represents a good example of the use of protecting groups to enhance regioselectivity in nucleoside synthesis.

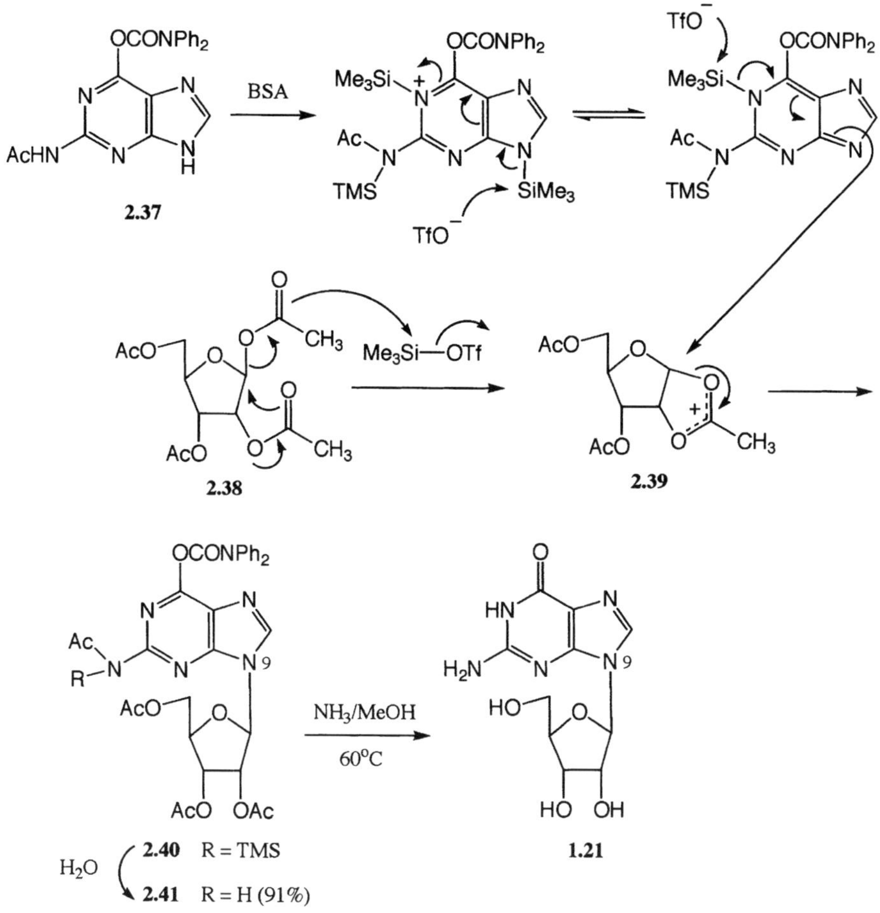

SCHEME 2.9 MECHANISM OF VORBRÜGGEN PURINE NUCLEOSIDE SYNTHESIS.

2.2.2.3 Synthesis of 2'-Deoxynucleosides

Application of the Vorbrüggen procedure to the synthesis of 2'-deoxynucleosides would be expected to result in anomeric mixtures (see Scheme 2.5, section 2.1.2). This can be prevented with the use of suitable protecting groups in the 3-position of the sugar, which are capable of 'directing' the approach of the heterocyclic base. The thiocarabamate protecting group has been found to act as a directing group, reaction of the thiocarbamate sugar (**2.42**) with TMSOTf results in formation of the intramolecular iminium ion intermediate (**2.43**).

Nucleophilic attack by the silylated heterocyclic base is then restricted to the top face of the sugar, with the β-nucleoside (**2.44**) being the major product formed (α/β = 4/96).[17] This reaction represents a good example of the use of protecting groups to influence the stereoselectivity of a glycosylation reaction (Scheme 2.10).

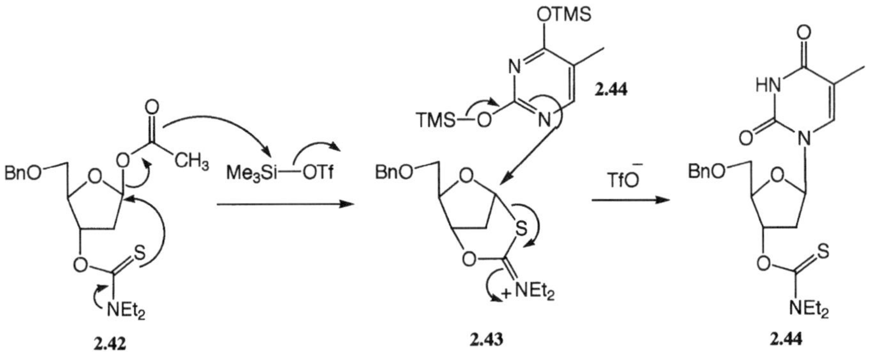

SYNTHESIS OF CONVENTIONAL D-NUCLEOSIDES

SCHEME 2.10 DIRECTING GROUPS IN 2'-DEOXYNUCLEOSIDE SYNTHESIS (Bn = $C_6H_5CH_2$).

2.2.3 Nucleoside Synthesis from Glycals

The use of the furanoid glycal procedure is of value in the convergent synthesis of 2'-deoxynucleosides as high regiospecificity and stereoselectvity often results with this method. This method relies on the addition of an electrophilic reagent across the C1–C2 double bond, producing an intermediate with the α-face blocked.

Compounds in which selenium and sulphur atoms are bound to more electrophilic elements can react with glycals to give addition products. The mechanism is similar to that in halogenation of alkenes, with attack on selenium or sulphur by the alkene π-electrons to form a bridged cationic intermediate.[18,19] N-Iodosuccinimide (NIS) reacts with glycals in a similar fashion forming an iodinium cationic intermediate complexed with succinimide anion[20] (Scheme 2.11).

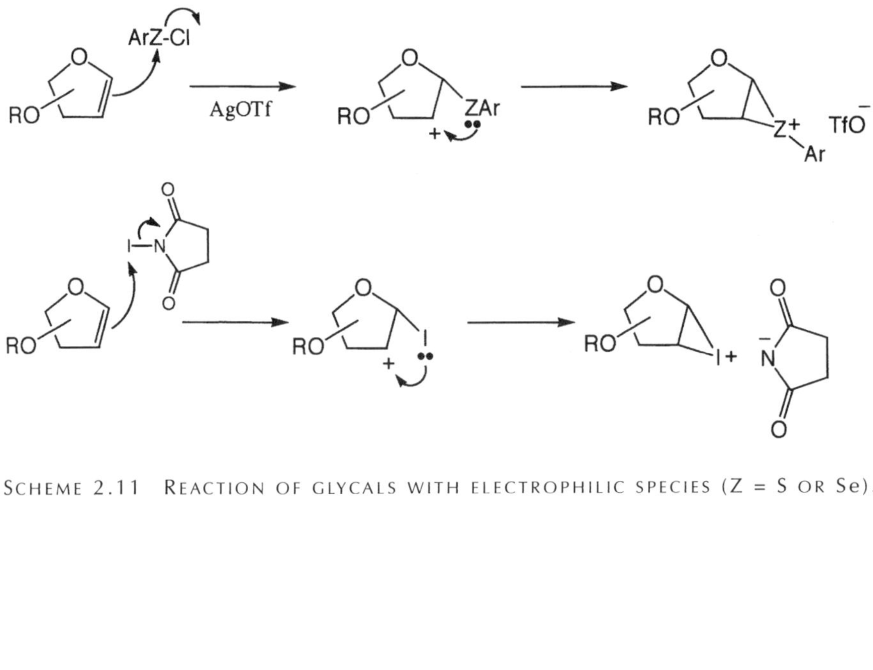

SCHEME 2.11 REACTION OF GLYCALS WITH ELECTROPHILIC SPECIES (Z = S OR Se).

2.2.3.1 Selenium-Mediated Reactions

The selenium mediated glycosylation reactions of glycals require that the hydroxyl group in the 3-position is 'ara', *i.e.* above the plane of the sugar ring. This is necessary to direct the formation of the selenium cation intermediate on the α-face of the sugar. Reaction of the glycal **2.45** with phenylselenyl chloride in the presence of the efficient chlorine activator silver triflate (AgOTf) leads to formation of the selenium cation intermediate (**2.46**) on the face opposite the substituent at position 3.[18] Nucleophilic attack by a silylated thymine produced predominantly the β-nucleoside (**2.47**, α : β = 1 : 9) with the phenylselenyl group removed with tributyltin hydride. The observed stereoselectivity of the nucleoside product **2.49** is owing to anchimeric assistance provided by the selenyl group in the glycosylation step (Scheme 2.12).

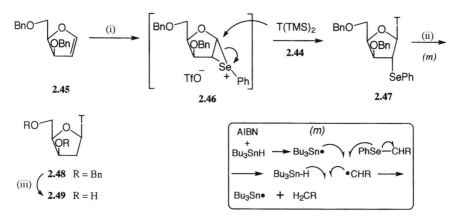

SCHEME 2.12 REAGENTS AND CONDITIONS: (I) PhSeCl, AgOTf, CH$_2$Cl$_2$, **2.44** (II) Bu$_3$SnH, AIBN, TOLUENE, REFLUX (III) H$_2$, Pd-C.

2.2.3.2 Sulphur-Mediated Reactions

Sulphur mediated glycosylation reactions of glycals rely on the formation of a S-Sn cation complex to control the direction of glycosylation. Reaction of the glycal **2.50** with phenylsulphenyl chloride (PhSCl) results in the formation of the cation intermediate (**2.51**), the use of SnCl$_4$ as the Lewis acid leads to the formation of a tin-sulphur complex (**2.52**) which amplifies the size of the group *in situ*, the bulk of this complex restricts the cation to the least hindered face, the α-face.[21] Glycosylation with silyl protected *N*-acetyl-cytosine (**2.53**) is then directed to the opposite face, giving predominantly the required β-nucleoside (**2.54**, α : β = 1 : 23), oxidation, elimination and subsequent deprotection of the phenylsulphenyl group gives the 2',3'-dideoxynucleoside **1.27** (Scheme 2.13).[19]

2.2.3.2 NIS-Mediated Reactions

The sheer bulk of the succinimide counter ion of the iodinium cationic intermediate is sufficient to result in its incorporation at the least hindered face, directing any subsequent attack by the heterocyclic base from the opposite face. This is shown in the reaction of the 5-pivaloyl-protected glycal (**2.57**) with *N*-iodosuccinimide

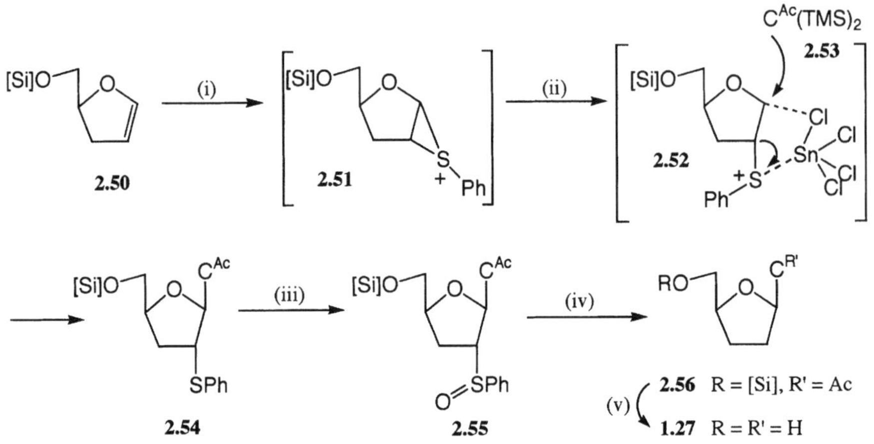

SYNTHESIS OF CONVENTIONAL D-NUCLEOSIDES

SCHEME 2.13 REAGENTS AND CONDITIONS: (I) PhSCl, CH_2Cl_2 (II) $SnCl_4$, **2.53** (III) $NaIO_4$ (IV) TOLUENE, PYRIDINE, REFLUX (V) DEPROTECTION. [Si] = TBDMS = TERT-BUTYLDIMETHYLSILYL.

followed by reaction with silylated thymine (**2.44**), resulting in exclusive formation of the β-nucleoside (**2.59**, B = thymine) *via* the intermediate **2.58**.[20] The 2-iodo group can then be removed by catalytic hydrogenation (Scheme 2.14).

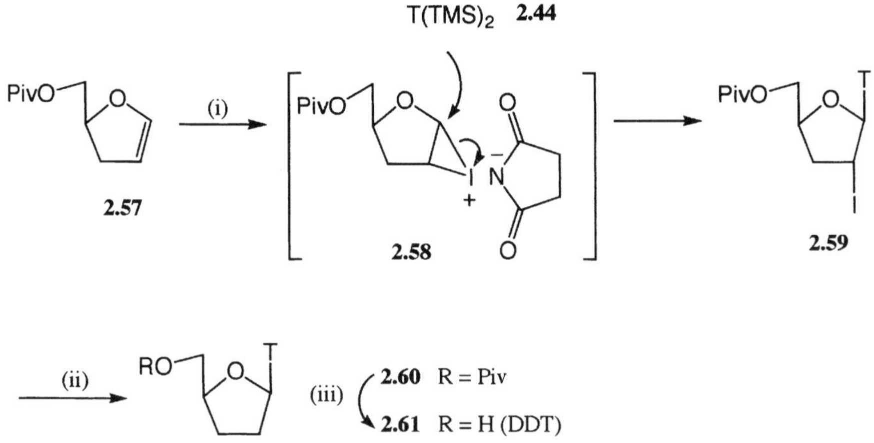

SCHEME 2.14 REAGENTS AND CONDITIONS: (I) NIS, **2.44**, THF (II) H_2, Pd-C, EtOH (III) NaOMe, MeOH. (Piv = PIVALOYL $(CH_3)_3C=O$).

2.2.4 Glycosylamine Procedure

This is a non-convergent method which involves building the pyrimidine and purine heterocycles directly onto a 1-aminosugar. This methodology has been used extensively in the synthesis of carbocyclic nucleosides (see Chapter 7).

2.2.4.1 Pyridine synthesis from 1-aminosugars

The amino sugar (**2.63**) can be prepared by treatment of the chlorosugar (**2.25**) with sodium azide followed by catalytic hydrogenation.[22] Reaction of **2.63** with 2-cyano-3-ethoxy-*N*-(ethoxycarbonyl)acrylamide (**2.64**)[23] gives the acylic intermediate (**2.65**), which cyclizes on acidification with acetic acid to give the 5-cyano-uridine derivative **2.66**[24] (Scheme 2.15).

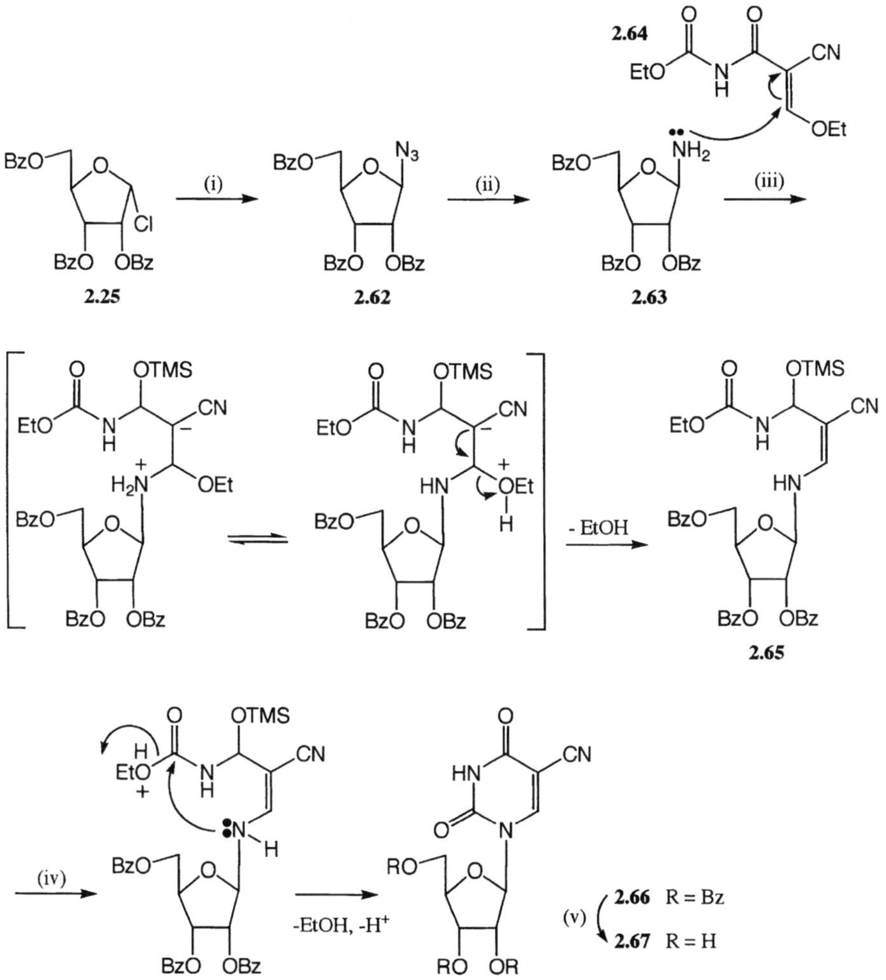

SCHEME 2.15 REAGENTS AND CONDITIONS: (I) NaN$_3$, DMF (II) H$_2$, Pd-C (III) EtOH, REFLUX (IV) AQUEOUS AcOH (V) NaOMe, MeOH.

2.2.4.2 Purine synthesis from 1-aminosugars

Purines can be prepared by reaction of an appropriate amino sugar (**2.68**, R = protected ribose) with 2-amino-4,6-dichloropyrimidine (**2.69**).[25] Coupling of 4-chlorobenzene diazonium chloride with the pyrimidine (**2.70**) in buffered aqueous acetic acid, results in electrophilic substitution at C-5 to give the azo compound (**2.71**) which is then reduced to the 5-aminopyrimidine (**2.72**). The 2-amino-6-chloropurine (**2.74**) is obtained *via* the intermediate (**2.73**) by acid catalyzed reaction of **2.72** with triethylorthoformate[26,27] (Scheme 2.16).

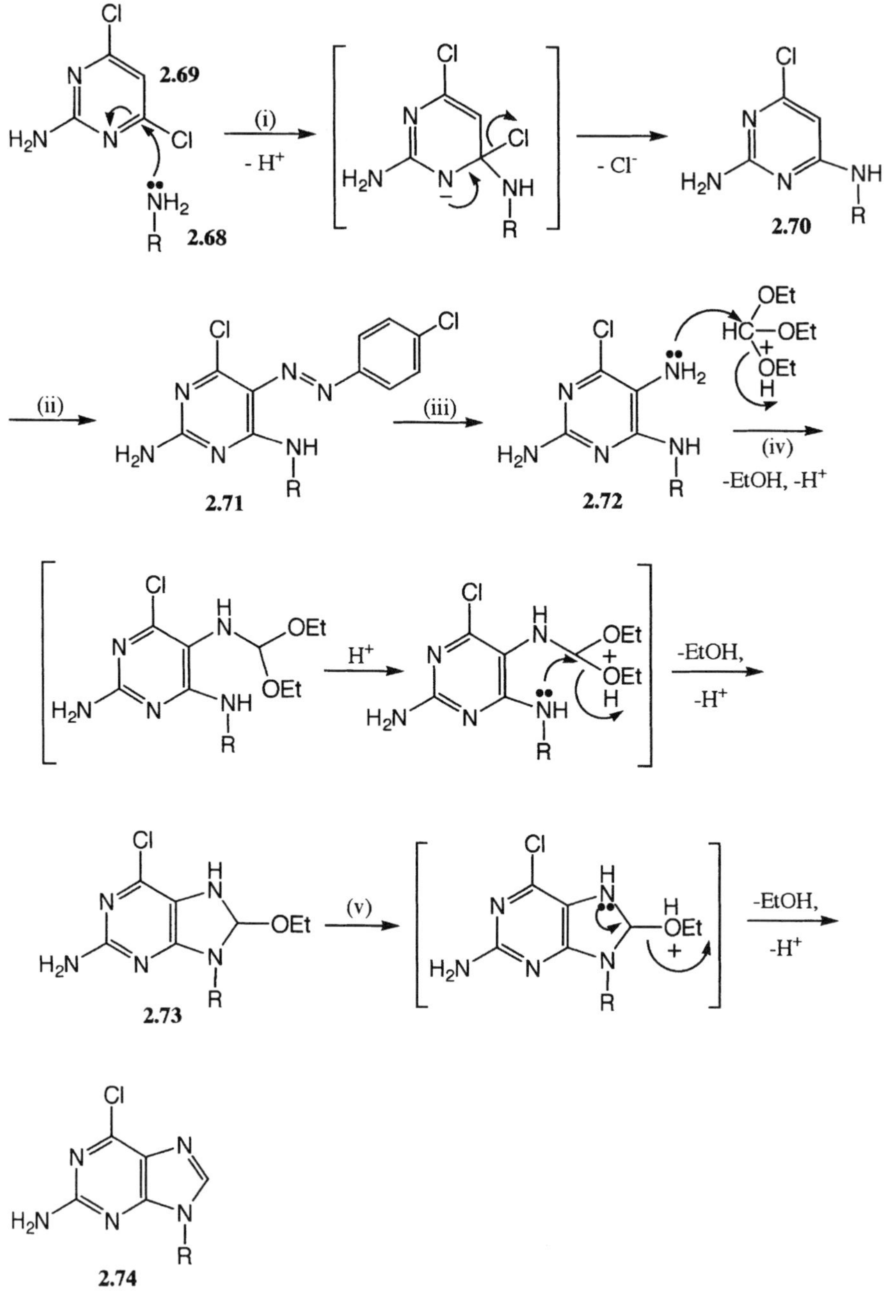

SYNTHESIS OF CONVENTIONAL D-NUCLEOSIDES

SCHEME 2.16 REAGENTS AND CONDITIONS: (I) Et_3N, BuOH, REFLUX (II) 4-Cl-C_6H_4-N_2^+ Cl^-, NaOAc, AcOH, H_2O (III) Zn, AcOH, EtOH, 70°C (IV) $(EtO)_3CH$, DMA, c.HCl (V) 50% AQUEOUS AcOH.

2.2.5 Enzymatic Synthesis

Enzymatic synthesis is primarily directed towards the synthesis of purine nucleosides and in particular 2'-deoxy or 2', 3'-dideoxypurine nucleosides. This method[28,29] employs two nucleoside phosphorylases, isolated from *Eschirichia coli*, as catalysts and pyrimidine nucleosides as pentofuranosyl donors.

The ribose-1-phosphate (**2.75**) is generated *in situ* from the thymidine nucleoside (**1.22**, R = OH, **2.61**, R = H) and inorganic phosphate by thymidine phosphorylase (TPase, EC 2.4.2.4). Purine nucleoside phosphorylase (PNP, EC 2.4.2.1) catalyses the formation of the new glycosyl bond between **2.75** and the purine base (**2.76**). The net result is the transfer of the sugar moiety from the thymidine nucleoside (**1.22/2.61**) to **2.77/2.78** with retention of the β-configuration in the nucleoside (Scheme 2.17).

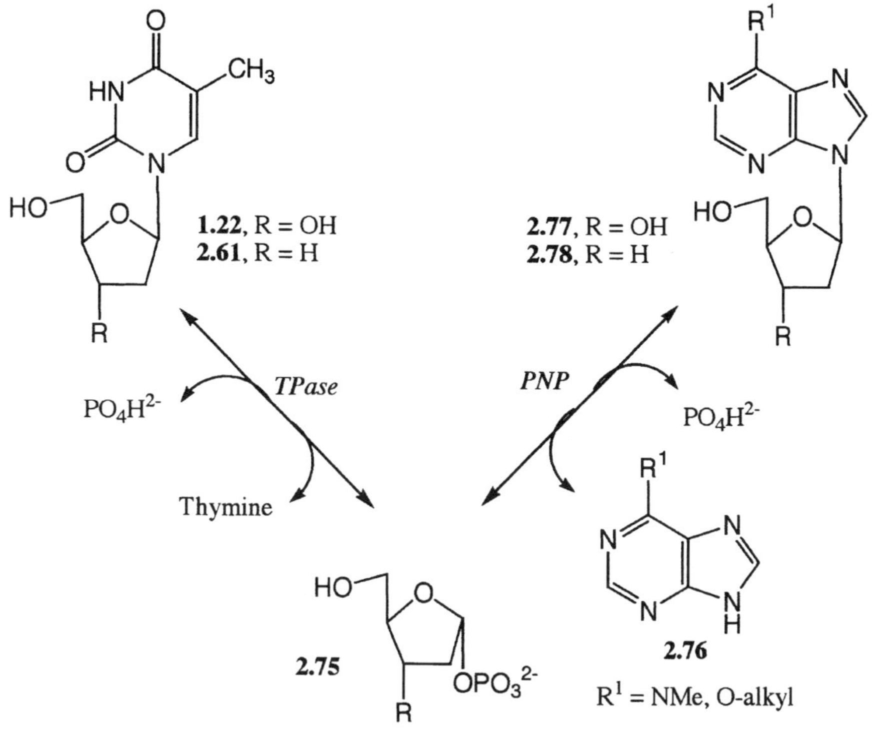

SCHEME 2.17 ENZYMATIC NUCLEOSIDE SYNTHESIS EMPLOYING TWO ENZYMES.

Hennen and Wong have described an enzymatic procedure which employs either N^7-methylated inosine or guanosine as activated ribosyl donors (Scheme 2.18).[30] This method is an improvement on many of the other enzymatic syntheses as the leaving group, the N^7-alkylated purine, undergoes isomerization rendering the reaction essentially irreversible. The reaction also only requires the use of one enzyme, purine nucleoside phosphorylase.

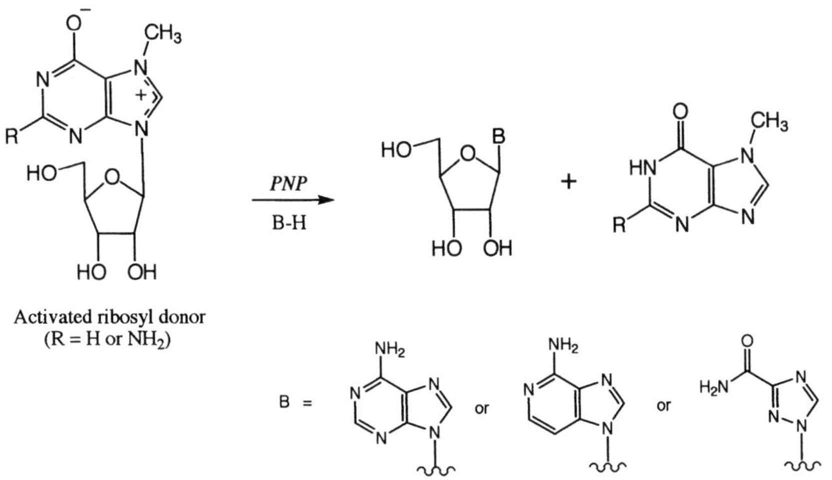

SCHEME 2.18 ENZYMATIC NUCLEOSIDE SYNTHESIS EMPLOYING ACTIVATED RIBOSYL DONORS.

2.3 SUGAR MODIFICATIONS

Modification of the sugar moiety of nucleosides may produce marked changes in their spectrum of biological activity and degree of selective toxicity, as well as in their chemical and physical properties. In particular, modification of the 2'- and 3'-positions has resulted in compounds with a broad range of biological activity.

Modification of the sugar moiety with either alteration or retention of the configuration of the sugar is readily achieved either *via* nucleophilic opening of anhydro-pyrimidine nucleosides or epoxides of purine nucleosides, as well as by a range of other methods.

2.3.1 Sugar Modifications via Anhydronucleosides

The anhydronucleosides are formed on reaction with a base, which results in deprotonation of NH-3 followed by displacement of a sulphonate group at either the 2'- or 3'-position. Treatment of **2.79** with sodium hydroxide gives the 2, 2'-anhydro derivative (**2.80**), which undergoes nucleophilic opening on reaction with a further equivalent of base. Detritylation of **2.81** by treatment with acetic acid gives the anticancer agent Ara-FU[31] (**2.82**) (Scheme 2.19).

Conversion to the lyxo-form (**2.88**) from the 2', 3'-dimesylate (**2.83**) proceeds through the 2, 2'-anhydro (**2.84**) with displacement of the 2'-OMs, followed by displacement of the 3'-OMs to form the 2, 3'-anhydro derivative (**2.86**). Attack of the 2'-hydroxy group on the carbonium ion generated at C2, results in the formation of the 2, 2'-anhydro (**2.87**) which is finally hydrolyzed to the lyxo-nucleoside[32] (**2.88**) (Scheme 2.20).

Anhydronucleosides are extremely versatile intermediates providing access to a wide variety of 2'- and/or 3'-substituted pyrimidines (Scheme 2.21).[33] Nucleophilic opening occurs with overall retention of configuration *via* a double Walden inversion from **2.89** to **2.91**.

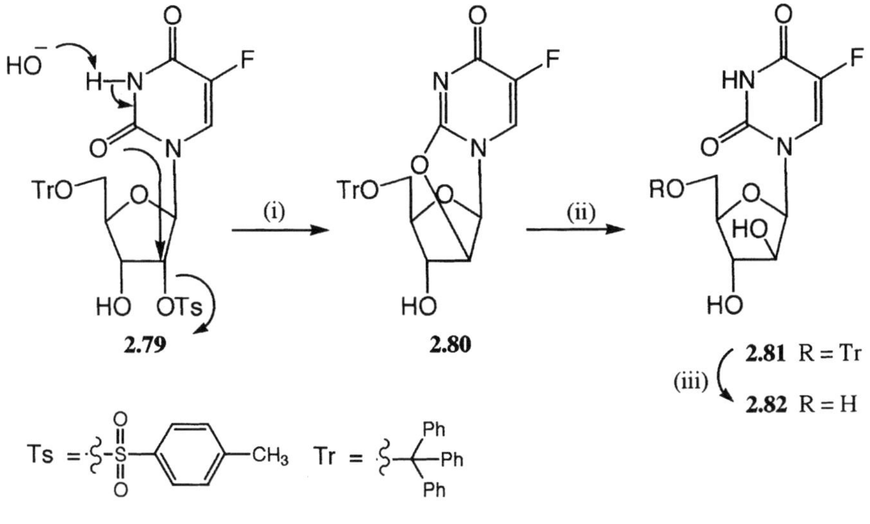

SYNTHESIS OF CONVENTIONAL D-NUCLEOSIDES

SCHEME 2.19 REAGENTS AND CONDITIONS: (I) NaOH, EtOH, REFLUX (II) 0.5N NaOH (III) 80% AcOH, REFLUX.

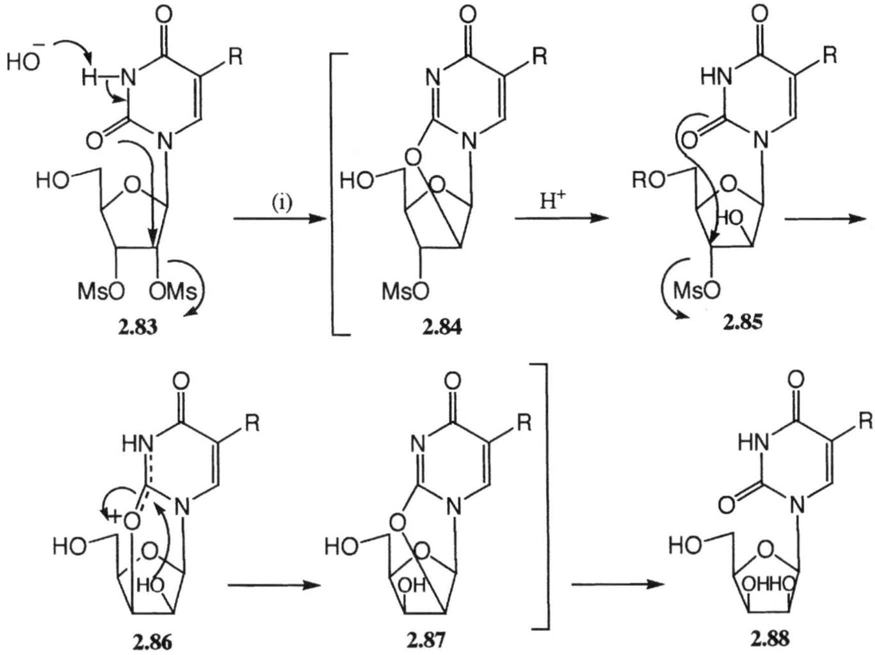

SCHEME 2.20 REAGENTS AND CONDITIONS: (I) H_2O, REFLUX. (Ms = $CH_3SO_2^-$).

NUCLEOSIDE MIMETICS 45

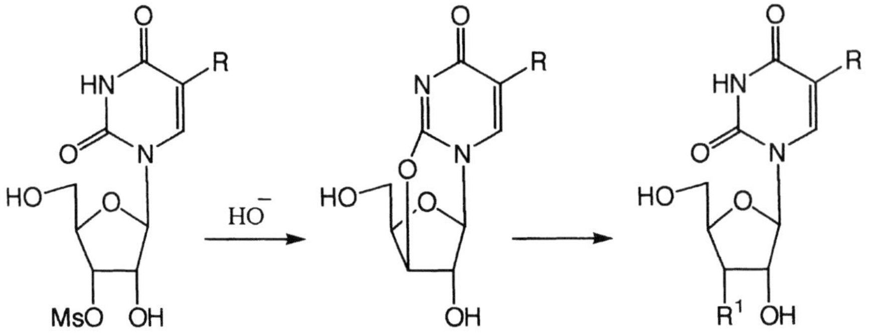

SYNTHESIS OF CONVENTIONAL D-NUCLEOSIDES

SCHEME 2.21 NUCLEOPHILIC OPENING OF A 2,3'-ANHYDRONUCLEOSIDE (R = H, F, CH$_3$; R^1 = N$_3$, NH$_2$, S-alkyl, Cl, Br, F).

2.3.2 Sugar Modifications via Epoxides

This method is readily applicable to purine nucleosides with the epoxides formed by reaction of a sulphonyl nucleoside with base, such as in the preparation of the anti-HSV and anticancer agent 1-β-D-arabinosyladenine (**1.10**, Ara-A, Vidarabine®)[34,35] from the corresponding 2'-O-mesyl-xylose derivative (**2.92**)[36] (Scheme 2.22).

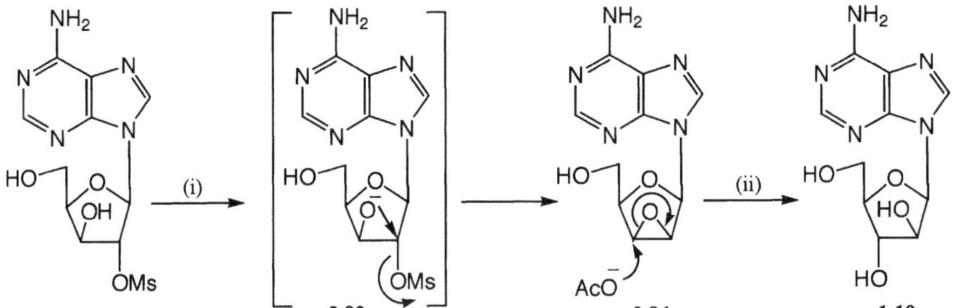

SCHEME 2.22 REAGENTS AND CONDITIONS: (I) NaOMe, MeOH, REFLUX (II) NaOAc, 95% AQUEOUS DMF, 156°C.

The epoxides, like the anhydronucleosides, are also very versatile intermediates, nucleophilic opening of either the 'up' or 'down' epoxide ring occurs predominantly at the 3'-position, however the down epoxide is far less reactive requiring strong nucleophiles to obtain ring opening (Scheme 2.23).[37]

2.3.3 Deoxygenation

Radical-induced reductive cleavage allows efficient deoxygenation of secondary alcohols, as shown for the conversion of adenosine (**1.20**) to 2'-deoxyadenosine (**1.24**).[38] The 3'- and 5'-hydroxy groups of **1.20** were first protected with Markiewicz

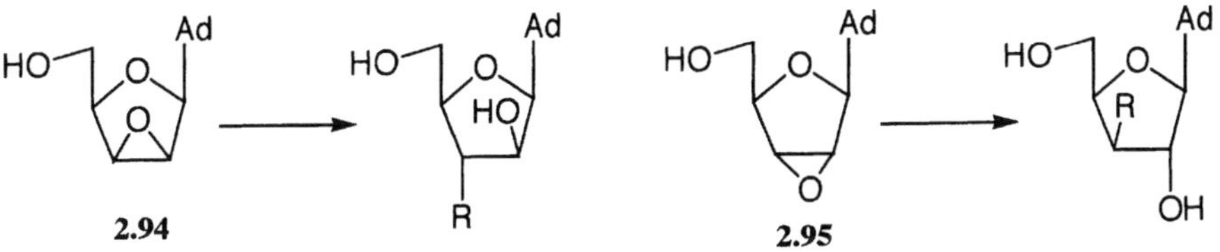

SCHEME 2.23 NUCLEOPHILIC EPOXIDE OPENING (R = N_3, NH_2, S-alkyl, Cl, Br, F).

reagent,[39] and the resulting silylated nucleoside **2.96** reacted with phenylchlorothionocarbonate in the presence of DMAP to give the thionoester **2.97**. Deoxygenation with Bu_3SnH following the Barton-McCombie procedure,[40] followed by removal of the silyl protecting group gives **1.24** in 78% overall yield from **1.20** (Scheme 2.24).

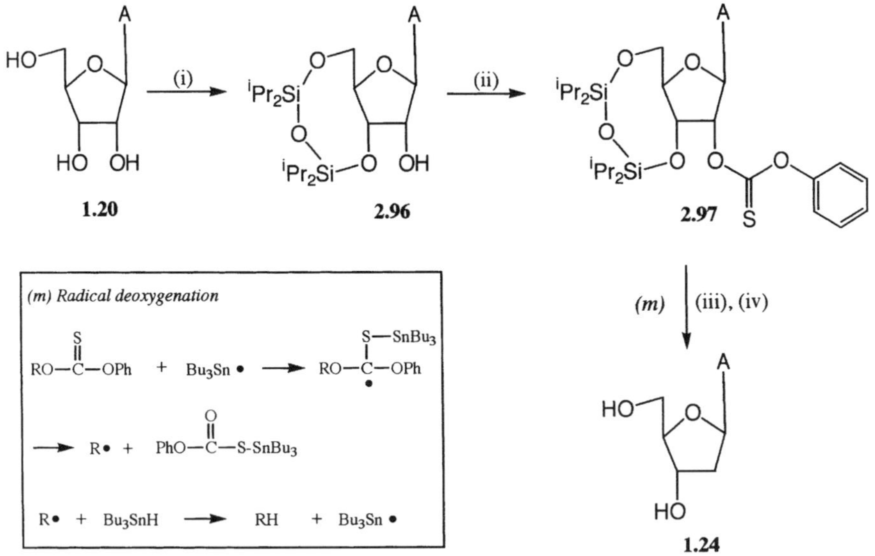

SCHEME 2.24 REAGENTS AND CONDITIONS: (I) [(^{i}Pr)$_2$SiCl]$_2$O (TIPS-Cl), PYRIDINE (II) ClC(S)OC$_6$H$_5$, DMAP, CH$_3$CN (III) Bu$_3$SnH, AIBN, TOLUENE, 75°C (IV) Bu$_4$NF, THF.

The combined reagents magnesium methoxide-sodium borohydride [Mg(OMe)$_2$ – NaBH$_4$] are effective in the deoxygenative rearrangement of a 3'-O-sulphonyl nucleoside (**2.98**) and the successive reduction of the 2'-keto intermediate (**2.100**) in a 1-pot procedure[41] (Scheme 2.25).

There has been considerable interest in 2',3'-dideoxy (DD) and 2',3'-didehydro-2',3'-dideoxy (D$_4$) ribonucleosides owing to their application in the treatment of AIDS.[42] DD- and D$_4$- pyrimidine nucleosides can be prepared via anhydro-derivatives (Scheme 2.26).[43,44]

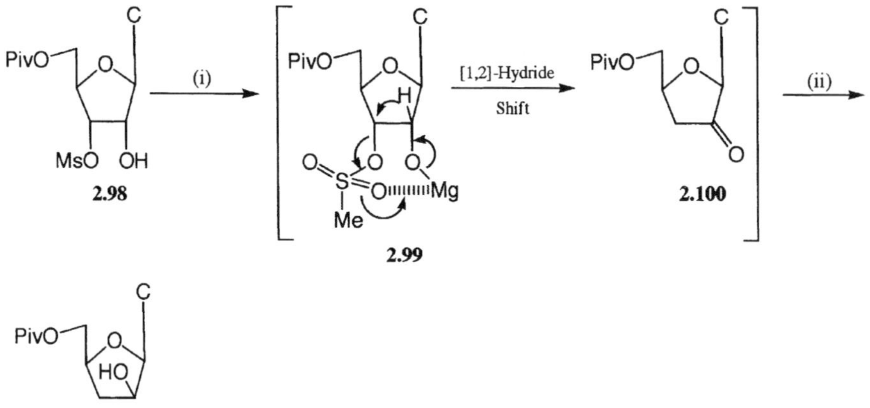

SCHEME 2.25 REAGENTS AND CONDITIONS: (I) Mg(OMe)$_2$ (II) NaBH$_4$.

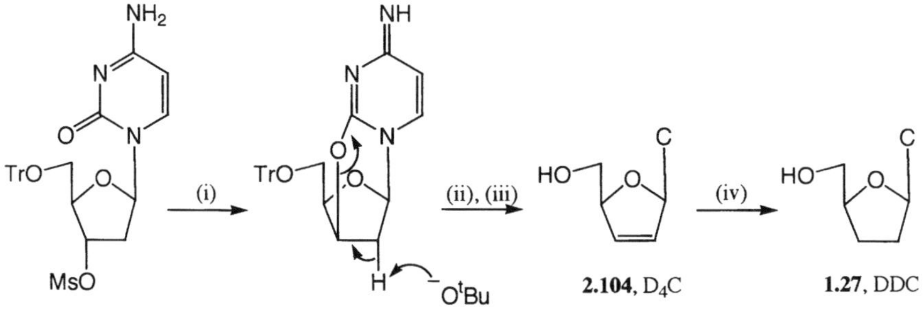

SCHEME 2.26 REAGENTS AND CONDITIONS: (I) NaOH, EtOH (II) KO^{t}Bu, DMSO (III) HCl, CHCl$_3$ (IV) H$_2$, 10% Pd-C, DIOXANE.

Treatment of the anhydronucleoside (**2.103**) with potassium tert-butoxide gives the 2', 3'-ene nucleoside (**2.104**, D$_4$C) by α-proton abstraction at C-2'. Catalytic hydrogenation of the 2', 3'-double bond affords 2',3'-dideoxycytidine (**1.27**, DDC, Zalcitabine®) a highly effective anti-HIV agent.[45] Alternative procedures for the synthesis of DD and D$_4$ nucleosides which are more efficient and of general application, involve either reduction of acetobromofuranosyl nucleosides with Zn-Cu couple (Scheme 2.27),[46,47] or treatment of 2',3'-O-thionocarbonates with trimethylphosphite (Scheme 2.28)[48] following the Corey-Winter procedure.[49]

Reaction of adenosine (**1.20**) with α-acetoxyisobutyryl bromide gives a 9:1 mixture of the xylo- and arabino-bromosugars (**2.106**) by the mechanism shown in Scheme 2.27. Reductive elimination of the bromoacetate with Zn/Cu couple gives D$_4$A (**2.107**). The 2',3'-O-thionocarbonate (**2.109**) is reacted with the thiophile trimethylphosphite to produce the unstable carbene, which readily undergoes elimination to give the alkene D$_4$U (**2.111**) and carbon dioxide (Scheme 2.28).

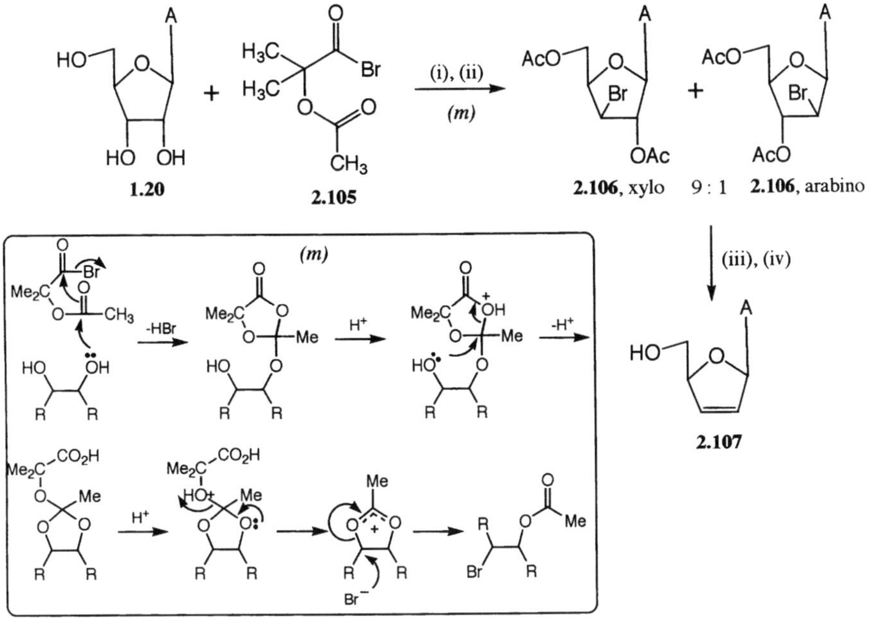

SCHEME 2.27 *REAGENTS AND CONDITIONS*: (I) CH$_3$CN, H$_2$O (II) Ac$_2$O, DMAP, PYRIDINE (III) Zn/Cu, DMF (IV) NH$_3$, MeOH.

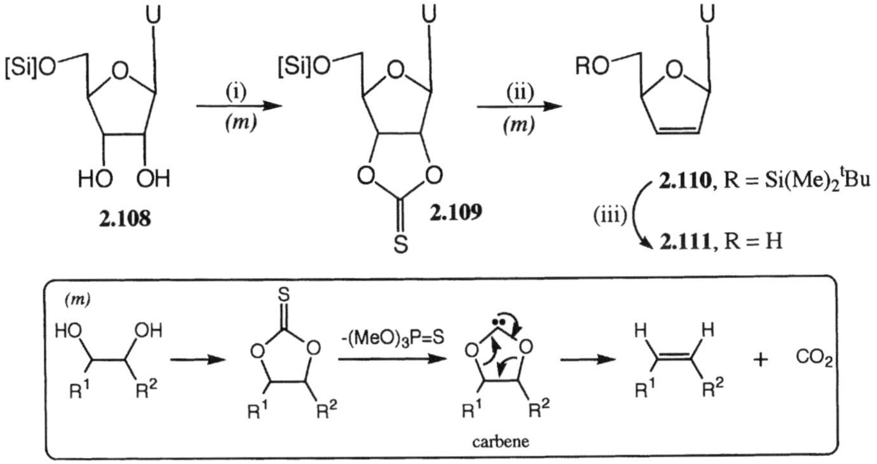

SCHEME 2.28 *REAGENTS AND CONDITIONS*: (I) 1,1'-THIOCARBONYLDIIMIDAZOLE, CHCl$_3$ (II) (MeO)$_3$P, 110°C (III) TBAF. [Si] = TBDMS.

2.3.4 Azido and Amino Sugars

Azido and amino sugars can be readily prepared *via* 2,2'- and 2,3'-anhydro-nucleosides.[50–52] An efficient method[53] for generating azido nucleosides, including 3'-azido-2',3'-dideoxythymidine (**1.26**, AZT) involves a 1-pot transformation of thymidine (**1.22**) into 2,3'-anhydro-5'-*O*-(4-methoxybenzoyl)-thymidine (**2.112**) by a tandem Mitsunobu reaction.[54] Subsequent ring opening with azide anion and 5'-*O*-deprotection gives the anti-HIV agent AZT (Scheme 2.29).

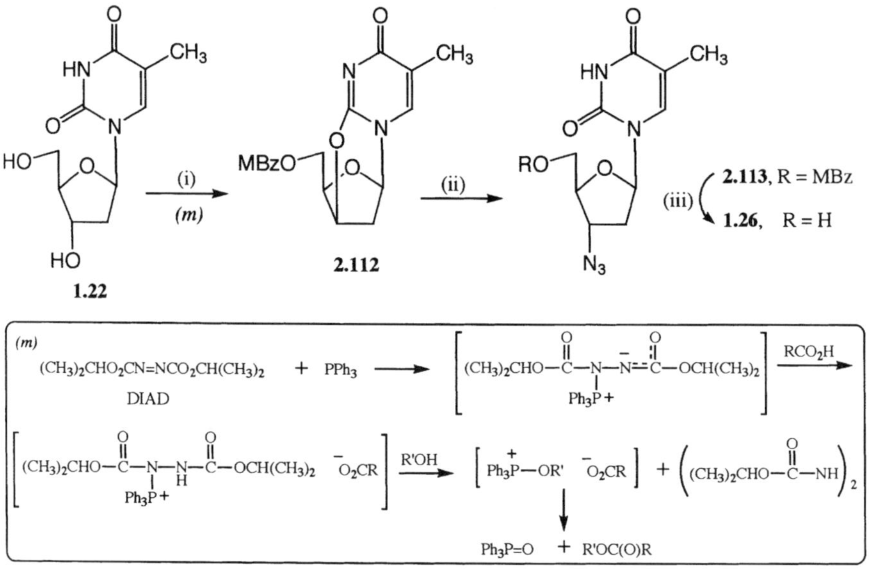

SCHEME 2.29 REAGENTS AND CONDITIONS: (I) DIAD, PPh$_3$, 4-MeOC$_6$H$_4$CO$_2$H, DMF (II) LiN$_3$, DMF, 125°C (III) NaOMe, MeOH. MBz = 4-MeOC$_6$H$_4$C(=O).

A modified Mitsunobu reaction can also be employed for the direct replacement of a hydroxy group with an azido group, using either hydrazoic acid[55] or diphenylphosphoryl azide (DPPA)[56,57] as the source of azide anion (Scheme 2.30). The 5'-azido-5'-deoxy-thymidine (**2.115**) is then converted to the 5'-amino compound (**2.116**) by reaction with triphenylphosphine, followed by hydrolysis of the resulting phopsphinimide.[58]

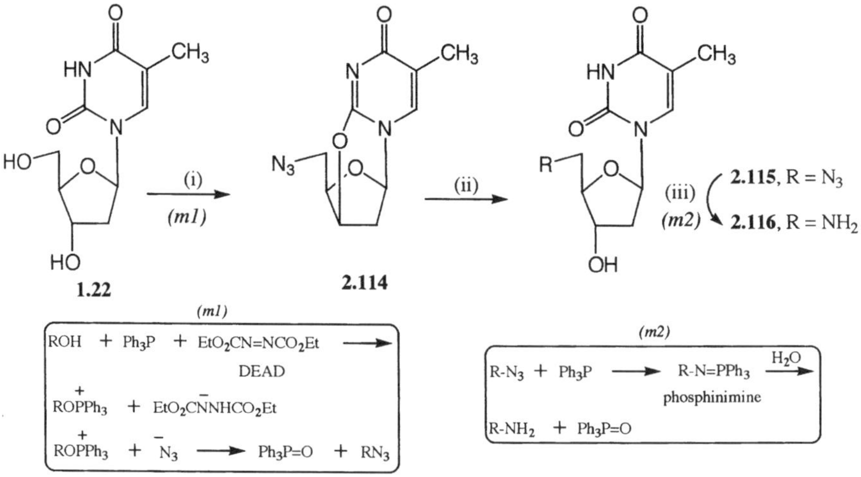

SCHEME 2.30 REAGENTS AND CONDITIONS: (I) DEAD, PPh$_3$, HN$_3$, DMF (II) aq. NaOH (III) PPh$_3$, PYRIDINE/NH$_4$OH.

2.3.5 Halogenation

5'-Halonucleosides are useful precursors for the synthesis of nucleotides, anhydronucleosides, sulphur analogues and deoxynucleosides. A convenient method for the synthesis of 5'-halo-5'-deoxy-nucleosides employs triphenylphosphine and carbon tetrahalides (chloride, bromide or iodide) in pyridine (Scheme 2.31).[59]

1.28, B = U
1.15, B = hypoxanthine

2.117, X = Cl, Br, I

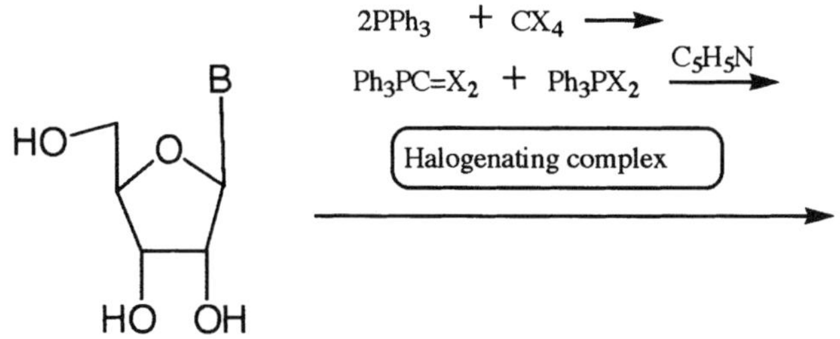

SCHEME 2.31 5′-HALOGENATION OF NUCLEOSIDES.

Fluorinated nucleosides have been of considerable interest since the discovery of 1-(2-fluoro-2-deoxy-β-D-arabino-furanosyl)-5-iodouracil (**2.118**, FIAU, Fialuridine®) and its cytidine counterpart (**2.119**, FIAC, Fiacitabine®),[60,61] as potent and selective antiherpesvirus agents, however development of both compounds were terminated in Phase II clinical trials owing to toxicity.

Difluorinated nucleosides are also of interest owing to compounds such as the anticancer agent 2', 2'-difluoro-2'-deoxycytidine (**2.120**, dFdC),[62] which inhibits DNA synthesis and ribonucleotide reductase[63] (Fig. 2.2). The fluorinating agent diethylaminosulphur trioxide (Et_2NSF_3, DAST)[64] has been used to prepare 2'- and 3'-fluorinated nucleosides.

2.118, FIAU **2.119**, FIAC **2.120**, dFdC

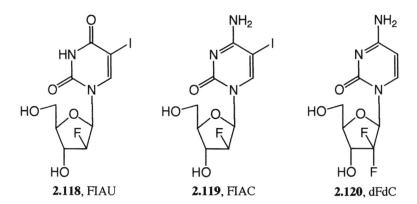

FIGURE 2.2 THERAPEUTIC FLURORINATED NUCLEOSIDES

SYNTHESIS OF CONVENTIONAL D-NUCLEOSIDES

Treatment of the 1-(3-deoxy-β-D-*threo*-pentofuranosyl)thymidine analogue (**2.121**) with 2 equivalents of DAST gives the 3'-fluoro-nucleoside (**2.122**) with inversion of configuration.[65] Reaction of the tritylated 3'-keto-thymidine (**2.123**) with DAST gives the 3', 3'-difluoro-DDT nucleoside (**2.124**)[66] which is converted to the 3'-fluoro-D$_4$T nucleoside (**2.125**) on treatment with sodium methanolate[65] (Scheme 2.32).

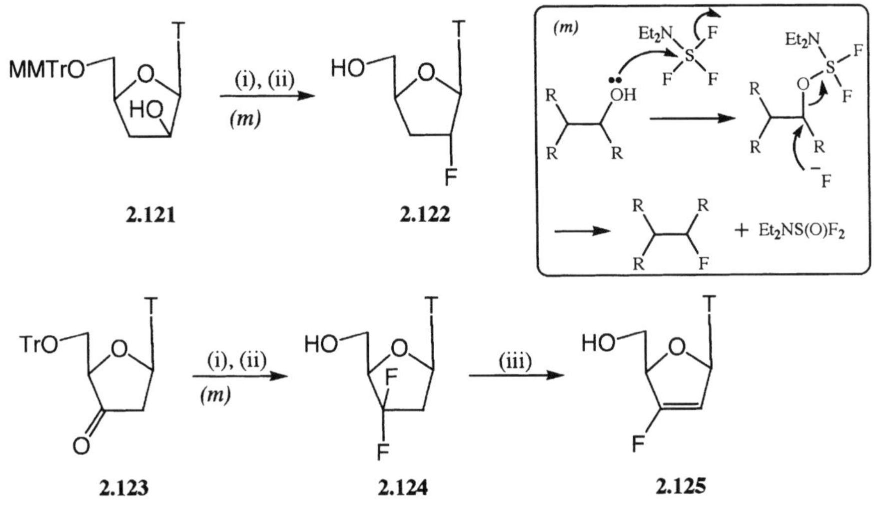

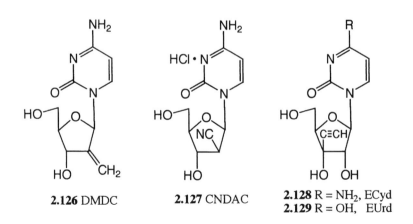

SCHEME 2.32 REAGENTS AND CONDITIONS: (I) DAST, DICHLOROETHANE (II) 80% AcOH, 60°C (III) SODIUM METHANOLATE, DMF. MMTr = MONOMETHOXYTRITYL.

2.126 DMDC

2.127 CNDAC

2.128 R = NH$_2$, ECyd
2.129 R = OH, EUrd

FIGURE 2.3 ANTITUMOUR 2'-C- AND 3'-C-NUCLEOSIDES

2.3.6 2'- and 3'-C-Nucleosides

2'-C- and 3'-C-nucleosides such as 1-(2-deoxy-2-methylene-β-D-*erythro*-pento furanosyl)cytosine (**2.126**, DMDC),[67] 1-(2-C-cyano-2-deoxy-β-D-*arabino*-pentofuranosyl)cytosine (**2.127**, CNDAC),[68] and 1-(3-C-ethynyl-β-D-*ribo*-pentofuranosyl)cytosine (**2.128**, ECyd) and its uracil derivative (**2.129**, EUrd)[69] (Fig. 2.3), have been developed as potent antitumour agents, active against leukaemias, lymphomas and solid tumours.

DMDC and CNDAC are synthesised from the 2'-keto-nucleoside (**2.232**) prepared by the Swern oxidation[70] of **2.130**. A Wittig reaction[71] of **2.131** with methylenetriphenyl phosphorane gives the 2'-methylene derivative (**2.132**) which on deprotection gives DMDC (**2.126**) (Scheme 2.33).[67] Reaction of **2.131** with sodium cyanide gives the 2'-cyano derivative (**2.133**) as an epimeric mixture which is deoxygenated (for mechanism, see Scheme 2.24) stereospecifically to give the arabinonucleoside (**2.134**). Deprotection and treatment with 3% HCl in methanol gives CNDAC (**2.127**) (Scheme 2.33).[68]

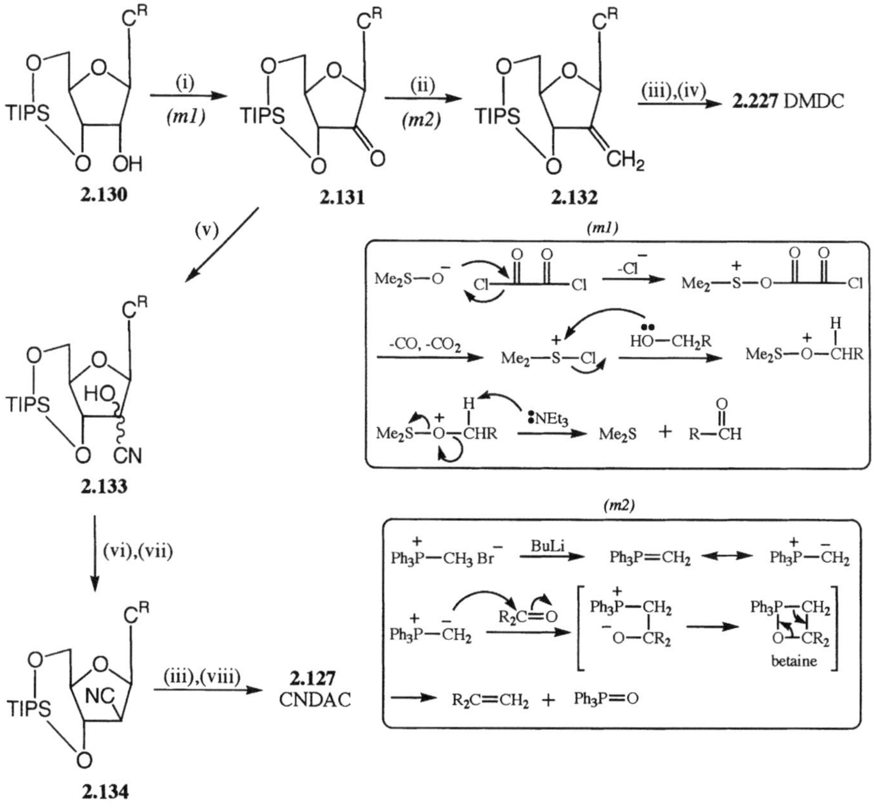

SCHEME 2.33 *REAGENTS AND CONDITIONS*: (I) OXALYL CHLORIDE, DMSO, Et$_3$N, CH$_2$Cl$_2$ (II) Ph$_3$P$^+$CH$_3$Br$^-$, BuLi, THF (III) Bu$_4$NF, AcOH, THF, 0°C (IV) NH$_3$/MeOH (V) NaCN, NaHCO$_3$, Et$_2$O-H$_2$O (VI) ClC(S)OC$_6$H$_5$, DMAP, Et$_3$N, CH$_3$CN, 0°C (VII) Bu$_3$SnH, AIBN, TOLUENE, 100°C (VIII) HCl/MeOH. R = Ac OR Bz.

ECyd (**2.128**) and EUrd (**2.129**) are prepared by the Vorbrüggen coupling[15] of the 3-*C*-ethynyl-ribofuranoside (**2.137**) and the appropriate silylated base. The introduction of the ethynyl moiety is achieved with high stereoselectivity on reaction of the 3-keto sugar (**2.135**) with LiC≡CTMS[69] (Scheme 2.34). The presence of a 2',3'-*cis* diol in ECyd and EUrd with the *ribo*-configuration is essential for toxicity.[72] ECyd strongly inhibits RNA synthesis by inhibiting RNA polymerase inducing apoptotic cell death.

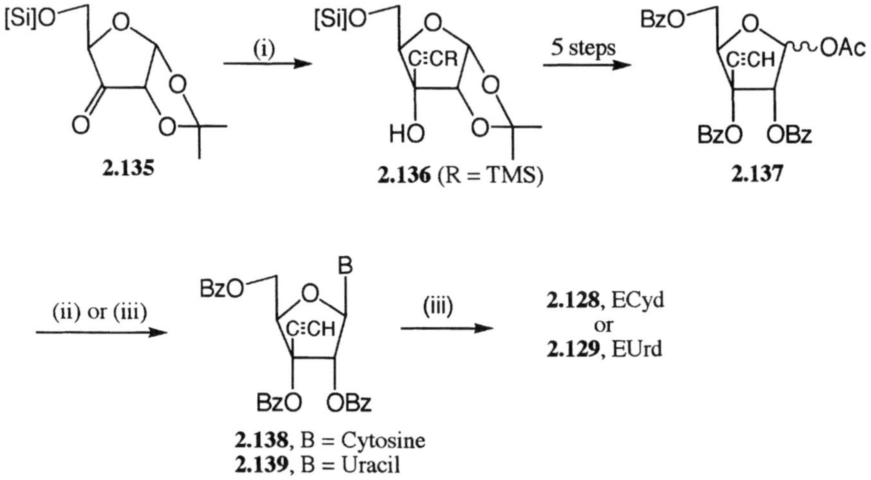

SCHEME 2.34 REAGENTS AND CONDITIONS: (I) TMSC≡CH, BuLi, THF, −78°C (II) C(TMS)$_2$, SnCl$_4$, CH$_3$CN (III) U(TMS)$_2$, TMSOTf, CH$_3$CN (IV) NH$_3$/MeOH. [Si] = TBDMS.

2.3.6.1 Mechanism of Action

The antitumour activity of these nucleosides is mainly owing to their inhibition of DNA synthesis (and in the case of ECyd, RNA synthesis) in tumour cells, in which they are metabolised to their corresponding 5'-diphosphates and 5'-triphosphates and inhibit ribonucleoside diphosphate reductase (RDPR) and/or DNA polymerase.

RDPR is one of the most important target enzymes for the inhibition of DNA synthesis.[72] RDPR catalyses the conversion of all the four main ribonucleoside 5'-diphosphates (rNDPs) to their corresponding 2'-deoxyribonucleoside 5'-diphosphates (dNDPs) in a *de novo* pathway (Fig. 2.4). The dNTPs are phosphorylated by nucleoside diphosphate kinase (NDPK) to the triphosphates (dNTPs) which are essential for DNA synthesis.

2.4 HETEROCYCLIC BASE MODIFICATIONS

Transformations of the heterocyclic base moiety have resulted in the preparation of nucleosides with a variety of therapeutic applications. The literature regarding base modifications of pyrimidine nucleosides and purine nucleosides up to early 1960s has been well documented,[7,37] the development of synthetic methodology and in particular the development in carbon-carbon coupling chemistry has resulted in efficient methods for the generation of many novel compounds.

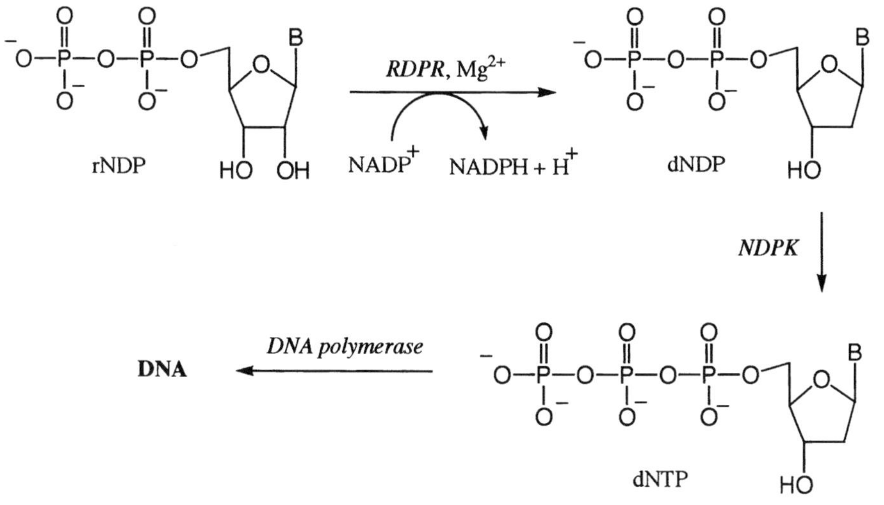

FIGURE 2.4 DE NOVO SYNTHESIS OF dNDPs.

2.4.1 Pyrimidine Modifications

Modification of the pyrimidine base of nucleosides has concentrated primarily on substitutions at C-4 and C-5. Modification at C-4 allows interconversion between nucleoside bases, whereas modification at C-5 has proven very effective in generating compounds with a broad spectrum of biological activity.

2.4.1.1 Modifications at C-4

Thiation of the C-4 position of the thymidine nucleoside (**2.140**) with phosphorus pentasulphide provides a useful intermediate (**2.141**) for the synthesis of cytosine nucleosides (**2.142**) (Scheme 2.35).[74] The use of P_2S_5 has been superceded for the preparation of 4-thiopyrimidine nucleosides[75] by the milder conditions of Lawesson's reagent (**2.143**, Ar = 4-methoxyphenyl),[76] the active thionating reagent is believed to be the highly reactive dithiophosphine ylide (**2.144**) which reacts with carbonyl groups *via* a mechanism involving Wittig-type intermediates[77] (*(m)* Scheme 2.35).

Direct conversion of uridine or thymidine nucleosides to the corresponding cytidine analogues is readily achieved, the anticancer agent 1-(β-D-arabinofuranosyl) cytosine (**2.147**, Ara-C) is prepared from protected AraU (**2.145**) *via* the 4-(triazol-1-yl)-intermediate (**2.146**) by the mechanism shown in Scheme 2.36.[78] AraC produces its biological effect in the same manner as DMDC and CNDAC, *i.e.* inhibition of RDPR (see section 2.3.6.1).

2.4.1.2 Modifications at C-5

A large variety of modification at C-5 can be achieved (Scheme 2.37) using a range of procedures which involve initial electrophilic addition across the C5–C6 double bond. These reactions have been performed on uridine, 2'-deoxyuridine, cytosine and other pyrimidine nucleoside anlogues.[79–89]

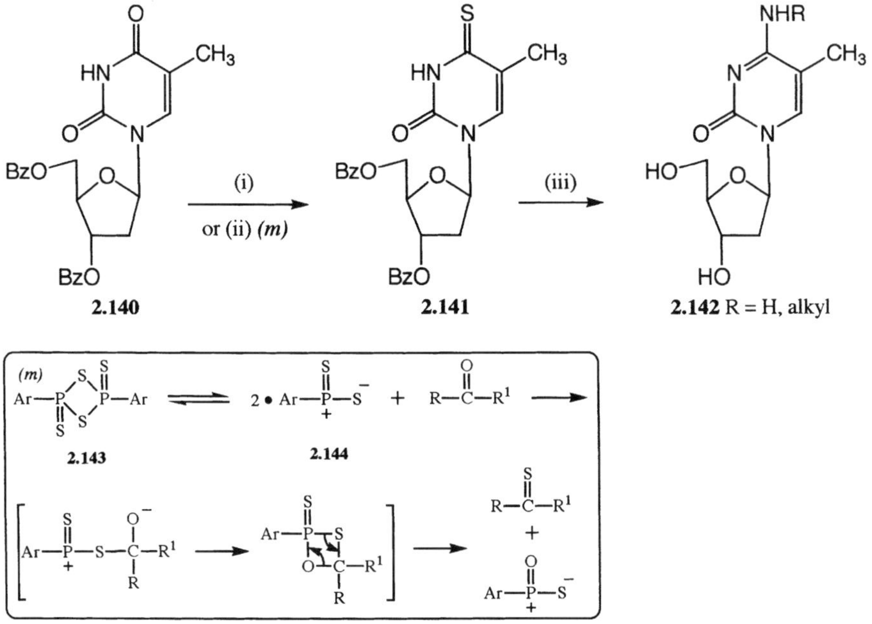

SYNTHESIS OF CONVENTIONAL D-NUCLEOSIDES

SCHEME 2.35 *REAGENTS AND CONDITIONS*: (I) P$_2$S$_5$, PYRIDINE, REFLUX (II) LAWESSON'S REAGENT (III) RNH$_2$, EtOH, 100°C.

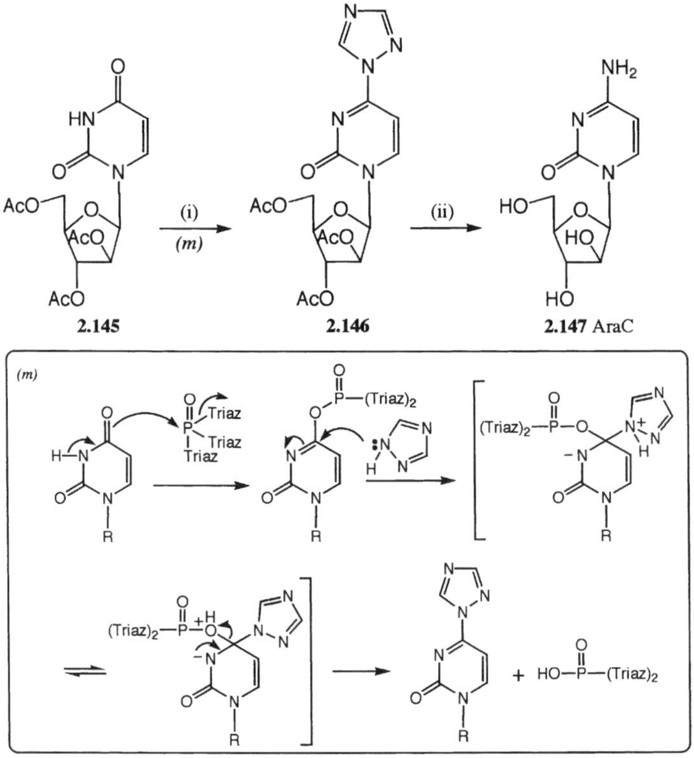

SCHEME 2.36 *REAGENTS AND CONDITIONS*: (I) POCl$_3$, 1,2,4-TRIAZOLE, Et$_3$N, CH$_3$CN (II) AQUEOUS NH$_3$, DIOXANE.

NUCLEOSIDE MIMETICS

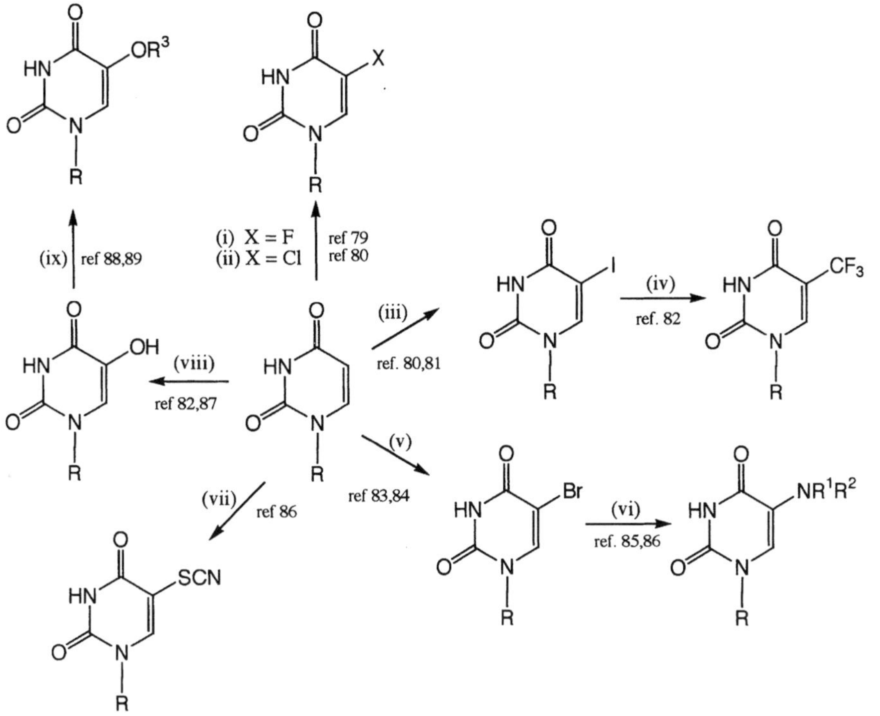

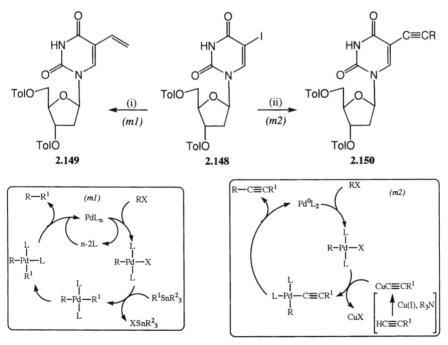

SCHEME 2.37 REAGENTS AND CONDITIONS: (I) CF₃OF, CH₂Cl₂, −78°C (II) N-CHLOROSUCCINIMIDE, CCl₄ (III) HIO₃, I₂, AcOH-CCl₄ OR I₂, NaHCO₃, EtOH (IV) CF₃I-Cu, HMPA, 40–50°C (V) Br₂, PYRIDINE, CCl₄ OR Br₂-H₂O, THF (VI) R¹R²NH, 50–80°C, STEEL-BOMB (VII) ClSCN, AcOH (VIII) Hg(OAc)₂, DIOXANE, 60°C (IX) ALKYL HALIDE (R³-X), NaOH, DMF-H₂O.

SCHEME 2.38 REAGENTS AND CONDITIONS: (I) N-METHYLPYRROLIDINONE, TRI(2-FURYL)PHOSPHINE, Pd₂(dba)₃, H₂C=CH-SnBu₃ (II) HC≡CR, Et₃N, (Ph₃P)₂PdCl₂, CuI, 50°C.

SYNTHESIS OF CONVENTIONAL D-NUCLEOSIDES

Carbon-carbon coupling methods have been extensively used for the introduction of alkyl, alkene and aryl substituents at C-5.[90–92] Stille coupling[93] of vinyltributyltin with protected 5-iodo-2-deoxyuridine (**2.148**) gives the 5-alkenyl-nucleoside (**2.149**),[94] (Scheme 2.38, *m1*), the 5-alkynyl-nucleoside (**2.150**)[95] can also be prepared from **2.148** using the methodology of Sonogashira *et al.*[96] (Scheme 2.38, *m2*).

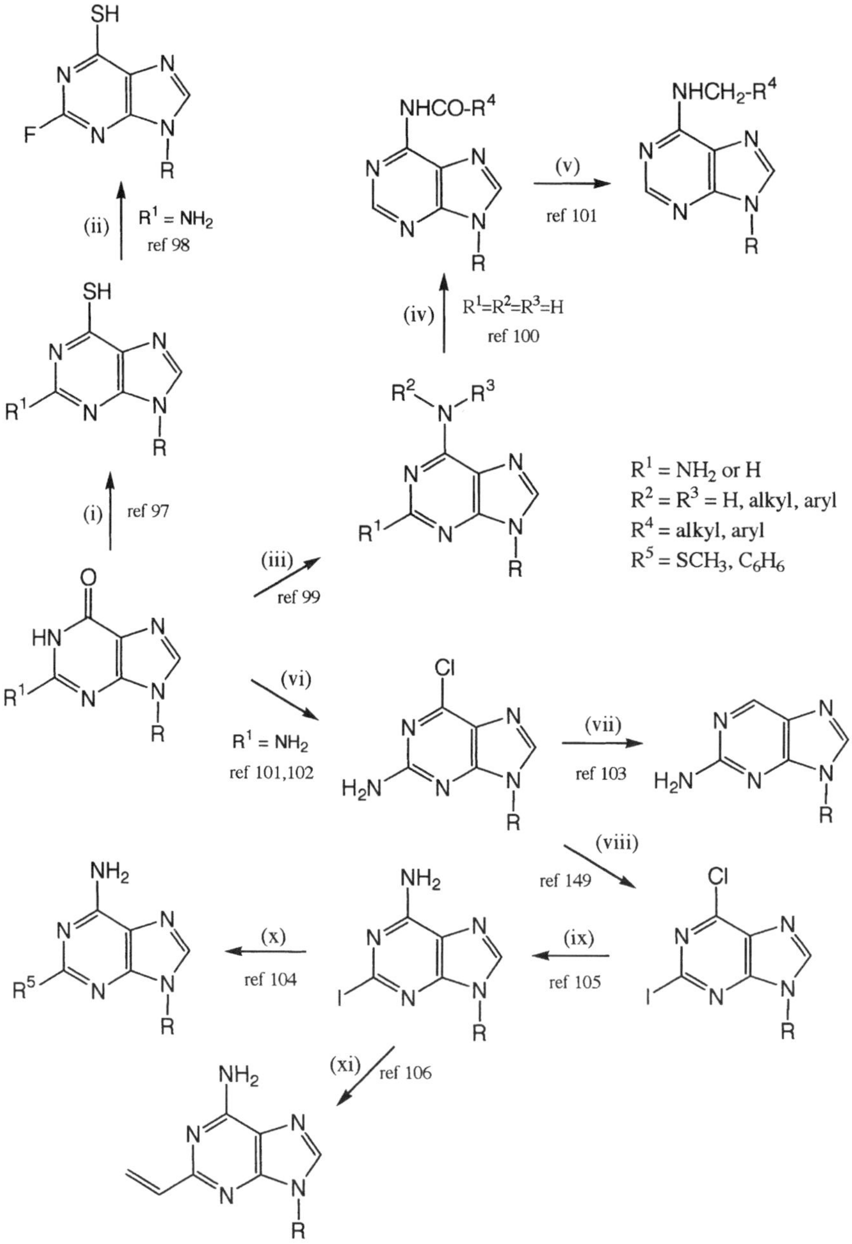

R^1 = NH_2 or H
R^2 = R^3 = H, alkyl, aryl
R^4 = alkyl, aryl
R^5 = SCH_3, C_6H_6

SCHEME 2.39 REAGENTS AND CONDITIONS: (I) P_2S_5, PYRIDINE (II) 48% HBF_4, aq. $NaNO_2$, $-10°C$ (III) 1° OR 2° AMINE, HMDS, $(NH_4)_2SO_4$, MeOH (IV) RCOCl, PYRIDINE (V) $LiAlH_4$, DIOXANE (VI) $POCl_3$, DIMETHYLANILINE, REFLUX OR PPh_3, CH_2Cl_2-CCl_4, DBU (VII) Et_3N, hv, 253.7NM (VIII) n-$C_5H_{11}ONO$, CH_2Cl_2, CH_3CN, REFLUX (IX) NH_3, EtOH (X) CH_3SH OR BENZENE, hv, 253.7NM (XI) $PdCl_2(CH_3CN)_2$, $Bu_3SnCH=CH_2$, DMF, REFLUX.

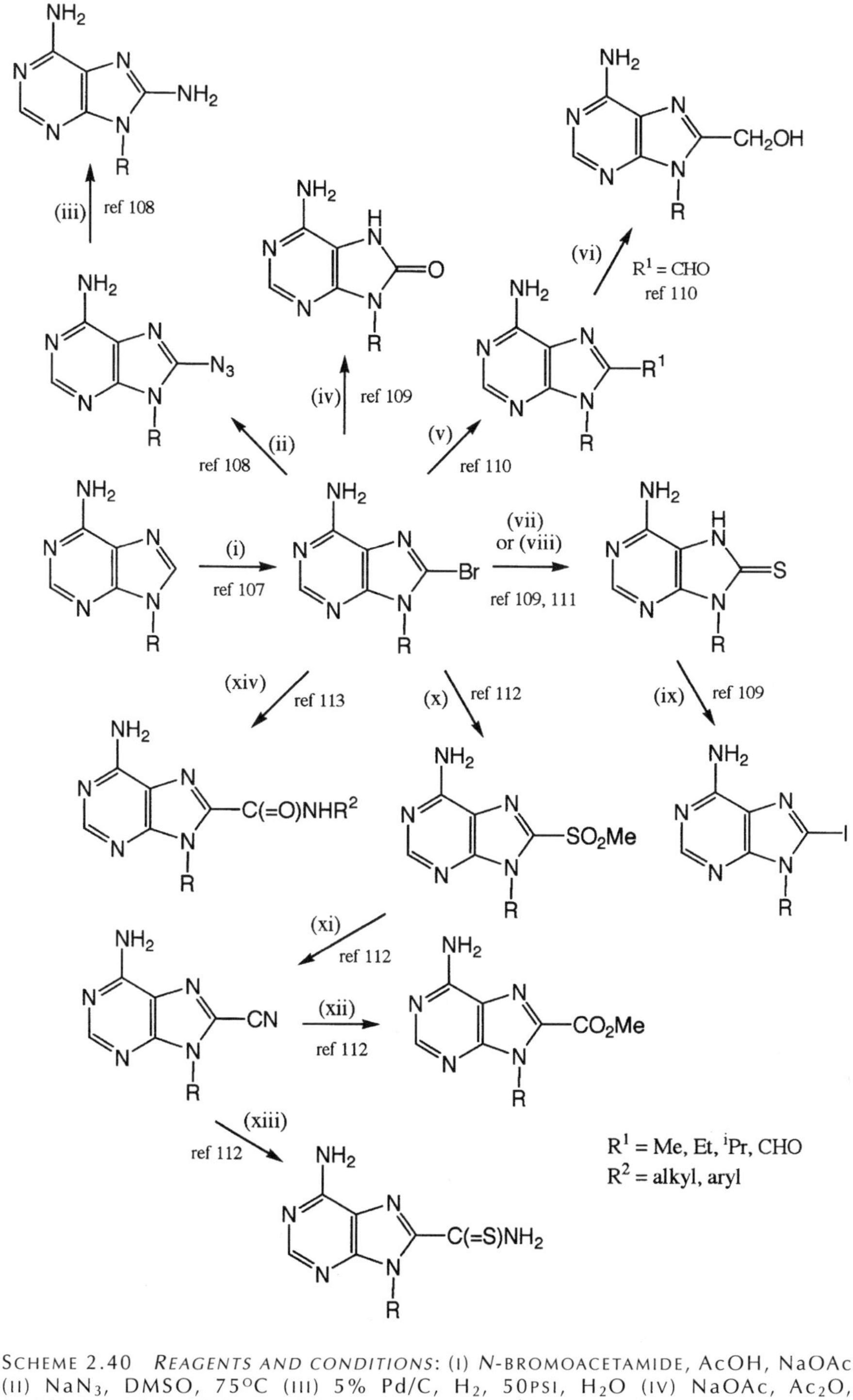

SCHEME 2.40 *REAGENTS AND CONDITIONS*: (I) N-BROMOACETAMIDE, AcOH, NaOAc (II) NaN₃, DMSO, 75°C (III) 5% Pd/C, H₂, 50PSI, H₂O (IV) NaOAc, Ac₂O, AcOH (V) LDA, THF, −70°C THEN MeI, EtI, ⁱPrI OR HCO₂Me (VI) NaBH₄, MeOH, THF (VII) NaSH, EtOH (VIII) (NH₂)₂C=S, EtOH (IX) I₂, KI, NaHCO₃, H₂O (X) NaSMe, DMF THEN 50% AcOH, KMnO₄ (XI) NaCN, DMF (XII) NaOMe, MeOH THEN 1N HCl, MeOH-H₂O (XIII) H₂S, PYRIDINE, 0°C (XIV) 1° AMINE, Et₃N, 50PSI CO, 3 MOL% Pd(PPh₃)₄, DMF, 80°C.

2.4.2 Purine Modifications

The difference in reactivity of the C-2, C-6 and C-8 positions of the purine moiety of purine nucleosides can allow selective modification. In terms of nucleophilic displacement of a chlorine atom from the purine moiety, the order of reactivity/rate of nucleophilic displacement is Cl-6>Cl-2 ~ Cl-8. Scheme 2.39 shows a number of modifications that can be achieved at C2 and C-6 from guanosine/inosine nucleosides (R = sugar moiety, R^1 = NH_2 or H respectively).

The 8-halo-purine nucleosides are versatile intermediates from which numerous derivatives can be prepared. The chemical modifications established in the 1960's are still the most commonly employed, more recent procedures exploit the acidity of the H-8 proton to achieve alkylation and carboxyamidation. Scheme 2.40 (R = sugar moiety) represents some of the modifications/reactions of 8-bromoadenosine, the methods described are applicable to other purine nucleosides including guanosine, xanthosine and inosine.

2.5 REFERENCES

1. G.E. Gilbert, T.B. Johnson, *J. Am. Chem. Soc.*, **1930**, *52*, 4489–4494.
2. G.E. Gilbert, T.B. Johnson, *J. Am. Chem. Soc.*, **1930**, *52*, 2001–2007.
3. G.A. Howard, B. Lythgoe, A.R. Todd, *J. Chem. Soc.*, **1947**,1052–1054.
4. E. Fischer, B. Helferich, *Ber.*, **1914**, *47*, 210–235.
5. J. Davoll, B. Lythgoe, A.R. Todd, *J. Chem. Soc.*, **1948**, 967–969.
6. J. Davoll, B. Lythgoe, A.R. Todd, *J. Chem. Soc.*, **1948**, 1685–1687.
7. J.J. Fox, I. Wempen, *Adv. Carbohydr. Chem.*, **1959**, *14*, 283–380 and references cited therein.
8. J. Davoll, B.A. Lowry, *J. Am. Chem. Soc.*, **1951**, *73*, 1650–1655.
9. B.R. Baker, K. Hewson, H.J. Thomas, J.A. Johnson, Jr., *J. Org. Chem.*,**1957**, *22*, 954–959.
10. M. Hoffer, *Chem. Ber.*, **1960**, *93*, 2777–2781 .
11. Z. Kazimierczuk, H.B. Cottam, G.R. Revankar, R.K. Robins, *J. Am. Chem. Soc.*, **1984**, *106*, 6379–6382.
12. U. Niedballa, H. Vorbrüggen, *J. Org. Chem.*, **1974**, *39*, 3654–3660.
13. H. Vorbrüggen, K. Krolikiewicz, B. Bennua, *Chem. Ber.*, **1981**, *114*, 1234–1235.
14. M. Lalonde, T.H. Chan, *Synthesis*, **1985**, 817–845.
15. H.Vorbrüggen, G. Höfle, *Chem. Ber.*, **1981**, *114*, 1256–1268.
16. R. Zou, M.J. Morris, *Can. J. Chem.*, **1987**, *65*, 1436–1437.
17. T. Mukaiyama, N. Hirano, M. Nishida, H. Uchiro, *Chem. Lett.*, **1996**, 99–100.
18. A. El–Laghdach, Y. Díaz, S. Castillón, *Tetrahedron Lett.*, **1993**, *34*, 2821–2822.
19. J. Wang, J.A. Burster, L.J. Wilson, D. Liotta, *Tetrahedron Lett.*, **1993**, *34*, 4881–4884.
20. C.U. Kim, P.F. Misco, *Tetrahedron Lett.*, **1992**, *33*, 5733–5736.
21. W.-B. Choi, L.J. Wilson, S. Yeola, D.C. Liotta, R.F. Schinazi, *J. Am. Chem. Soc.*, **1991**, *113*, 9377–9379.
22. J. Baddiley, J.G. Buchanan, R. Hodges, J.F. Prescott, *J. Chem. Soc.*, **1957**, 4769–4774.
23. G. Shaw, *J. Chem. Soc.*, **1955**, 1834–1840.
24. G. Shaw, R.N. Warrener, M.H. Maguire, R.K. Ralph, *J. Chem. Soc.*, **1958**, 2294–2299.
25. G.W. Kenner, C.W. Taylor, A.R. Todd, *J. Chem. Soc.*, **1949**, 1620–1624.
26. J.A. Montgomery, S.J. Clayton, H.J. Thomas, W.M. Shannon, G. Arnett, A.J. Bodner, I.–K., Kim, G.L. Cantoni, P.K. Chiang, *J. Med. Chem.*, **1982**, *25*, 626–629.
27. Y.F. Shealy, J.D. Clayton, *J. Pharm. Sci.*, **1973**, *62*, 1432–1434.
28. T.A. Krenitsky, G.W. Koszalka, J.V. Tuttle, *Biochemistry*, **1981**, *20*, 3615–3621.
29. C.L. Burns, M.H. St. Clair, L.W. Frick, T. Spector, D.R. Averett, M.L. English, T.J. Holmes, T.A. Krenitsky, G.W. Koszalka, *J. Med. Chem.*, **1983**, *36*, 378–384.
30. W.J. Hennen, C–H. Wong, *J. Org. Chem.*, **1989**, *54*, 4692–4695.
31. N.C. Yung, J.H. Burchenal, R. Fecher, R. Duschinsky, J.J. Fox, *J. Am. Chem. Soc.*, **1961**, *83*, 4060–4065.
32. J.F. Codington, I.L. Doerr, J.J. Fox, *J. Org. Chem.*, **1965**, *30*, 476–481.
33. J.J. Fox, *Pure Appl. Chem.*, **1969**, *18*, 223–255.
34. J.J. Furth, *Cancer Res.*, **1968**, *28*, 2061–2067.
35. S.S. Cohen, *Med. Biol.*, **1976**, *54*, 299–326.
36. E.J. Reist, A. Benitez, L. Goodman, B.R. Baker, W.W. Lee, *J. Org. Chem.*, **1962**, *27*, 3274–3270.

37. J.A. Montgomery, H.J. Thomas, *Adv. Carbohydr. Chem.*, **1962**, *17*, 301–369.
38. M.J. Robins, J.S. Wilson, *J. Am. Chem. Soc.*, **1981**, *103*, 933–934.
39. W.T. Markiewicz, *J. Chem. Res. (S)*, **1979**, 24–25.
40. D.H.R. Barton, S.W. McCombie, *J. Chem. Soc. Perkin Trans 1*, **1975**, 1574–1585.
41. M. Kawana, N. Yamasaki, M. Nishikawa, H. Kuzuhara, *Chem. Lett.*, **1987**, 2419–2422.
42. E. De Clercq, A. Van Aerschot, P. Herdewijn, M. Baba, R. Pauwels, J. Balzarini, *Nucleosides & Nucleotides*, **1989**, *8*, 659–671.
43. J.P. Horwitz, J. Chua, M.A. Da Rooge, M. Noel, I.L. Klundt, *J. Org. Chem.*, **1966**, *31*, 205–211.
44. J.P. Horwitz, J. Chua, M. Noel, J.T. Donatti, *J. Org. Chem.*, **1967**, *32*, 817–818.
45. J.C. Adkins, D.H. Peters, D. Faulds, *Drugs*, **1997**, *53*, 1054–1080.
46. M.J. Robins, F. Hansske, N.H. Low, J.–I. Park, *Tetrahedron Lett.*, **1984**, *25*, 367–370.
47. S. Greenberg, J.G. Moffatt, , *J. Am. Chem. Soc.*,**1973**, *95*, 4016–4025.
48. L.W. Dudycz, *Nucleosides & Nucleotides*, **1989**, *8*, 35–41.
49. E.J. Corey, R.A.E. Winter, *J. Am. Chem. Soc.*, **1963**, *85*, 2677– 2678.
50. T.–S. Lin, W.R. Mancini, *J. Med. Chem.*, **1983**, *26*, 544–548.
51. R.T. Sudhakar, C.B. Reese, *J. Chem. Soc. Chem. Commun.*, **1989**, 997–998.
52. J.P. Verheyden, D. Wagner, J.G. Moffatt, *J. Org. Chem.*, **1971**, *36*, 250–254.
53. S. Czernecki, J.–M. Valéry, *Synthesis*, **1991**, 239–240.
54. O. Mitsunobu, *Synthesis*, **1981**, 1–28.
55. H. Loibner, E. Zbiral, *Liebigs Ann. Chem.*, **1978**, 78–86.
56. G. Gosselin, V. Boudou, J.–F. Griffon, G. Pavia, C. Pierra, J.–L. Imbach, A. Faraj, J.–P. Sommadossi, *Nucleosides & Nucleotides*, **1998**, *17*, 1731–1738.
57. A. Matsuda, J. Yasuoka, T. Sasaki, T. Ueda, *J. Med. Chem.*, **1991**, *34*, 999–1002.
58. W.S. Mungall, G.L. Greene, G.A. Heavner, R.L. Letsinger, *J. Org. Chem*, **1975**, *40*, 1659–1662.
59. A.K.M. Anisuzzaman, R.L. Whistler, *Carbohydr. Res.*, **1978**, *61*, 511–518.
60. K.A. Watanabe, U. Reichman, K. Hirota, C. Lopez, J.J. Fox, *J. Med. Chem.*, **1979**, *22*, 21–24.
61. G. Lopez, K.A. Watanabe, J.J. Fox, *Antimicrob.Agents Chemother.*, **1980**, *17*, 803–806.
62. L.W. Hertel, J.S. Kroin, J.W. Misner, J.M. Tustin, *J. Org. Chem.*, **1988**, *53*, 2406–2409.
63. W. Plunkett, V. Gandhi, P. Chubb, B. Nowak, V. Heinemann, S. Mineishi, A. Sen, L.W. Hertel, G.B. Grindley, *Nucleosides & Nucleotides*, **1989**, *8*, 775–785.
64. W.J. Middleton, *J. Org. Chem.*, **1975**, *40*, 574–578.
65. A. Van Aerschot, P. Herdewijn, J. Balzarini, R. Pauwels, R. Pauwels, E. De Clercq, *J. Med. Chem.*, **1989**, *32*, 1743–1749.
66. D. Bergstrom, E. Romo, P. Shum, *Nucleosides & Nucleotides*, **1987**, *6*, 53–63.
67. A. Matsuda, K. Takenuki, M. Tanaka, T. Sasaki, T. Ueda, *J. Med. Chem.*, **1991**, *34*, 812–819.
68. A. Matsuda, Y. Nakajima, A. Azuma, M. Tanaka, T. Sasaki, *J. Med. Chem.*, **1991**, *34*, 2917–2919.
69. H. Hattori, M. Tanaka, M. Fukushima, T. Sasaki, A. Matsuda, *J. Med. Chem.*, **1996**, *39*, 5005–5011.
70. A.J. Mancuso, S.–L. Huang, D. Swern, *J. Org. Chem.*, **1978**, *43*, 2480–2482.
71. For a review of the Wittig reaction see B.E. Maryanoff, A.B. Reitz, *Chem. Rev.*, **1989**, *89*, 863–927.
72. H. Hattori, E. Nozawa, T. Iino, Y. Yoshimura, S. Shuto, Y. Shimamoto, M. Nomura, M. Fukushima, M. Tanaka, T. Sasaki, A. Matsuda, *J. Med. Chem.*, **1998**, *41*, 2892–2902.
73. J. Stubbe, *Adv. Enzymol.*, **1989**, *63*, 349–420.
74. J.J. Fox, D. Van Praag, I. Wempen, I.L. Doerr, L. Cheong, J.E. Knoll, M.L. Eidinoff, A. Bendich, G.B. Brown, *J. Am. Chem. Soc.*, **1959**, *81*, 178–187.
75. K. Kaneko, H. Katayama, T. Wakabayashi, T. Kumonaka, *Synthesis*, **1988**, 152–154.
76. For a review of the reactions of Lawesson's reagent see M.P. Cava, M.I. Levinson, *Tetrahedron*, **1985**, *41*, 5061–5087.
77. J. Perregaard, I. Thomsen, S.–O. Lawesson, *Bull. Soc. Chim. Belg.*, **1977**, *86*, 321.
78. K.J. Divakar, C.B. Reese, *J. Chem. Soc. Perkin Trans. 1*, **1982**, 1171–1176.
79. R.A. Sharma, M. Bobek, A. Bloch, *J. Med. Chem.*, **1974**, *17*, 466–468.
80. B-C. Pan, Z-H. Chen, E. Chu, M-Y. Wang Chu, S-H. Chu, *Nucleosides & Nucleotides*, **1998**, *17*, 2367–2382.
81. P.K. Chang, A.D. Welch, *J. Med. Chem.*, **1963**, *6*, 428–430.
82. T-S. Lin, Y-S. Gao, W.R. Mancini, *J. Med. Chem.*, **1983**, *26*, 1691–1696.
83. P.K. Chang, A.D. Welch, *Biochem. Pharmacol.*, **1961**, *6*, 50–52.
84. T-S. Lin, M.S. Chen, C. McLaren, Y–S. Gao, I. Ghazzouli, W.H. Prusoff, *J. Med. Chem.*, **1987**, *30*, 440–444.
85. M. Roberts, D.W. Visser, *J. Am. Chem. Soc.*, **1952**, *74*, 668–669.
86. T-S. Lin, Y-S. Gao, R.F. Schinazi, C.K. Chu, J-N. Chiang, W.H. Prusoff, *J. Med. Chem.*, **1988**, *31*, 336–340.
87. E.G. Podrebarac, C.C. Cheng in "*Synthetic Procedures in Nucleic Acid Chemistry*", W.W. Zorbach, R.S. Tipson (eds.), Interscience, New York, N.Y., **1968**, pp 412–413.

88. P.F. Torrence, J.W. Spencer, A.M. Bobst, *J. Med. Chem.*, **1978**, *21*, 228–231.
89. G-F. Huang, M. Okada, E. De Clercq, P.F. Torrence, *J. Med. Chem.*, **1981**, *24*, 390–393.
90. D.E. Bergstrom, *Nucleosides & Nucleotides*, **1982**, *1*, 1–34.
91. P. Wigerinck, L. Kerremans, P. Claes, R. Snoeck, P. Maudgal, E. De Clercq, P. Herdewijn, *J. Med. Chem.*, **1993**, *36*, 538–543 and references cited therein.
92. B.L. Flynn, V. Macolino, G.T. Crisp, *Nucleosides & Nucleotides*, **1991**, *10*, 763–779 and references cited therein.
93. J.K. Stille, *Angew. Chem. Int. Ed. Engl.*, **1986**, *25*, 508–524.
94. V. Farina, S.I. Hauck, *Synlett*, **1991**, 157–159.
95. M.J. Robins, P.J. Barr, *J. Org. Chem.*, **1983**, *48*, 1854–1862.
96. K. Sonogashira, Y. Tohda, N. Hagihara, *Tetrahedron Lett.*, **1975**, 4467–4470.
97. J.J. Fox, I. Wempen, A. Hampton, I.L. Doerr, *J. Am. Chem. Soc.*, **1958**, *80*, 1669–1675.
98. J.A. Montgomery, K. Hewson, *J. Am. Chem. Soc.*, **1960**, *82*, 463–468.
99. H. Vorbrüggen, K. Krolikiewicz, *Liebigs Ann. Chem.*, **1976**, 745–761.
100. G.S. Ti, B.L. Gaffney, R.A. Jones, *J. Am. Chem. Soc.*, **1982**, *104*, 1316–1319.
101. E. Lescrinier, C. Pannecouque, J. Rozenski, A. Van Aerschot, L. Kerremans, P. Herdewijn, *Nucleosides & Nucleotides*, **1996**, *15*, 1863–1869.
102. M.J. Robins, B. Uznanski, *Can. J. Chem.*, **1981**, *59*, 2601–2607.
103. L. De Napoli, A. Messere, D. Montesarchio, G. Picciali, C. Santacroce, M. Varra, *J. Chem. Soc., Perkin Trans. 1*, **1994**, 923–925.
104. V. Nair, D.A. Young, R. DeSilvia, Jr., *J. Org. Chem.*, **1987**, *52*, 1344–1347.
105. V. Nair, D.A. Young, *J. Org. Chem.*, **1985**, *50*, 406–408.
106. V. Nair, *Nucleosides & Nucleotides*, **1989**, *8*, 699–708.
107. M. Ikehara, H. Tada, K. Muneyama, M. Kaneko, *J. Am. Chem. Soc.*, **1966**, *88*, 3165–3167.
108. R.E. Holmes, R.K. Robins, *J. Am. Chem. Soc.*, **1965**, *87*, 1772–1776.
109. T. Maruyama, Y. Sato, T. Sakamoto, *Nucleosides & Nucleotides*, **1998**, *17*, 115–122.
110. H. Hayakawa, K. Haraguchi, H. Tanaka, T. Miyasaka, *Chem. Pharm. Bull.*, **1987**, *35*, 72–79.
111. R.E. Holmes, R.K. Robins, *J. Am. Chem. Soc.*, **1964**, *86*, 1242–1245.
112. A. Matsuda, Y. Nomoto, T. Ueda, *Chem. Pharm. Bull.*, **1979**, *27*, 183–192.
113. C. Tu, C. Keane, B. Eaton, *Nucleosides & Nucleotides*, **1997**, *16*, 227–237.

CHAPTER 3

SUGAR MODIFIED NUCLEOSIDES

3.1 INTRODUCTION

Several classes of nucleoside derivatives containing significantly altered carbohydrate moieties (Fig. 3.1) possess antiviral and anticancer properties, and are often associated with increased enzymatic and chemical stability and increased target specificity. Ideally these sugar modified nucleosides should have a spatial orientation of the 5'-hydroxy and nucleobase pharmacophores which allows them to mimic that found in natural nucleosides.

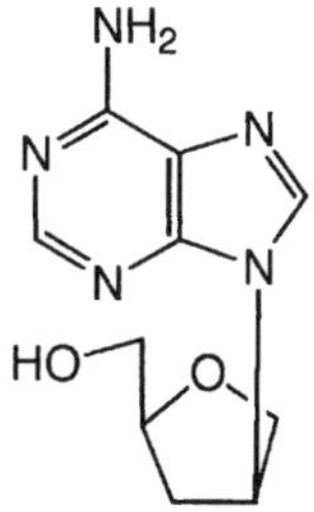

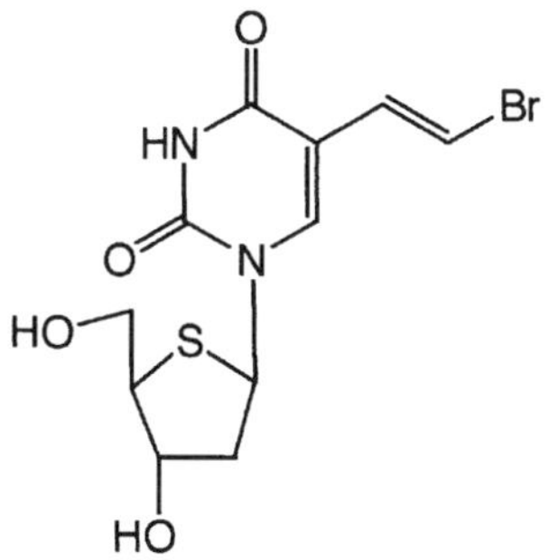

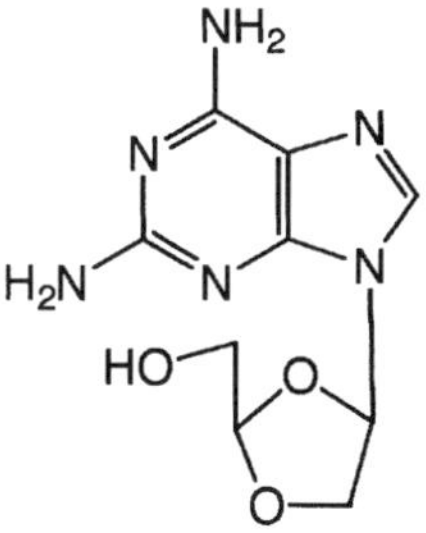

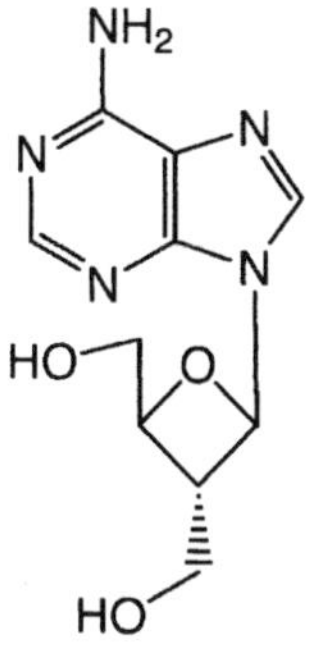

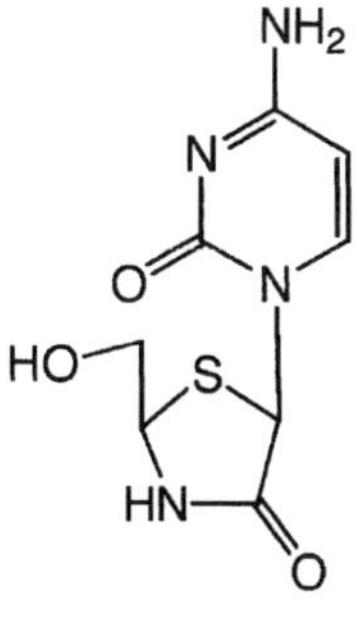

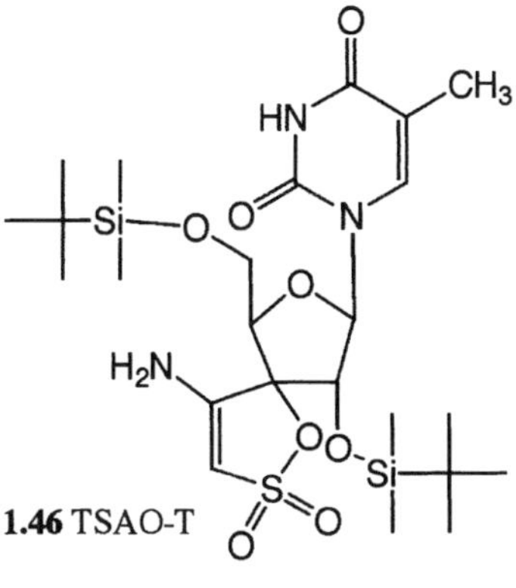

FIGURE 3.1 SUGAR MODIFIED NUCLEOSIDES.

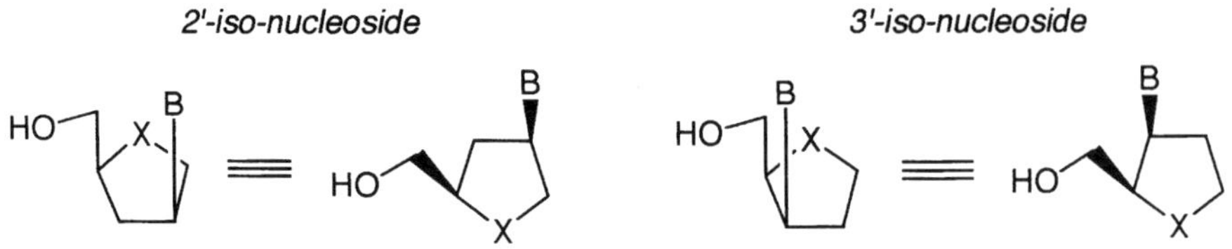

FIGURE 3.2 2'- AND 3'-ISONUCLEOSIDE STRUCTURE (B=NUCLEOBASE, X = O,S,NH,Se,PH).

3.2 ISONUCLEOSIDES

In isonucleosides the base moiety is located at either the 2'- or 3'-sugar carbon (Fig. 3.2). The rationale behind the synthesis of isonucleosides is that even with the transposition of the heterocyclic base to the 2'- or 3'-position, the spatial arrangement between the base and the 5'-hydroxy group is maintained, moreover the 'new' glycosidic bond is more stable towards enzymatic hydrolysis than its conventional counterpart.[1]

3.2.1 2'-Isonucleosides

2'-Isonucleosides are prepared by direct coupling of a 2-isosugar with an appropriate heterocyclic base. 2-Isosugars bearing oxygen,[2–4] sulphur,[5] NH,[6] N-OH,[6] selenium[7] or PH[7] as the ring heteroatom have been reported.

In this series 2'-iso-DDA (**3.1**) and 2'-iso-DDG exhibit significant and selective anti-HIV activity as well as enhanced hydrolytic stability. Synthesis of the required 2-isosugar (**3.9**) involves an acid catalysed rearrangement of (**3.6**) to (**3.7**) as the key step.[2] This is suggested to occur as a result of acid-catalysed ring opening by attack of methanol at C-1 followed by recyclization *via* nucleophilic displacement of the tosyl group by the 2-hydroxy (Scheme 3.1).

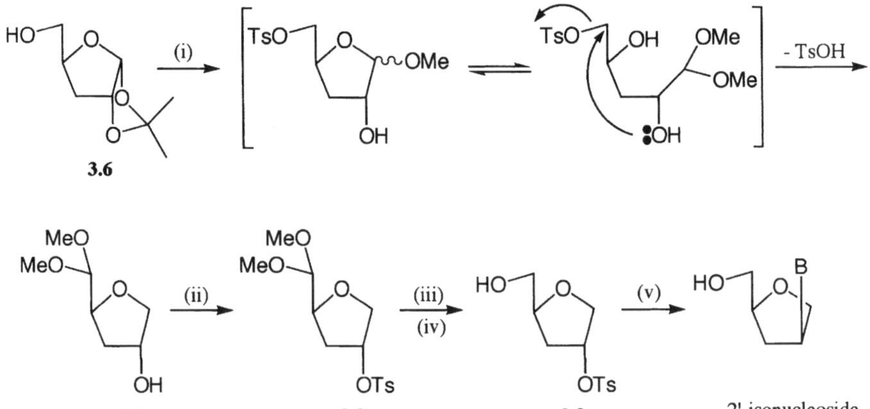

SCHEME 3.1 *REAGENTS AND CONDITIONS*: (I) 1% AcOH, MeOH, 70°C (II) TsCl, PYRIDINE, 0°C (III) 1% CF_3CO_2H, DIOXANE, 80°C (IV) $NaBH_4$, DIOXANE (V) B (A, G, C OR T), K_2CO_3, 18-CROWN-6, DMF, 80°C.

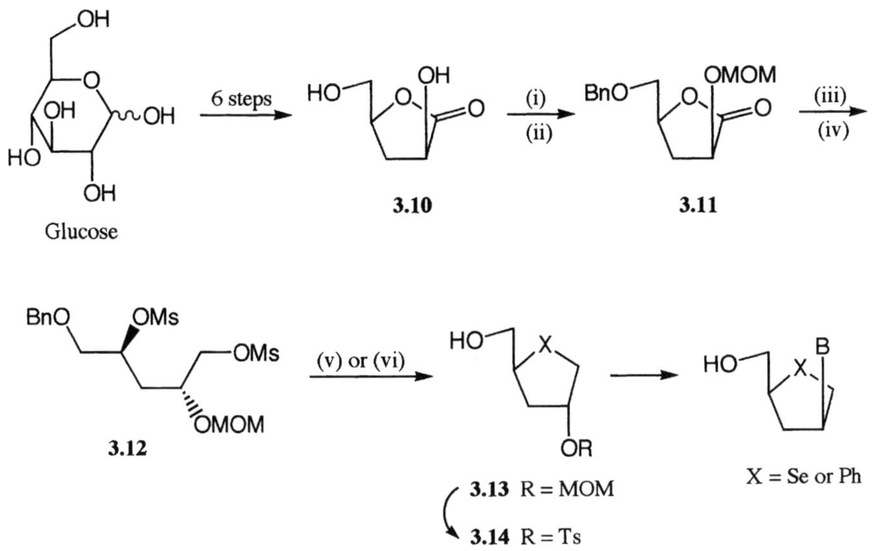

SCHEME 3.2 REAGENTS AND CONDITIONS: (I) BnBr, NaH, DMF (II) (MeO)$_2$CH$_2$P$_2$O$_5$ (III) LiAlH$_4$, THF (IV) MsCl, PYRIDINE, CH$_2$Cl$_2$ (V) Na$_2$Se, ACETONE/H$_2$O (VI) PH$_3$, 2NaH, DMSO.

Synthesis of 2'-isonucleosides containing a ring heteroatom other than oxygen can be prepared from glucose using a general procedure (Scheme 3.2).[7] In this reaction scheme the bis-mesylate (**3.12**) is the key/common intermediate.[4,7]

3.2.2 3'-Isonucleosides

The synthesis of 3'-isonucleosides can involve prior synthesis of an appropriate 3-isosugar or intramolecular glycosylation.[8,9] The 3-isosugar (**3.17**) was prepared from 1-deoxyribose (**3.15**) via a series of protection/deprotection reactions. Radical de-oxygenation (see mechanism, Scheme 2.24) provides **3.17** which is displaced by adenine to give 3'-iso-DDA (**3.18**) (Scheme 3.3).[8]

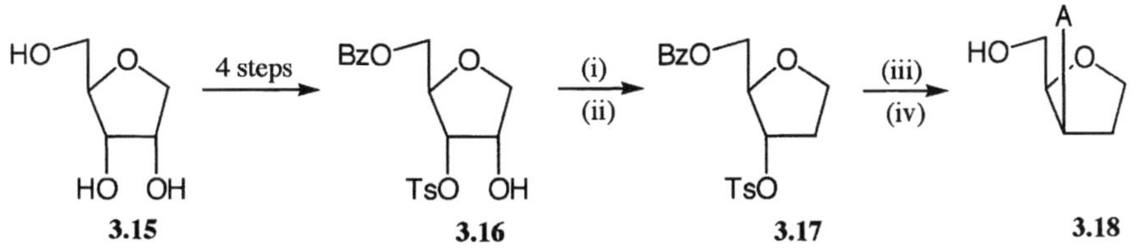

SCHEME 3.3 REAGENTS AND CONDITIONS: (I) (Imid)$_2$CS, (CHCl)$_2$, 85°C (II) nBu$_3$SnH, AIBN, TOLUENE, 120°C (III) ADENINE, K$_2$CO$_3$, 18-CROWN-6, DMF, 90°C (IV) NaOMe, MeOH.

Intramolecular glycosylation is achieved on reaction of the 2,3-unsaturated glycoside (**3.20**), prepared by β-elimination of the thioglycoside (**3.19**), with Me$_2$S(SMe)BF$_4$ which activates (**3.20**) to give the oxonium cation (**3.21**). Intramolecular attack by the pyrimidine at C-3' followed by hydrolysis gives the 3'-isonucleoside (**3.22**) (Scheme 3.4).[9]

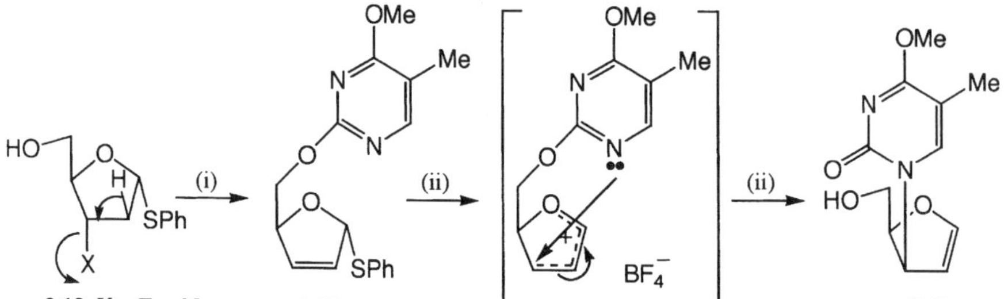

3.19 X = F or N₃ **3.20** **3.21** **3.22**

SCHEME 3.4 *REAGENTS AND CONDITIONS*: (I) 2-CHLORO-4,5-DIMETHOXY-PYRIMIDINE, NaH, DMF (II) Me₂S(SMe)BF₄, MeCN, −20°C (III) Aq. NaOH, 0°C.

3.3 4'-SUBSTITUTED NUCLEOSIDES

Substitution of the 4'-ring oxygen with other heteroatoms, such as sulphur[10] and nitrogen,[11,12] affects both the conformation and biological properties of the nucleoside. With regards to biological activity the most promising of this class of nucleoside has been the 4'-thionucleosides which display a broad spectrum of biological activity and enhanced chemical and enzymatic stability.[13] The unprotected 4'-azanucleosides are very unstable and to date have not been shown to exhibit any significant biological activity whereas the 4'-selenium nucleosides are associated with increased toxicity. Sulphur does not appear to have a great impact on the sugar conformation[14] whereas replacing the ring oxygen with N-R (R = BOC, Ts) or the larger selenium atom would be expected to result in a distortion of the sugar moiety and C4'–C5' conformation.

3.3.1 4'-Thionucleosides

Synthesis of 4-thioribose,[15–18] 2-deoxy-4-thioribose,[19–21] 2,3-dideoxy-4-thioribose[22,23] and 2',3'-dideoxy-3-*C*-(hydroxymethyl)-4-thioribose sugars[24,25] for incorporation into 4'-thionucleosides have been described. The more recent syntheses[17,18] of 4-thio-D-ribose use D-ribose as the starting material rather than the more expensive L-lyxose used in the earlier methods.[15,16] The synthesis in Scheme 3.5 starts with the protected ribose sugar (**3.23**) which is readily prepared from D-ribose.[26]

Dithioacetalization of (**3.23**) provides the source of sulphur, the next step requires inversion of configuration at C-4 which is achieved by a Mitsunobu reaction[18] (see Scheme 2.30 for mechanism) to give the key L-lyxose intermediate (**3.27**). This intermediate is perfectly set up to undergo an iodide-mediated intramolecular S_N2 cyclization, to give the required 4-thio-D-ribose (**3.29**).

2-Deoxy-4-thio-D-ribose can also be conveniently prepared from D-ribose and it was this synthesis on which the preparations of 4-thioribose from D-ribose were based.[19] One of the main difficulties encountered with the synthesis of 2'-deoxy-4'-thionucleosides is the poor anomeric ratios obtained on coupling with the heterocyclic base, which is a result of the inability to use 1-halo-4-thiosugars in a direct displacement reaction (see Scheme 2.5).

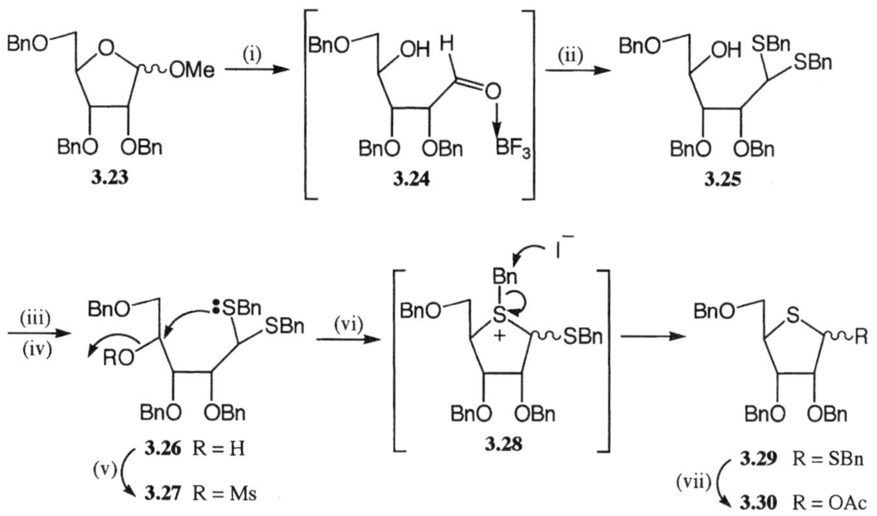

SCHEME 3.5 REAGENTS AND CONDITIONS: (I) BF₃, Et₂O (II) BnSH (III) Ph₃P, DEAD, PARA-NO₂-C₆H₄CO₂H, THF (IV) K₂CO₃, MeOH (V) MsCl, PYRIDINE (VI) Bu₄NI, BaCO₃, PYRIDINE (VII) Hg(OAc)₂, AcOH, Ac₂O.

Unlike 1-halo-4-O-sugars, the 1-halo-4-thiosugars are unstable and rapidly undergo α-elimination to produce thiophene derivatives.[27] The main method for circumventing the problem of unfavourable anomeric rations is the use of a suitable directing group in the C3-sugar position.[28,29] With this in mind Mann et al.[21] incorporated this requirement into their synthesis of 2-deoxy-4-thioribose (Scheme 3.6), the key step of which involved an acid-catalysed pyranose-furanose rearrangement (**3.34**) → (**3.35**).

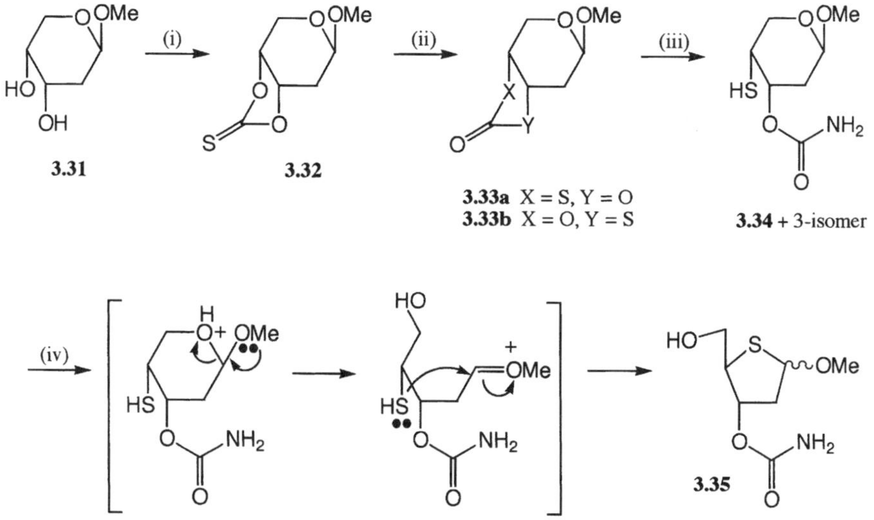

SCHEME 3.6 REAGENTS AND CONDITIONS: (I) (Imid)₂C=S, THF, REFLUX (II) ᵗBu₄N⁺Nr⁻, DIGLYME, REFLUX (III) NH₃, MeOH, 50°C (IV) DOWEX H⁺, MeOH.

SUGAR MODIFIED NUCLEOSIDES

The 4'-thio-pyrimidine nucleosides have been the most extensively studied, primarily as a result of ease of synthesis. 4'-S-BVDU (**3.2**)[30] exhibits antiherpes activity comparable with its 4'-O-counterpart BVDU,[31] however the stability of 4'-S-BVDU against nucleoside phosphorylase and its reduced toxicity *in vivo* makes it a more superior antiviral agent than BVDU.[30]

4'-S-BVDU is prepared by Vorbrüggen coupling of the bis-silylated BVDU (**3.37**) with the thiosugar (**3.36**), followed by anomeric separation and deprotection (Scheme 3.7).[30] Synthesis of 2'-deoxy-4'-thio-purine nucleosides is best achieved by enzymatic methods owing to the extremely poor anomeric ratios, typically α:β = 9:1, obtained. In this procedure the 2'-deoxy-4'-thio-pyrimidine nucleoside (**3.38**) is mixed with a purine base (**3.39**) in the presence of *trans*-N-deoxyribosylase (E.C. 2.4.2.6), and as only the β-pyrimidine nucleoside is a substrate for the enzyme, the reaction yields selectively the β-purine nucleoside (**3.40**) (Scheme 3.37).[32]

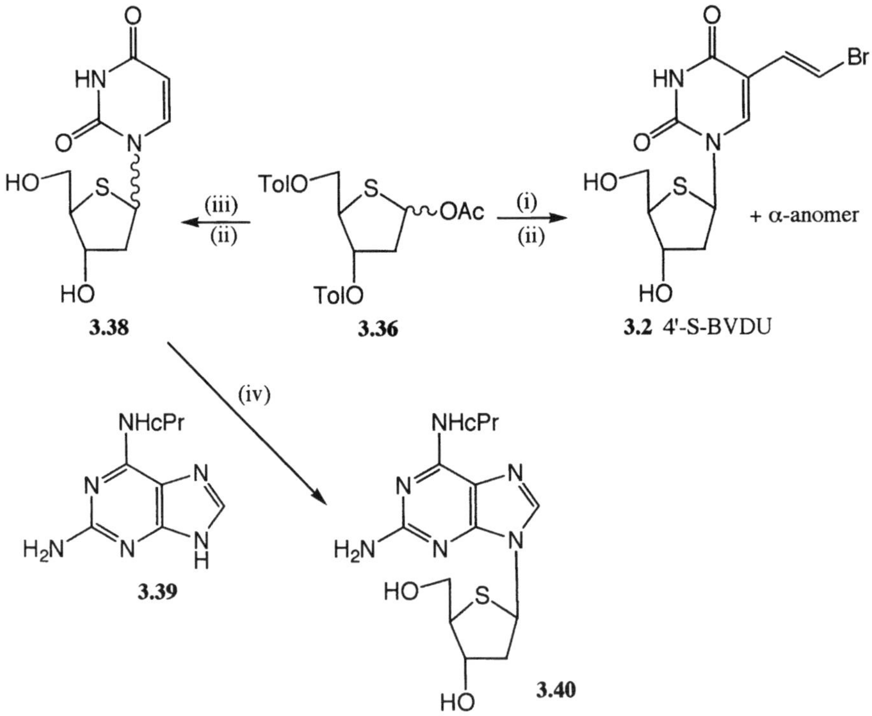

SCHEME 3.7 REAGENTS AND CONDITIONS: (I) $(TMS)_2$BVU (**3.37**), MeCN, $SnCl_4$ (II) NaOMe, MeOH (III) $(TMS)_2$U (**2.35**), MeCN, $SnCl_4$ (IV) TRANS-N-DEOXYRIBOSYLASE, PH 6.0 CITRATE BUFFER.

3.3.2 4'-Aza-nucleosides

4'-Aza-ribose and 4'-aza-xylose nucleosides were first prepared in the 1960s,[33,34] more efficient syntheses of 4'-aza-ribose and 2',3'-dideoxy-4'-azanucleosides have been described.[11,12,23] 4-Deoxy-4-(trifluoroacetamido)-β-D-ribofuranose (**3.41**) can be prepared from methyl 2,3-O-isopropylidene-α-L-lyxopyranoside (**3.42**).[12] Synthesis of the triflate derivative (**3.43**) allows the introduction of the azido group (**3.44**) which is reduced to the amine and subsequently protected to give the trifluoracetamido sugar (**3.45**). Acid hydrolysis of the 1-methoxy and isopropylidene group opens the ring, resulting in intramolecular nucleophilic attack by the amine at C-1, with rearrangement to give the key sugar (**3.41**) (Scheme 3.8).

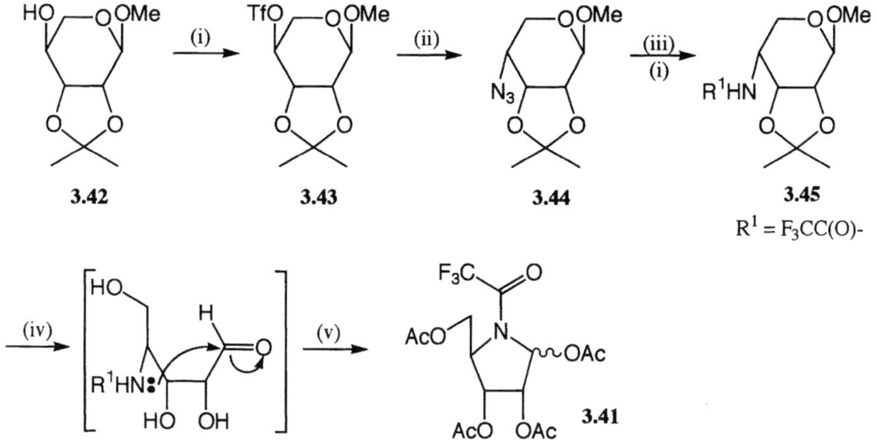

SCHEME 3.8 REAGENTS AND CONDITIONS: (I) Tf$_2$O, DMAP, PYRIDINE, CH$_2$Cl$_2$ (II) NaN$_3$, DMF (III) H$_2$, Pd/C, EtOH (IV) AcOH, H$_2$O (V) Ac$_2$O, AcOH, H$_2$SO$_4$.

The elegant methodology of Rassu et al.[23] allows the preparation of 2,3-dideoxy-4-aza, 4-thio and 4-oxo-sugars of both the D- and L-series for incorporation into the corresponding 2′,3′-dideoxynucleosides (Scheme 3.9).

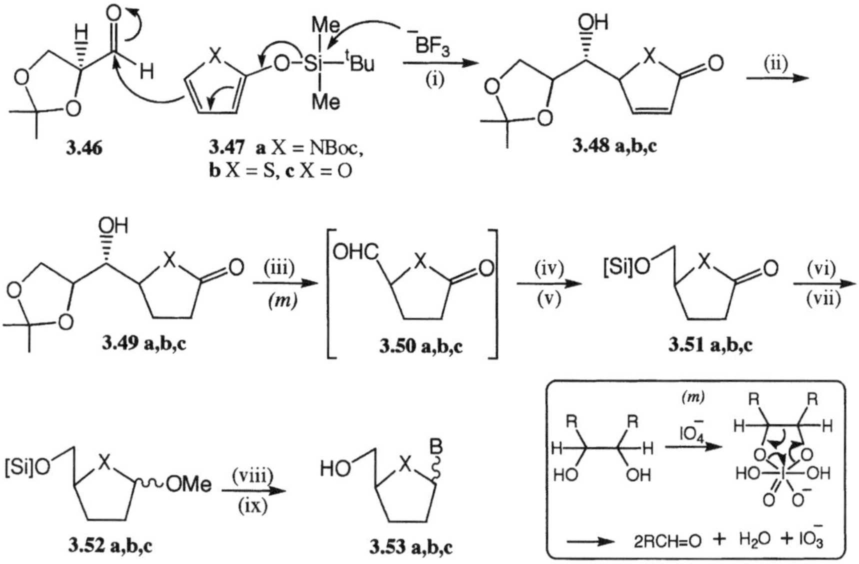

SCHEME 3.9 REAGENTS AND CONDITIONS: (I) BF$_3$•OEt$_2$, CH$_2$Cl$_2$, −90°C (II) H$_2$, Pd/C, THF (III) Aq. AcOH THEN NaIO$_4$, SiO$_2$, CH$_2$Cl$_2$ (IV) NaBH$_4$, MeOH (V) ^{t}Bu(Me)$_2$SiCl, IMIDAZOLE, CH$_2$Cl$_2$ (VI) DIBAL-H, CH$_2$Cl$_2$, −90°C (VII) CH(OMe)$_3$, BF$_3$•OEt$_2$, THF (VIII) (TMS)$_2$B, 1,2-DICHLOROETHANE, TMSOTf/SnCl$_4$ (IX) Bu$_4$NF, THF. [Si] = TBDMS.

Lewis acid promoted condensation of the siloxy diene reagents (**3.47 a-c**) with the L-glyceraldehyde acetonide (**3.46**) provides the lactam, thiolactone and lactone (**3.48 a-c**) respectively, which then undergo catalytic hydrogenation to give the 2,3-dideoxy derivatives (**3.49**). Removal of the acetonide group followed by oxidative cleavage of the resulting triol with sodium periodate (Scheme 3.9 [*m*]) gives the aldehydes (**3.50**). Four more steps provides the key sugars (**3.52**) which are reacted with silyl protected pyrimidines under Vorbrüggen conditions to give the 2',3'-dideoxy-D-nucleosides (**3.53 a-c**). Reaction of the siloxy diene reagents with the corresponding D-glyceraldehyde acetonide would have resulted in the formation of the 2',3'-dideoxy-L-nucleosides.

3.4 DIOXO-, OXATHIO- AND DITHIO-NUCLEOSIDES

Nucleosides containing sugar moieties with two heteroatoms in the sugar ring have proved to be quite potent antiviral agents, especially the dioxo- and oxathio-nucleosides.[35–37] The most challenging aspect in the preparation of the 1,3-dioxolanyl-nucleosides has been the development of asymmetric syntheses of the sugar moieties,[38,39] which have allowed in-depth structure-activity relationship studies.[38,40,41]

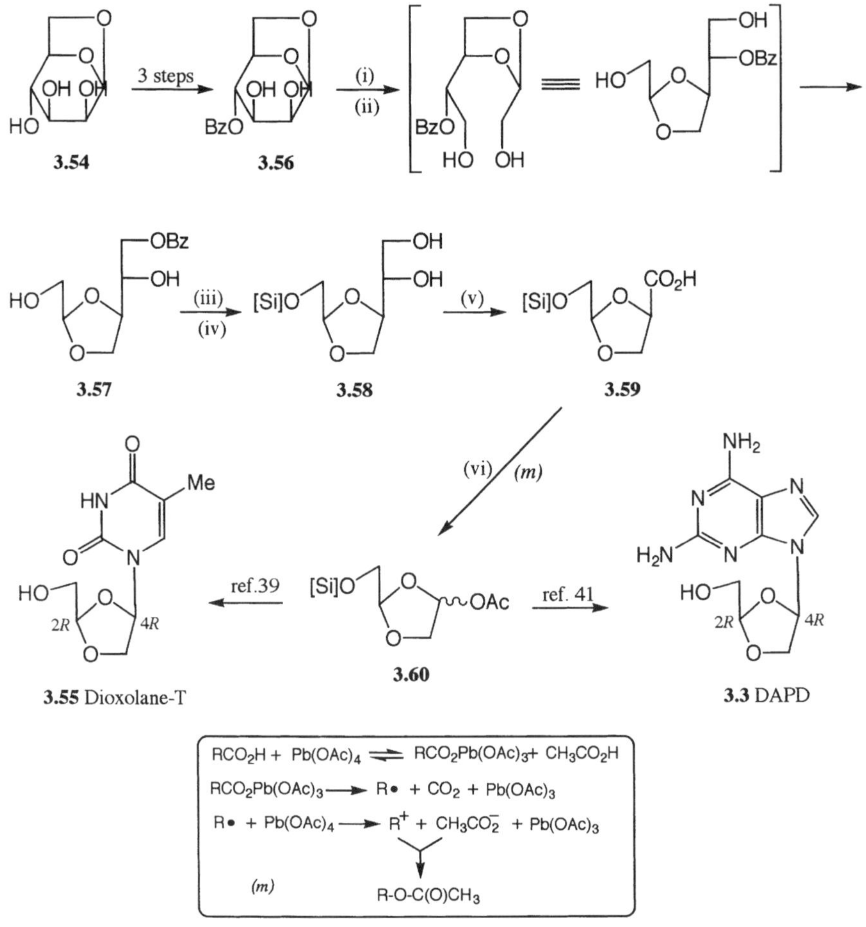

SCHEME 3.10 *REAGENTS AND CONDITIONS*: (I) NaIO$_4$, EtOH (II) NaBH$_4$, EtOH (III) ^{t}Bu(Ph)$_2$SiCl, IMIDAZOLE, DMF (IV) NaOMe, MeOH (V) NaIO$_4$, RuO$_2$, CH$_3$CN, CCl$_4$, H$_2$O (VI) Pb(OAc)$_4$, EtOAc, PYRIDINE. [Si] = ^{t}Bu(Ph)$_2$Si.

Using 1,6-anhydro-D-mannopyranose (**3.54**) as the chiral template, the enantiomerically pure dioxolane-T (**3.55**) and DAPD (**3.3**) can be obtained (Scheme 3.10).[38] The key step involves the rearrangement of (**3.56**) after periodate oxidative cleavage and reduction to the dioxolane sugar (**3.57**). Oxidation of the diol (**3.58**) to the carboxylic acid (**3.59**) followed by oxidative decarboxylation with Pb(OAc)$_4$ (Scheme 3.10 [m]),[42] gives the required 2R, cis and trans isomers (**3.60**). Vorbrüggen coupling with either silyl-protected cytosine or 2,6-diaminopurine gives the respective (2R,4R)-nucleosides (**3.55**) and (**3.3**) after deprotection and anomeric separation.

A similar preparation as that described in Scheme 3.10, though using 1,6-thioanhydro-D-mannopyranose as the chiral template, allows the preparation of the enantiomerically pure 1,3-oxathiolanyl-nucleosides.[43]

Using D-mannitol (**3.61**) as the starting material Scheme 3.11 describes an efficient divergent asymmetric synthesis of the dioxolanyl-sugar moiety that allows access to all four enantiomers.[39] Reaction of (**3.61**) with (**3.62**) gives the bis-acetal (**3.63**) which undergoes RuO$_4$ Wolfe oxidation[44] to give the acids (**3.64** and **3.65**). After separation by column chromatography the acids are then converted into the acetyl-sugars (**3.66** and **3.67**).

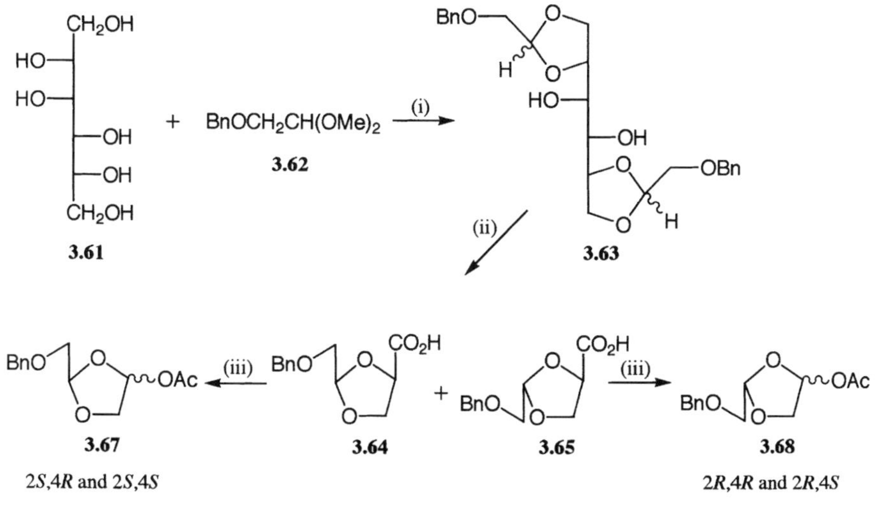

SCHEME 3.11 *REAGENTS AND CONDITIONS*: (I) SnCl$_2$, DME (II) RuCl$_3$ HYDRATE, NaOCl, H$_2$O, CH$_3$CN, DICHLOROETHANE (III) Pb(OAc)$_4$, CH$_3$CN, PYRIDINE.

Several synthetic procedures for the asymmetric synthesis of the 1,3-oxathiolanyl nucleosides have been reported,[43,45–48] two facile procedures are described in Scheme 3.12.[49,50] Reaction of benzoyloxyacetaldehyde (**3.69**) with mercaptoacetaldehyde dimethylacetal (**3.70**) in the presence of ZnCl$_2$ gave the oxathiolane (**3.71**) as a 1:1 mixture of anomers.[49] However reaction of (**3.69**) with either 2-sulphanylethanol (**3.72**, X = O) or ethane-1,2-dithiol (**3.72**, X = S) followed by a Pummerer rearrangement with benzoyl peroxide gave the 'reverse' oxathiolane sugar (**3.73**, X = O) and the dithiolane sugar (**3.73**, X = S) respectively.[50] After coupling **3.71** and **3.73**

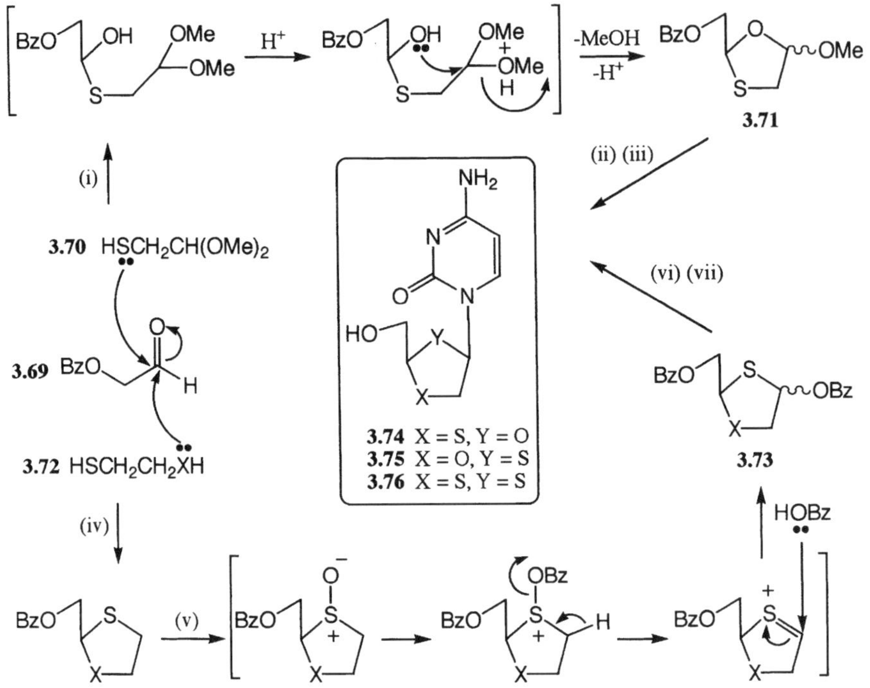

SCHEME 3.12 REAGENTS AND CONDITIONS: (I) ZnCl₂, TOLUENE (II) (TMS)₂C, TMSOTf, CH₃CN (III) ANOMERIC SEPARATION (aq. EtOH RECRYSTALLIZATION) (IV) *p*-TsOH, C₆H₆ (V) (C₆H₅CO)₂, C₆H₆ (VI) TMSI, CH₂Cl₂, (TMS)₂C (VII) MeOH, NH₃ THEN ANOMERIC SEPARATION (CHROMATOGRAPHY).

(X=O and X=S) with silylated cytosine and subsequent deprotection, the nucleosides **3.74-3.76**) were obtained.

2',3'-Dideoxy-3'-thiacytidine (**3.74**, BCH-189) and its 5-fluoro-analogue (+)-FTC are the most active compounds against HIV-1 and HIV-2 in the D-series.[36] In the 'reverse' oxathiolane series 2'-deoxy-3'-oxa-4'-thiocytidine (**3.75**, (−)-dOTC)[50,51] and its 5-fluoro-analogue display potent and selective anti-HIV-1 activity.[37]

3.5 OXAZA-, ISOXA- AND THIAZA-NUCLEOSIDES

The oxazolidine nucleosides are, like the 4'-azanucleosides, unstable without protection of the 3'-NH, and have not shown any significant antiviral activity. The key oxazolidine intermediate (**3.78**) was prepared in 7 steps from (±)-glycidol (**3.77**) and on coupling with silylated thymine and subsequent deprotection with methanolic ammonia, only the β-nucleosides (**3.79** and **3.80**) were isolated (Scheme 3.13).[52] *cis*-1-(2-Hydroxymethyl-3-methoxycarbonyl-1,3-oxazolidin-5-yl)thymine (**3.80**) results from a transesterification reaction during the deprotection step.

The preparation of the 1,3-oxazolidine systems has proved difficult owing to the unstable nature of the aminal (N-R) moiety. However the positioning of the nitrogen group in the isoxazolidine systems is such that the resulting isoxazolidine nucleosides have shown greater stability[53] and more promise with regards to

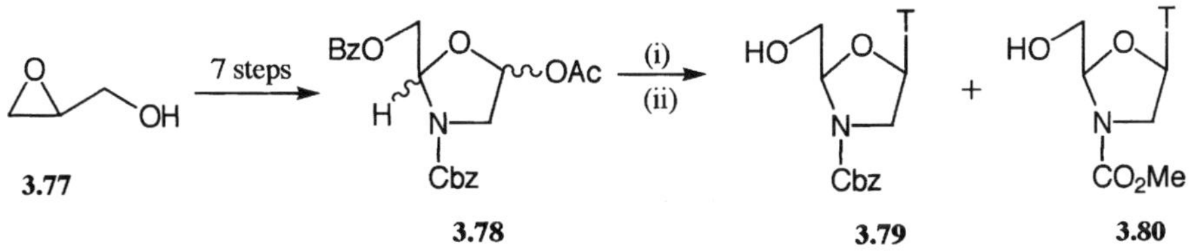

SCHEME 3.13 *REAGENTS AND CONDITIONS*: (I) (TMS)$_2$T, TMSOTf, DICHLOROETHANE (II) MeOH, NH$_3$.

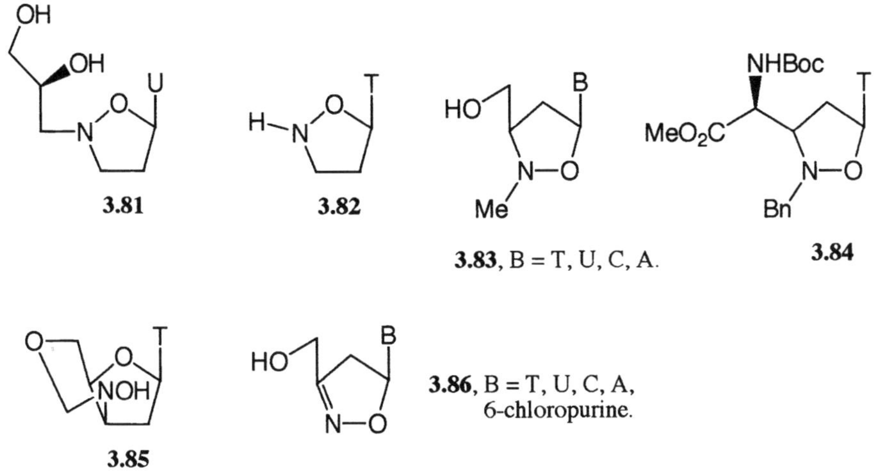

FIGURE 3.3 ISOXAZOLE-NUCLEOSIDES.

biological activity.[54] 1,2- and 3,4-isoxazolidine nucleosides (Fig. 3.3, **3.81–3.84**) have been prepared,[53,55–57] as well as the bicyclo- and dihydro-isoxazole nucleosides (Fig. 3.3, **3.85**, **3.86**),[58–60] the latter being the most interesting biologically with activity against HIV.

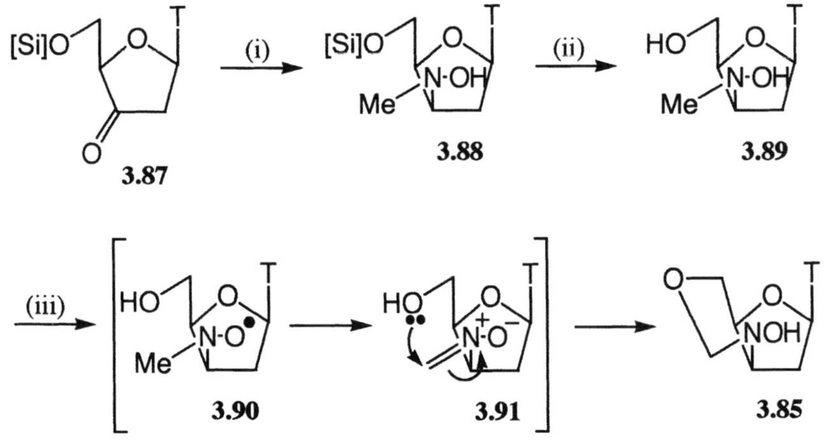

SCHEME 3.14 *REAGENTS AND CONDITIONS*: (I) MeNHOH•HCl, MeOH, PYRIDINE THEN NaBH$_3$CN (II) 85% ACETIC ACID (III) DDQ, CH$_2$Cl$_2$, THF. [Si] = TBDMS.

Synthesis of the bicyclo-isoxazole nucleoside (**3.85**) involves initial stereospecific formation of the *N*-methylnitrone followed by reductive deoxygenation with sodium cyanoborohydride to give the stable methylhydroxyl amine (**3.88**). On oxidation with DDQ the methylenic nitrone (**3.91**) is formed which undergoes a spontaneous cyclization to (**3.85**) (Scheme 3.14).[58]

The dihydro-isoxazole nucleosides (**3.86**) are prepared by a 1,3-dipolar cycloaddition of either a N^1-vinyl-pyrimidine or N^9-vinyl-purine (**3.92**) with the nitrile oxide (**E**, Scheme 3.15),[59–61] generated by the reaction of protected 2-nitroethanol (**A**, Scheme 3.15, R = tetrahydropyran) with phenyl isocyanate (**C**), according to the described mechanism.[62]

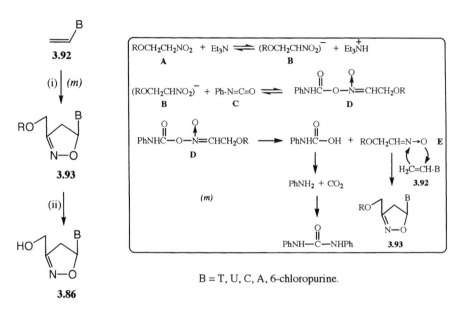

SCHEME 3.15 *REAGENTS AND CONDITIONS*: (I) PhNCO, Et$_3$N, THPO(CH$_2$)$_2$NO$_2$, C$_6$H$_6$ (II) DOWEX 50 (H$^+$), MeOH. [R = THP].

The thiazolidine nucleosides with either the 3'-thio-4'-aza- or 3'-aza-4'-thio-substitution (**3.94** and **3.95** respectively) have been prepared (Fig. 3.4),[63] however, both classes are devoid of biological activity probably owing to the lack of a suitable functional group at which phosphorylation could occur. The thiazolidinone nucleosides (Fig. 3.4, **3.96**, **3.97**) have also been described,[64,65] with the unusual nucleosides **3.97** displaying activity against both HIV-1 and duck hepatitis B virus.[65]

The key precursor of the thiazolidinone nucleosides **3.96** and **3.97**, is 2-benzoyloxy methyl-1,3-thiazolidin-4-one (**3.100**),[66] readily prepared from the reaction of the imine (**3.98**) with mercaptoacetic acid (**3.99**) (Scheme 3.16). *N*-alkylation gave the intermediate **3.101** which was condensed with various nucleobases using Cs$_2$CO$_3$ as the coupling agent to give the thiazolidinone **3.97** after deprotection.[65]

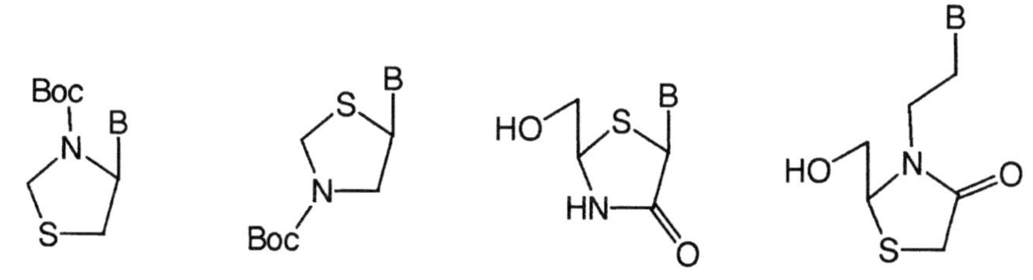

3.94, B = T, C.   **3.95**, B = T, C.   **3.96**, B = T, C, 5-Fluoro-C.   **3.97**, B = T, A, 5-Fluoro-C.

FIGURE 3.4 THIAZOLIDINE AND THIAZOLIDINONE NUCLEOSIDES.

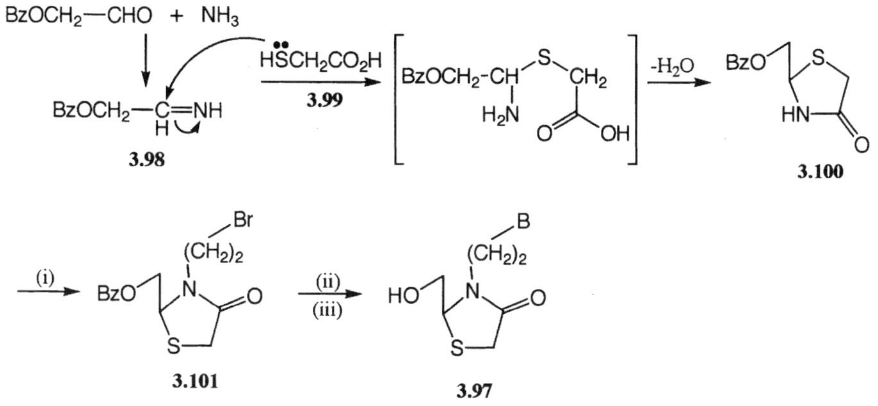

SCHEME 3.16 REAGENTS AND CONDITIONS: (I) KOH, Br(CH$_2$)$_2$Br, DMF (II) NUCLEOBASE (B), Cs$_2$CO$_3$, DMF, 100°C (III) NH$_3$, MeOH.

3.6 RING SIZE

The isolation of oxetanocin-A (**3.4**),[67] a nucleoside antibiotic which contains an oxetane ring in place of a furanose moiety, from *Bacillus megaterium* NK84-0218 led to an interest in analogues of oxetanocin-A and other nucleosides containing a 4-ring moiety in place of the normal furanose sugar (Fig. 3.5). In recent years pyranose nucleosides have emerged as potential antiviral agents, with the 1,5-anhydrohexitol nucleosides (**3.106**) showing particular promise.[68]

3.6.1 Oxetanocin-A and Derivatives

Oxetanocin-A is a broad spectrum antibiotic which also inhibits HIV.[69] Several synthetic procedures have been described for the preparation of the oxetane moiety, including ring contractions of ribose[70] or glucose,[71] [2 + 2] photocycloadditions,[72] cationic iodocyclization,[73] or *via* internal S$_N$2 displacement of an epoxide[74] or mesylate.[75]

Photocycloaddition of 2-methylfuran (**3.108**) with propionyloxyacetaldehyde (**3.107**) provides a short synthesis of oxetanocin, with the resulting photoadduct (**3.109**) of this reaction converted into the required oxetane (**3.110**) in a 1-pot procedure involving ozonolysis of the double bond followed by reduction and

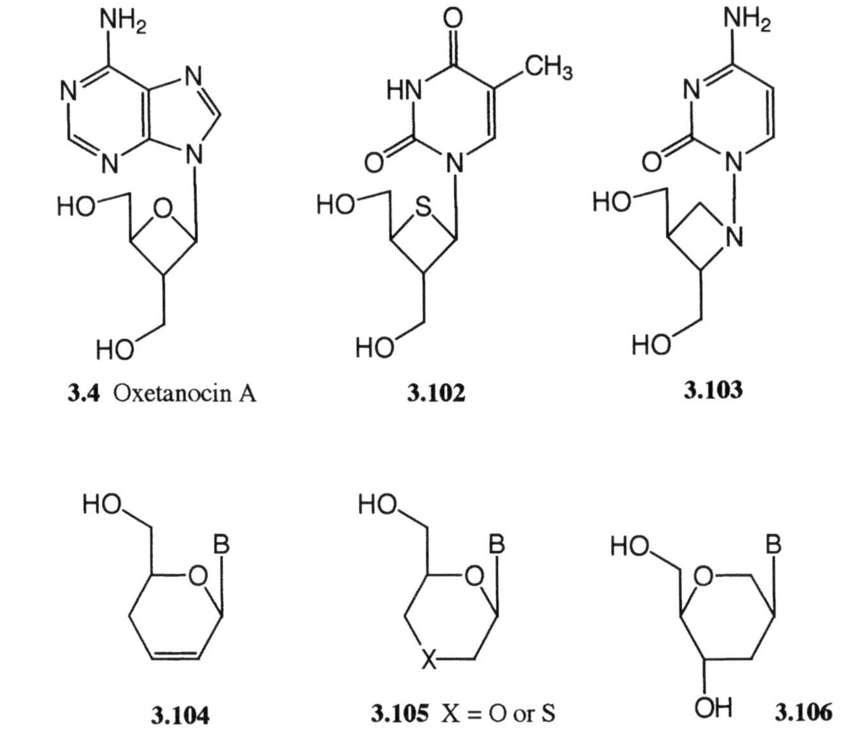

SUGAR MODIFIED NUCLEOSIDES

3.4 Oxetanocin A **3.102** **3.103**

3.104 **3.105** X = O or S **3.106**

FIGURE 3.5 4- AND 6-RING NUCLEOSIDES.

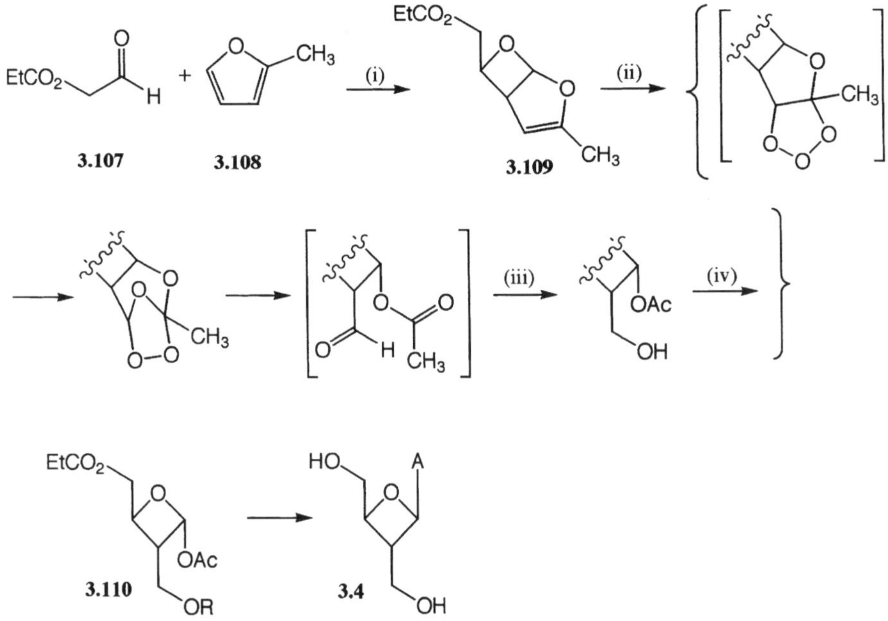

SCHEME 3.17 *REAGENTS AND CONDITIONS*: (I) C_6H_6, 8°C THEN IRRADIATION (W HANOVIA LAMP) (II) O_3, CH_2Cl_2, −78°C, DMS (III) $NaBH_4$-ALUMINA GEL (IV) MeOC(O)C(O)Cl, Et_3N, DMAP. R = CO_2Me.

acylation of the resulting aldehyde (Scheme 3.17). Coupling of the oxetane with adenine under Vorbrüggen conditions gives oxetanocin-A as a 1:3 α:β mixture.[72]

The thietane (**3.102**)[76] and azetidine (**3.103**)[77] analogues have been prepared though neither class displayed any biological activity in the assays used. Both require multistep syntheses with the thietane nucleoside proving particularly difficult to obtain, owing to a competing elimination reaction in the final coupling step which results in the thiete (**3.111**) being the major product isolated (Scheme 3.18).

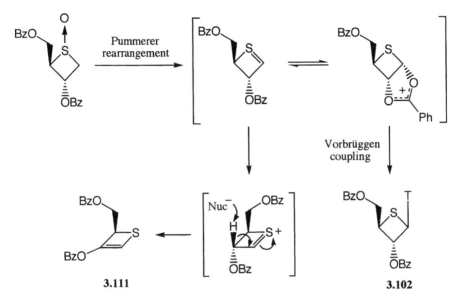

SCHEME 3.18   SYNTHESIS OF THIETANE NUCLEOSIDE **3.102**.

3.6.2 Pyranose Nucleosides

The six-membered 'pyranose' nucleosides are conformationally less flexible than conventional furanose nucleosides, therefore for activity to be observed there must be an exact steric and electronic fit between the enzyme and pyranose nucleoside substrate. The structural requirements of nucleosides with a six-membered carbohydrate moiety have been reviewed by Herdewijn.[68] Several pyranose nucleosides have been prepared (Fig. 3.5),[78–81] of these the 1,5-anhydrohexitols (**3.106**) are the most promising therapeutically, achieving optimal binding with the enzyme active site.[82–84]

The key intermediate, 1,5-anhydro-4,6-*O*-benzylidene-3-deoxy-D-glucitol (**3.116**),[82] can be prepared in multigram quantities from 3-deoxy-diacetone-D-glucose (**3.112**) (Scheme 3.19). The 5-iodouracil derivative (**3.117**), a highly selective, viral TK-dependent inhibitor of HSV-1 and HSV-2,[85] is synthesised from the alcohol **3.116** using Mitsunobu reaction conditions with subsequent deprotection.

Structural requirements for activity include a 1,4-substitution pattern between the hydroxymethyl group and base moiety, and a 1,3-relationship between the heterocyclic moiety, which adopts an axial conformation, and the ring oxygen.[68]

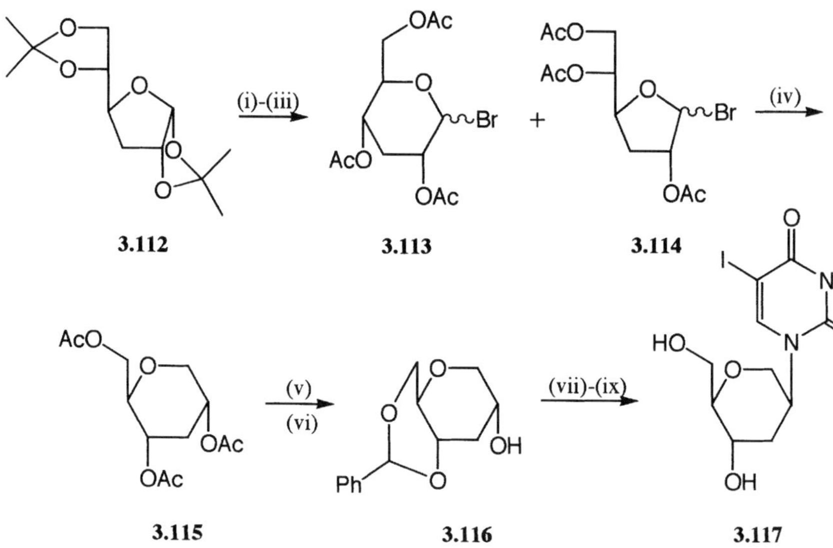

SCHEME 3.19 REAGENTS AND CONDITIONS: (I) IRA-120 (H⁺) RESIN, aq. EtOH, REFLUX (II) Ac$_2$O, PYRIDINE (III) HBr, AcOH (IV) Bu$_3$SnH, Et$_2$O (V) NaOMe, MeOH (VI) PhCH(OMe)$_2$, DIOXANE (VII) Ph$_3$P, DEAD, THF THEN N^3-BENZOYL-5-IODOURACIL (VIII) NH$_3$, MeOH (IX) 80% AeOH, REFLUX.

3.7 SPIRO-NUCLEOSIDES

The discovery of TSAO-T ([1-[2',5'-bis-*O*-(*tert*-butyldimethylsilyl)-β-D-ribofuranosyl]thymine]-3'-spiro-5"-(4"-amino-1",2"-oxathiole-2",2"-dioxide) (**1.46**) as a specific and potent inhibitor of HIV-1[86,87] led to considerable research into the preparation of spironucleosides. TSAO-T differs from the DDN's in that rather than competitively inhibiting the incorporation of natural substrates into DNA, it acts as a non-nucleoside HIV specific RT inhibitors by interacting with an allosteric non-substrate binding site of RT located near the polymerase active site.[88] The essential pharmacophore of the TSAO compounds is the 4"-amine group of the 3'-spiro moiety which interacts with a well defined carboxylic acid residue of the RT target enzyme.[89]

The key cyano mesylate (**3.119**) can be obtained stereoselectively from 5-*O*-benzoyl-1,2-*O*-isopropylidene-β-D-*erythro*-pentofuranose-3-ulose (**3.118**).[90] The key transformation of the cyano mesylate nucleoside (**3.120**) into the spiro-nucleoside (**3.121**) involves initial proton abstraction from the methyl group of the mesylate with subsequent attack of the resulting carbanion at the nitrile carbon atom (*m*, Scheme 3.20).[91] Structure-activity relationship studies have been carried out on the TSAO compounds, an overview of which has been published by Camarasa *et al*.[92]

1",3"-Thiazolidine-2"-spiro-3'-deoxyuridine derivatives (*e.g.* **3.122**, **3.123**),[93] and tricyclic *cis*-fused-spiro-isoxazolidine nucleosides (*e.g.* **3.124**) have also been prepared,[94] but were devoid of antiviral activity (Fig. 3.6).

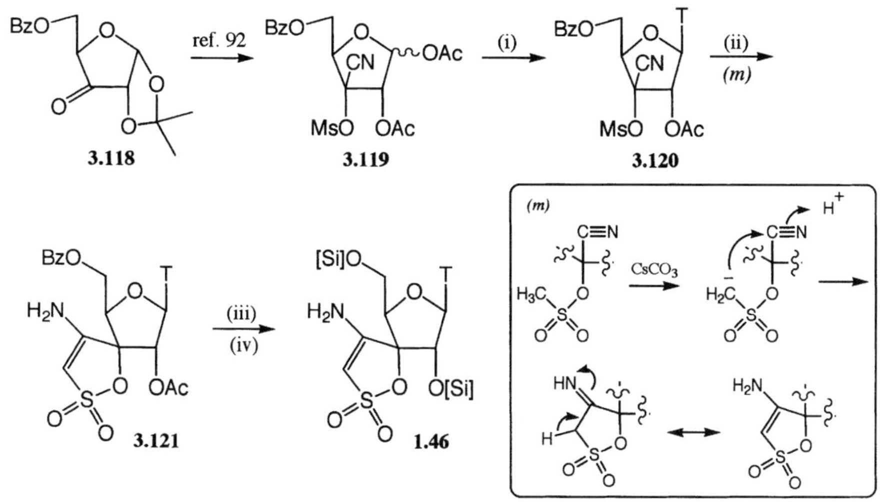

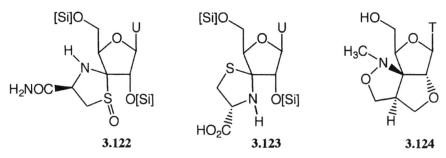

SCHEME 3.20 REAGENTS AND CONDITIONS: (I) (TMS)$_2$T, MeCN, TMSOTf (II) Cs$_2$CO$_3$, MeCN (III) NH$_3$, MeOH (IV) TBDMSCl, DMAP, MeCN, 80°C. [Si] = TBDMS.

FIGURE 3.6 THIAZOLIDINE AND ISOXAZOLIDINE SPIRO-NUCLEOSIDES ([Si] = TBDMS).

3.8 REFERENCES

1. V. Nair in "*Nucleosides and Nucleotides as Antitumor and Antiviral Agents*", C.K. Chu, D.C. Baker (eds.), Plenum Press, New York, New York, **1996**, pp 127–140.
2. D.M. Huryn, B.C. Sluboski, S.Y. Tam, L.J. Todaro, M. Weigle, *Tetrahedron Lett.*, **1989**, *30*, 6259–6262.
3. D.M. Huryn, B.C. Sluboski, M. Weigle, I. Sim, B,D, Anderson, H. Mitsuya, S. Broder, *J. Med. Chem.*, **1992**, *35*, 2347–2354.
4. J.A. Tino, J.M. Clark, A.K. Field, G.A. Jacobs, K.A. Lis, T.L. Michalik, B. McGeever–Rubin, W.A. Slusarchyk, S.H. Spergel, J.E. Sundeen, A.V. Tuomari, E.R. Weaver, M.G. Young, R. Zahler, *J. Med. Chem.*, **1993**, *36*, 1221–1229.
5. M.F. Jones, S.A. Noble, C.A. Robertson, R. Storer, *Tetrahedron Lett.*, **1991**, *32*, 247–250.
6. K–M.E. Ng, L.E. Orgel, *J. Med. Chem.*, **1989**, *32*, 1754–1757.
7. K. Schürrle, W. Piepersberg, *J. Carbohydr. Chem.*, **1996**, *15*, 435–447.
8. Z.M. Nuesca, V. Nair, *Tetrahedron Lett.*, **1994**, *35*, 2485–2488.
9. K. Sujino, H. Sugimura, *J. Chem. Soc., Chem. Commun.*, **1994**, 2541–2542.
10. R.T. Walker in "*Anti–infectives. Recent Advances in Chemistry and Structure–Activity Relationships*", P.H. Bentley, P.J. O'Hanlon (eds), The Royal Society of Chemistry, Cambridge, U.K., **1997**, pp 203–237.
11. L. Pickering, B.S. Malhi, P.L. Coe, R.T. Walker, *Nucleosides & Nucleotides*, **1994**, *13*, 1493–1506.
12. G.D. Kini, W.J. Hennen, *J. Org. Chem.*, **1986**, *51*, 4436–4439.

13. R.E. Parks Jr., J.D. Stoeckler, C. Cambor, T.M. Savarese, G.W. Crabtree, S.–H. Chu, in "*Molecular Actions and Targets for Cancer Chemotherapeutic Agents*", A.C. Sartorelli, J.S. Lazo, J.R. Bertino (eds.), Academic Press, New York, **1981**, pp 229–252.
14. L.H. Koole, J. Plavec, H. Liu, B.R. Vincent, M.R. Dyson, P.L. Coe, R.T. Walker, G.W. Hardy, S.G. Rahim, J. Chattopadhyaya, *J. Am. Chem. Soc.*, **1992**, *114*, 9936–9934.
15. R.L. Whistler, W.E. Dick, T.R. Ingle, R.M. Rowell, B. Urbas, *J. Org. Chem.*, **1964**, *29*, 3723–3725.
16. E.J. Reist, D.E. Gueffroy, L. Goodman, *J. Am. Chem. Soc.*, **1964**, *86*, 5658–5663.
17. K.N. Tiwari, J.A. Secrist III, J.A. Montgomery, *Nucleosides & Nucleotides*, **1994**, *13*, 1819–1828.
18. C. Leydier, L. Bellon, J-L. Barascut, J. Deydier, G. Maury, H. Pelicano, M. El Aloaoui, J–L. Imbach, *Nucleosides & Nucleotides*, **1994**, *13*, 2035–2050.
19. M.R. Dyson, P.L. Coe, R.T. Walker, *Carbohydr. Res.*, **1991**, *216*, 237–248.
20. K.S. Jandu, D.L. Selwood, *J. Org. Chem.*, **1995**, *60*, 5170–5173.
21. J. Mann, A.J. Tench, A.C. Weymouth–Wilson, S. Shaw–Ponter, R.J. Young, *J. Chem. Soc., Perkin Trans. 1*, **1995**, 677–681.
22. J.A. Secrist III, R.M. Riggs, K.N. Tiwari, J.A. Montgomery, *J. Med. Chem.*, **1992**, *35*, 533–538.
23. G. Rassu, F. Zanardi, L. Battistini, E. Gaetani, G. Casiraghi, *J. Med. Chem.*, **1997**, *40*, 168–180.
24. J. Branalt, I. Kvarnström, G. Niklasson, S.C.T. Svensson, B. Classon, B. Samuelsson, *J. Org. Chem.*, **1994**, *59*, 1783–1788.
25. E. Ichikawa, S. Yamamura, K. Kato, *Bioorg. Med. Chem. Lett.*, **1999**, *9*, 1113–1114.
26. J. Barker, H.G. Fletcher Jr., *J. Org. Chem.*, **1961**, *26*, 4605–4609.
27. Y. Wang, G. Inguaggiato, M. Jasamai, M. Shah, D. Hughes, M.J. Slater, C. Simons, *Bioorg. Med. Chem.*, **1999**, *7*, 481–487.
28. R.J. Young, S. Shaw–Ponter, G.W. Hardy, G. Mills, *Tetrahedron Lett.*, **1994**, *35*, 8687–8690.
29. G. Inguaggiato, M. Jasamai, J.E. Smith, M.J. Slater, C. Simons, *Nucleosides & Nucleotides*, **1999**, *18*, 457–467.
30. I. Basnak, G.P. Otter, R.J. Duncombe, N.B. Westwood, M. Pietrarelli, G.W. Hardy, G. Mills, S.G. Rahim, R.T. Walker, *Nucleosides & Nucleotides*, **1998**, *17*, 29–38.
31. E. De Clercq, J. Descamps, P. De Somer, P.J. Barr, A.S. Jones, R.T. Walker, *Proc. Natl. Acad. Sci. USA*, **1979**, *76*, 2947–2591.
32. N.A. Van Draanen, G.A. Freeman, S.A. Short, R. Harvey, R. Jansen, G. Szczech, G.W. Koszalka, *J. Med. Chem.*, **1996**, *39*, 538–542.
33. E.J. Reist, D.E. Gueffroy, R.W. Blackford, L. Goodman, *J. Org. Chem.*, **1966**, *31*, 4025–4030.
34. E.J. Reist, L.V. Fisher, L. Goodman, *J. Org. Chem.*, **1967**, *32*, 2541–2544.
35. T.S. Mansour, B. Belleau, K. Bednarski, T. Breining, W.L. Brown, A. Cimpoia, M. DiMarcio, D.M. Dixit, C.A. Evans, H. Jin, J.L. Kraus, D. Lafleur, N. Lee, N. Nguyen–Ba, M.A. Siddiqui, H.L.A. Tse, B. Zacharie in "*Recent Advances in the Chemistry of Anti–Infective Agents*", P.H. Bentley, R. Ponsford (eds), The Royal Society of Chemistry, Cambridge, U.K., **1993**, pp 336–347.
36. L. Cui, R.F. Schinazi, G. Gosselin, J-L. Imbach, C.K. Chu, R.F. Rando, G.R. Revankar, J–P. Sommadossi, *Biochem. Pharmacol.*, **1996**, *52*, 1577–1584.
37. T.S. Mansour, H. Jin, W. Wang, E.U. Hooker, C. Ashman, N. Cammack, H. Salomon, A.R. Belmonte, M.A. Wainberg, *J. Med. Chem*, **1995**, *38*, 1–4.
38. H.O. Kim, S.K. Ahn, A.J. Alves, J.W. Beach, L.S. Jeong, B.G. Choi, P. Van Roey, R.F. Schinazi, C.K. Chu, *J. Med. Chem.*, **1992**, *35*, 1987–1995.
39. C.A. Evans, D.M. Dixit, M.A. Siddiqui, H. Jin, H.L.A. Tse, A. Cimpoia, K. Bednarski, T. Breining, T.S. Mansour, *Tetrahedron: Asymmetry*, **1993**, *4*, 2319–2322.
40. H.O. Kim, R.F. Schinazi, S. Nampelli, K. Shanmuganathan, D.L. Cannon, A.J. Alves, L.S. Jeong, J.W. Beach, C.K. Chu, *J. Med. Chem.*, **1993**, *36*, 30–36.
41. T.S. Mansour, H. Jin, W. Wang, D.M. Dixit, C.A. Evans, H.L.A. Tse, B. Belleau, J.W. Gillard, E.U. Hooker, C. Ashman, N. Cammack, H. Salomon, A.R. Belmonte, M.A. Wainberg, *Nucleosides & Nucleotides*, **1995**, *14*, 627–635.
42. R.A. Sheldon, J.K. Kochi, *Org. React.*, **1972**, *19*, 279–421.
43. L.S. Jeong, R.F. Schinazi, J.W. Beach, H.O. Kim, K. Shanmuganathan, S. Nampelli, M.W. Chun, W–K. Chung, B.G. Choi, C.K. Chu, *J. Med. Chem.*, **1993**, *36*, 2627–2638.
44. S. Wolfe, S.K. Hasan, J.R. Campbell, *J. Chem. Soc. D*, **1970**, 1420–1421.
45. C.K. Chu, J.W. Beach, L.S. Leong, B.G. Choi, F.I. Comer, A.J. Alves, R.F. Schinazi, *J. Org. Chem.*, **1991**, *56*, 6503–6505.
46. L.S. Jeong, A.J. Alves, S.W. Carrigan, H.O. Kim, J.W. Beach, C.K. Chu, *Tetrahedron Lett.*, **1992**, *33*, 595–598.
47. J.W. Beach, L.S. Jeong, A.J. Alves, D. Pohl, H.O. Kim, C–N. Chang, S–L. Doong, R.F. Schinazi, Y–C. Cheng, C.K. Chu, *J. Org. Chem.*, **1992**, *57*, 2217–2219.

48. J. Branalt, I. Kvarnström, B. Classon, B. Samuelsson, *J. Org. Chem.*, **1996**, *61*, 3604–3610.
49. J.A.V. Coates, I.M. Mutton, C.R. Penn, R. Storer, C. Williamson, International Patent Application Publication No. WO 91/17159, **1991**.
50. N. Nguyen-Ba, W. Brown, N. Lee, B. Zacharie, *Synthesis : Stuttgart*, **1998**, 759–762.
51. R. Caputo, A. Guaragna, G. Palumbo, S. Pedatella, *Nucleosides & Nucleotides*, **1998**, *17*, 1739–1745.
52. J. Du, C.K. Chu, *Nucleosides & Nucleotides*, **1998**, *17*, 1–13.
53. A. Leggio, A. Liguori, A. Procopcio, C. Siciliano, G. Sindona, *Tetrahedron Lett.*, **1996**, *37*, 1277–1280.
54. J.M.J Tronchet, M. Iznaden, F. Barbalat–Rey, I. Komaromi, N. Dolatshahi, G. Bernardinelli, *Nucleosides & Nucleotides*, **1995**, *14*, 1737–1758.
55. J.M.J Tronchet, M. Iznaden, F. Barbalat–Rey, H. Dhimane, A. Ricca, J. Balzarini, E. De Clercq, *Eur. J. Med. Chem.*, **1992**, *27*, 555–560.
56. Y. Xiang, H–J. Gi, D. Niu, R.F. Schinazi, K. Zhao, *J. Org. Chem.*, **1997**, *62*, 7430–7434.
57. P. Merino, S. Franco, F.L. Merchan, T. Tejero, *Tetrahedron Lett.*, **1998**, *39*, 6411–6411.
58. J.M.J Tronchet, M. Zsély, K. Capek, I. Komaromi, M. Geoffroy, E. De Clercq, J. Balzarini, *Nucleosides & Nucleotides*, **1994**, *13*, 1871–1889.
59. Y.J. Xiang, J. Chen, R.F. Schinazi, K. Zhao, *Bioorg. Med. Chem. Lett.*, **1996**, *6*, 1051–1054.
60. H–J. Gi, Y. Xiang, R.F. Schinazi, K. Zhao, *J. Org. Chem.*, **1997**, *62*, 88–92.
61. P. Li, H–J. Gi, L. Sun, K. Zhao, *J. Org. Chem.*, **1998**, *63*, 366–369.
62. T. Mukaiyama, T. Hoshino, *J. Am. Chem. Soc.*, **1960**, *82*, 5339–5342.
63. G. Rassu, F. Zanardi, L. Cornia, G. Casiraghi, *Nucleosides & Nucleotides*, **1996**, *15*, 1113–1120.
64. J.C. Graciet, P. Faury, M. Camplo, A.S. Charvet, N. Mourier, C. Trabaud, V. Niddam, V. Simon, J.L. Kraus, *Nucleosides & Nucleotides*, **1995**, *14*, 1379–1392.
65. J.C. Graciet, V. Niddam, M. Gamberoni, C. Trabaud, J. Dessolin, M. Medou, N. Mourier, F. Zoulim, C. Borel, O. Hantz, M. Camplo, J.C. Chermann, J.L. Kraus, *Bioorg. Med. Chem. Lett.*, **1996**, *6*, 1775–1780.
66. A.R. Surrey, R.A. Cutler, *J. Am. Chem. Soc.*, **1954**, *76*, 578–580.
67. N. Shimada, S. Hasegawa, T. Harada, T. Tonisawa, T. Fujii, *J. Antibiot.*, **1986**, *39*, 1623–1625.
68. P. Herdewijn in "*Anti-infectives. Recent Advances in Chemistry and Structure–Activity Relationships*", P.H. Bentley, P.J. O'Hanlon (eds), The Royal Society of Chemistry, Cambridge, U.K., **1997**, pp 316–327.
69. H. Hoshino, N. Shimizu, N. Shimada, T. Takita, T. Takeuchi, *J. Antibiot.*, **1987**, *40*, 1077–1078.
70. F.X. Wilson, G.W.J. Fleet, K. Vogt, Y. Wang, D.R. Witty, S. Choi, R. Storer, P.L. Meyers, C.J. Wallis, *Tetrahedron Lett.*, **1990**, *31*, 6931–6934.
71. D.W. Norbeck, J.B. Kramer, *J. Am. Chem. Soc.*, **1988**, *110*, 7217–7218.
72. R. Hambalek, G. Just, *Tetrahedron Lett.*, **1990**, *31*, 5445–5448.
73. M.E. Jung, C.J. Nichols, *Tetrahedron Lett.*, **1996**, *37*, 7667–7670.
74. S. Niitsuma, Y–I. Ichikawa, K. Kato, T. Takita, *Tetrahedron Lett.*, **1987**, *28*, 3967–3968.
75. S. Nishiyama, S. Yamamura, K. Kato, T. Takita, *Tetrahedron Lett.*, **1988**, *29*, 4739–4742.
76. N. Nishizono, N. Koike, Y. Yamagata, S. Fujii, A. Matsuda, *Tetrahedron Lett.*, **1996**, *37*, 7569–7572.
77. F. Hosono, S. Nishiyama, S. Yamamura, T. Izawa, K. Kato, Y. Terada, *Tetrahedron*, **1994**, *50*, 13335–13346.
78. P. Herdewijn, A. Van Aerschot, *Bull. Soc. Chim. Belg.*, **1990**, *99*, 895–901.
79. H. Hansen, E. Pedersen, *Arch. Pharm.* (Weinheim), **1992**, *325*, 491–497.
80. P. Herdewijn, A. Van Aerschot, J. Balzarini, E. De Clercq, *Nucleosides & Nucleotides*, **1991**, *10*, 119–127.
81. A. Van Aerschot, G. Janssen, P. Herdewijn, *Bull. Soc. Chim. Belg.*, **1990**, *99*, 769–777.
82. I. Verheggen, A. Van Aerschot, S. Toppet, R. Snoeck, G. Janssen, J. Balzarini, E. De Clercq, P. Herdewijn, *J. Med. Chem.*, **1993**, *36*, 2033–2040.
83. N. Hossain, J. Rozenski, E. De Clercq, P. Herdewijn, *J. Org. Chem.*, **1997**, *62*, 2442–2447.
84. N. Hossain, P. Herdewijn, *Nucleosides & Nucleotides*, **1998**, *17*, 1781–1786.
85. I. Verheggen, A. Van Aerschot, L. Van Meervelt, J. Rozenski, L. Wiebe, R. Snoeck, G. Andrei, J. Balzarini, P. Claes, E. De Clercq, P. Herdewijn, *J. Med. Chem.*, **1995**, *38*, 826–835.
86. J. Balzarini, M.J. Pérez-Pérez, A. San-Félix, D. Schols, C.F. Perno, A.M. Vandamme, M.J. Camarasa, E. De Clercq, *Proc. Natl. Acad. Sci. USA*, **1992**, *89*, 4392–4396.
87. M.J. Camarasa, M.J. Pérez-Pérez, A. San-Félix, J. Balzarini, E. De Clercq, *J. Med. Chem.*, **1992**, *35*, 2721–2727.
88. J. Balzarini, M.J. Pérez–Pérez, A. San-Félix, M.J. Camarasa, I.C. Bathurst, P.J. Barr, E. De Clercq, *J. Biol. Chem.*, **1992**, *267*, 11831–11838.
89. S. Veláquez, M.L. Jimeno, M.J. Camarasa, J. Balzarini, *Tetrahedron*, **1994**, *37*, 11013–11022.
90. M.J. Pérez–Pérez, A. San-Félix, J. Balzarini, E. De Clercq, M.J. Camarasa, *J. Med. Chem.*, **1992**, *35*, 2988–2995.
91. M-J. Camarasa, M-J. Pérez, S. Veláquez, A. San-Félix, R. Alvarez, S. Ingate, M-L. Jimeno, E. De Clercq, J. Balzarini in "*Anti-infectives. Recent Advances in Chemistry and Structure–Activity Relationships*", P.H. Bentley, P.J. O'Hanlon (eds), The Royal Society of Chemistry, Cambridge, U.K., **1997**, pp 259–268.

92. A. Calvo-Mateo, M-J. Camarasa, A. Diaz-Ortiz, F.G. de las Heras, *J. Chem. Soc., Chem. Commun.*, **1988**, 1114–1115.
93. J.M.J. Tronchet, I. Kovács, F. Barbalat-Rey, M. Vega Holm, *Nucleosides & Nucleotides*, **1998**, *17*, 1115–1123.
94. J. Rong, J. Plavec, J. Chattopadhyaya, *Tetrahedron*, **1994**, *50*, 4921–4936.

CHAPTER 4

HETEROCYCLIC BASE MODIFIED NUCLEOSIDES

4.1 INTRODUCTION

Modification of the heterocyclic base moiety of nucleosides has been less extensively studied than sugar modifications, however research in this area has led to many compounds with potent activity. 5-Ring heterocyclic nucleosides structurally derived from 5-amino-1-β-D-ribofuranosylimidazole-4-carboxamide (AICAR, **1.29**) by modification at the 5-position, such as Bredinin (**4.1**) and EICAR (**1.55**), or by modification of the ring skeleton, such as Ribavirin (**1.8**) and TCNR (**4.2**), are known to exhibit a broad spectrum of activity against a range of viruses including hepatitis C, influenza and respiratory syncytial virus,[1] in addition to a variety of other chemotherapeutic applications.

The deaza-purine modified nucleosides, such as the naturally occurring tubercidin (**4.3**), have been extensively studied as anticancer agents,[2] benzimidazole nucleosides are of more recent interest with the 2,5,6-trichloro compound TCRB (**4.4**) showing activity against CMV.[3] Of the pyridine modified nucleosides, the 5-aza-compounds such as the anticancer agent 5-azacytidine (**4.5**),[4] and those containing a fused furan moiety (*e.g.* **4.6**) which displays inhibitory activity against VZV,[5] are of particular interest (Fig. 4.1).

4.2 5-RING HETEROCYCLIC BASES

The 5'-monophosphates of the synthetic 5-ring heterocyclic nucleosides mimic the biosynthetic nucleoside monophosphate 5-amino-1-β-D-ribofuranosylimidazole-4-carboxamide (AICAR, **1.29**) and as such inhibit inosine 5'-monophosphate (IMP) dehydrogenase, one of the key rate-controlling enzymes in the *de novo* biosynthesis of purine nucleotides (see section 1.7.3.1). Inhibitors of IMP dehydrogenase have a broad spectrum of biological activity including antiviral, anticancer and immunosuppresive activities.[6] AICAR (Acadenosine®) itself is a beneficial chemotherapeutic for the treatment of stroke owing to its ability to act as an adenosine-regulating agent.[7]

4.2.1 Imidazole Nucleosides

The most promising compounds in this class of nucleosides are bredinin (Mizoribine®) and the 5-substituted derivatives such as EICAR. Bredinin, used clinically as an

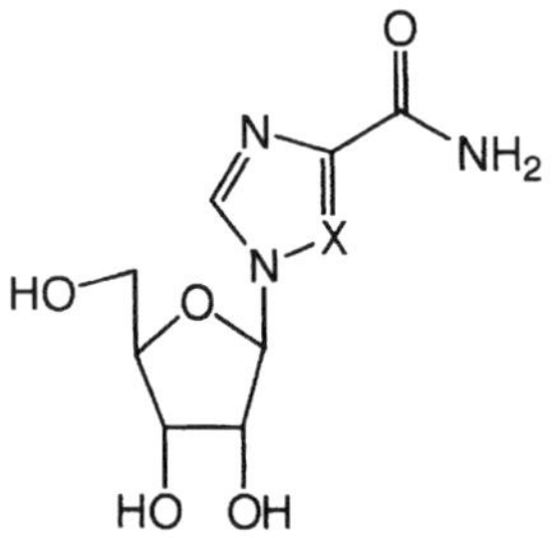

HETEROCYCLIC BASE MODIFIED NUCLEOSIDES

Bredinin 4.1, X = C-OH
Ribavirin 1.8, X = N
EICAR 1.55, X = C-C≡CH

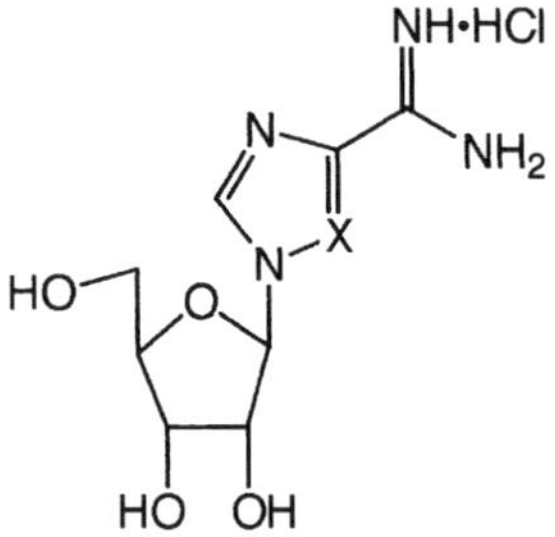

4.2 TCNR

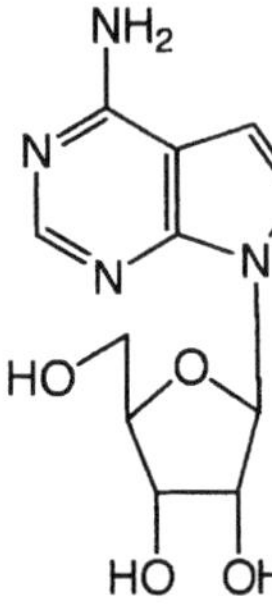

4.3 Tubercidin

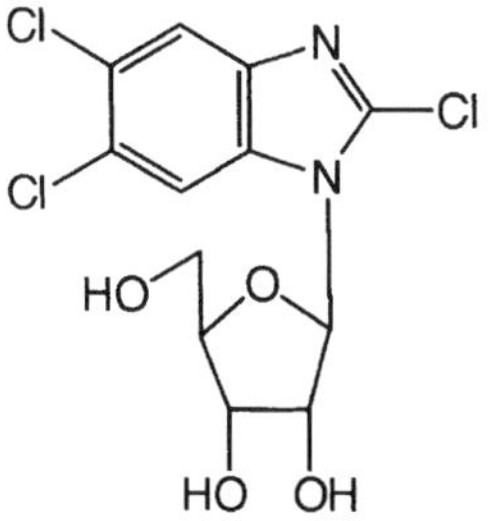

4.4 TCRB

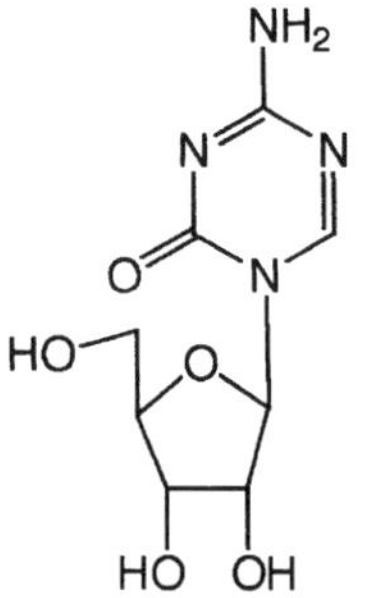

4.5 5-Azacytidine

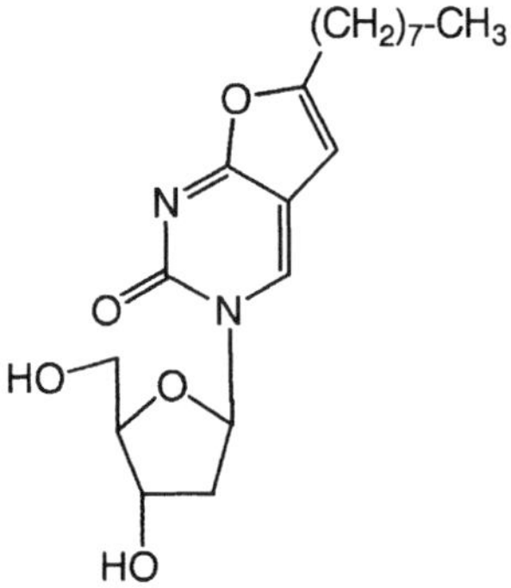

4.6

FIGURE 4.1 HETEROCYCLIC BASE MODIFIED NUCLEOSIDES.

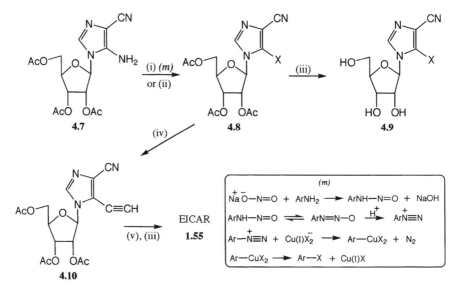

SCHEME 4.1 *REAGENTS AND CONDITIONS*: (I) NaNO$_2$, MeOH, (A) CuI$_2$, 6N HCl, X = I (B) CuCl$_2$, 6N HCl, X = Cl (C) CuBr$_2$, HBr, X = Br (II) (CH$_3$)$_2$CH(CH$_2$)$_2$ONO, CH$_2$I$_2$ (III) NH$_4$OH, MeOH, SEALED VESSEL, 4–5 DAYS (IV) HC≡CH, (PhCN)$_2$PdCl$_2$, Et$_3$N, CH$_3$CN (V) 30% H$_2$O$_2$.

immunosuppressant for post transplant patients,[8] was first isolated from *Eupencillium brefeldianum* M-2166.[9]

Of the syntheses reported,[10-12] the direct Vorbrüggen coupling of silylated 4(5)-carbamoyl imidazolium-5(4)-olate with an acylated ribose resulted in the best overall yield.[11] EICAR is generated from a palladium cross-coupling reaction of the 5-iodo-imidazole nucleoside (**4.8**) with acetylene and catalytic bis(benzonitrile)palladium dichloride followed by reaction of (**4.10**) with H_2O_2 and deprotection (Scheme 4.1).[13,14]

The 5-halo derivatives (**4.9**, X = I, Cl, Br), which display significant antiviral activity, are prepared by diazotization of **4.7** involving an electron transfer process between the diazonium ion and copper (I), a process known as the Sandmeyer reaction (Scheme 4.1, *m*).[15] In the case of the synthesis of EICAR the diazonium ion, produced by reaction of **4.7** with isoamyl nitrite, is captured by the nucleophile I⁻ generated from diiodomethane (see Scheme 2.2).

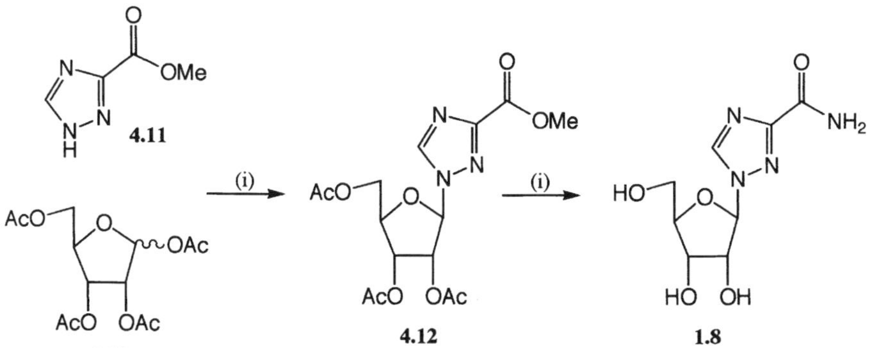

SCHEME 4.2 *REAGENTS AND CONDITIONS*: (I) BIS(*P*-NITROPHENYL)PHOSPHATE, 165°C (II) NH_3, MeOH.

4.2.2 Triazole Nucleosides

Ribavirin (**1.8**, Virazole®, 1-β-D-ribofuranosyl-1*H*-1,2,4-triazole-3-carboxamide), which inhibits both IMPDH and PNP, is used in the treatment of RSV infections, Lassa fever, Hepatitis (A, B and C), measles and mumps.[16,17] Ribavirin was first synthesised using the acid-catalysed fusion procedure by heating a mixture of the 1,2,4-triazole (**4.11**) with the peracetylated sugar (**2.38**) and catalytic bis-(*p*-nitrophenyl)phosphate at 165°C (Scheme 4.2).[18]

Ribavirin derivatives have also been prepared by enzymatic synthesis[19] and Vorbrüggen coupling.[20] Of the derivatives prepared, TCNR (**4.2**, 1-β-D-ribofuranosyl-1*H*-1,2,4-triazole-3-carboxamidine) and 5-amino-TCNR (**4.13**) have been the most promising,[21,22] and act as competitive reversible inhibitors of human lymphoblast purine nucleoside phosphorylase.[23] TCNR and its derivatives are better substrates for human PNP than Ribavirin owing to the interaction of the protonated carboxamidine with the Asn 243 and Glu 201 residues in the PNP active site.[24]

5-Amino-TCNR is prepared by dehydration of the carboxamide (**4.14**) with phosgene to give the carbonitrile (**4.15**), followed by transformation of the carbonitrile function to the amidine unit with concomitant nucleophilic displacement of the 5-chloro group (Scheme 4.3).

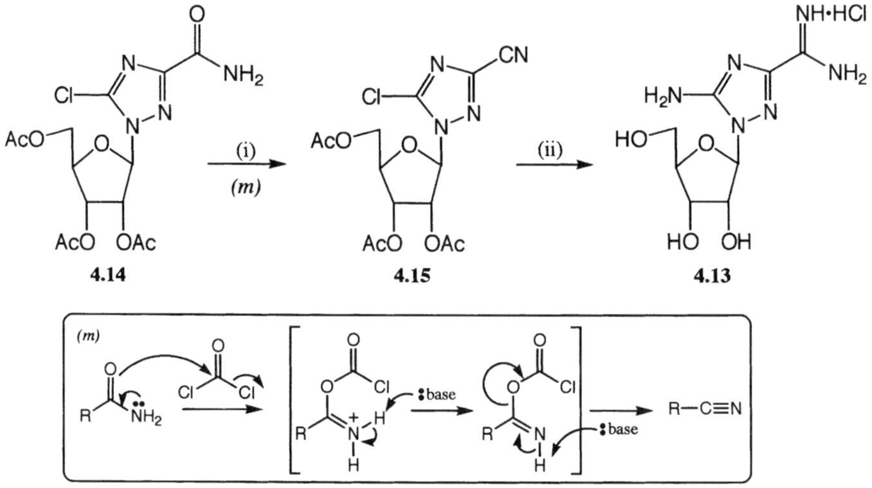

SCHEME 4.3 REAGENTS AND CONDITIONS: (I) 20% PHOSGENE, CH_2Cl_2, PYRIDINE (II) NH_4Cl, lq. NH_3, STEEL BOMB, 110°C, 24H.

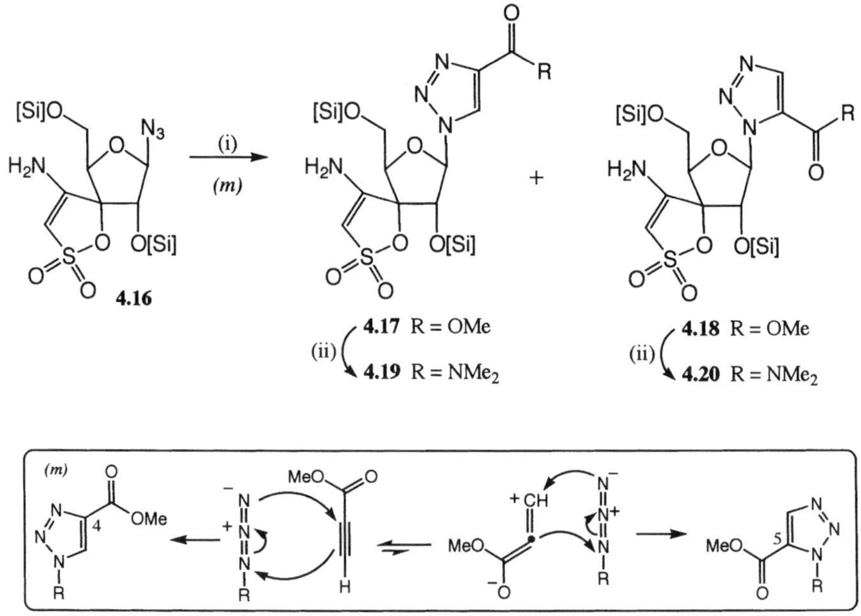

SCHEME 4.4 REAGENTS AND CONDITIONS: (I) METHYL PROPIOLATE, TOLUENE, REFLUX, 18H (II) Me_2NH, EtOH, 8H. [Si] = TBDMS.

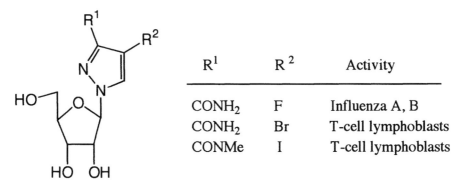

FIGURE 4.2 ACTIVE PYRAZOLE NUCLEOSIDES.

Several publications have described the synthesis of 1,2,3-triazole nucleosides *via* 1,3-dipolar cycloaddition of an azido sugar with an acetylene,[25–27] the most active compounds being the 1,2,3-triazole-TSAO compounds which are potent inhibitors of HIV-1. Reaction of the azido sugar (**4.16**) with the unsymmetrical acetylene, methyl propiolate, yields the 4- and 5-substituted triazoles (**4.17** and **4.18**) in a ratio of approximately 3:1, which are converted to the dimethylamido derivatives (**4.19** and **4.20**) by reaction with dimethylamine (Scheme 4.4).[25]

The pyrazole-3-carboxamide derivatives (Fig. 4.2), which can be formally considered as 4-deaza-analogues of ribavirin, have also shown activity against T-cell lymphocytes[28,29] and Influenza A and B *in vitro*.[30] What is noticeable with all the nucleosides which contain a 5-ring heterocyclic base moiety, is that modification of either the AICAR or ribavirin structure is very limited, with loss of activity generally observed with fairly minor modifications.

4.3 PURINE-MODIFIED NUCLEOSIDES

A wide range of fused heterocyclic base systems which mimic the conventional purine moiety have been prepared, or in some cases isolated from natural sources, many of which exhibit significant biological activity. Synthesis of these nucleosides requires either preparation of the heterocyclic base prior to coupling with the sugar moiety or construction of the base from a suitable sugar precursor.

4.3.1 Benzimidazole Nucleosides

Benzimidazole nucleosides have been of interest for some time,[31] however it is only recently that the value of this class of nucleosides as antiviral agents has emerged. Many benzimidazole nucleosides have been prepared,[32–35] of these 2,5,6-trichloro-1-(β-ribofuranosyl)benzimidazole (**4.4**, TCRB) and the 6-bromo analogue (**4.21**, BDCRB) are selective against CMV and non-cytotoxic.[3] Both compounds act *via* a novel mechanism involving inhibition of viral processing and packaging, inhibition of CMV DNA processing has been shown to be mediated through the specified protein kinase UL89 gene product.[36]

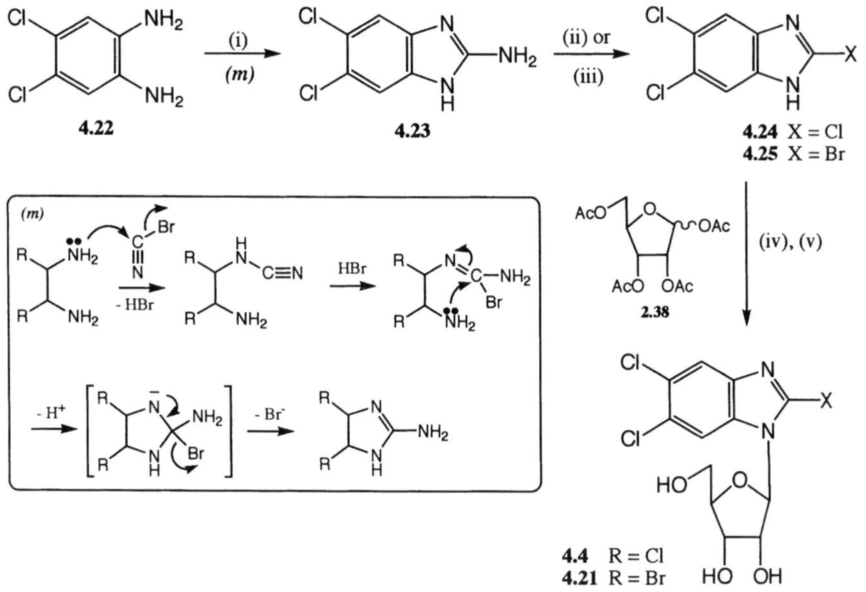

SCHEME 4.5 *REAGENTS AND CONDITIONS*: (I) CNBr, MeOH (II) HCl, NaONO, CuCl$_2$ (III) HBr, NaONO, CuBr$_2$ (IV) **2.38**, BSA, CH$_3$CN, TMSOTf, 16 H (V) NH$_3$, MeOH.

Synthesis of the benzimidazole components of TCRB and BDCRB can be achieved by reaction of commercially available 4,5-dichloro-*o*-phenylenediamine (**4.22**) with cyanogen bromide. Diazotization of the resulting 2-amino-5,6-dichlorobenzimidazole (**4.23**),[37] gives the required benzimidazoles bases (**4.24, 4.25**), which after Vorbrüggen coupling and deprotection of the sugar component produces TCRB and BDCRB (Scheme 4.5).

4.3.2 7-Deazapurine Nucleosides

Isolation of the 7-deazanucleosides, also described as pyrrolo[2,3-*d*]pyrimidine nucleosides, such as tubercidin (**4.4**), toyocamycin (**4.26**) and sangivamycin (**4.27**), which exhibit significant antitumour activity *in vivo*,[38] resulted in the synthesis of a series of both sugar and base modified derivatives.[2] The 7-deazanucleosides have potential advantages over conventional purine nucleosides, in the case of tubercidin these advantages include stability towards adenosine deaminase and purine nucleoside phosphorylase.[39,40] Tubercidin and its derivatives, such as the 2'-deoxy-2'-*ara*-fluoro-tubercidin (**4.28**), can be prepared by the direct sodium-salt glycosylation of 4-chloropyrrolo[2,3-*d*]pyrimidine (**4.29**) and an appropriate sugar (**4.30**) (Scheme 4.6).[41]

The synthesis of the base moiety involves initial formation of the pyrimidine (**4.32**) by reaction of the mononitrile (**4.31**) with urea in the presence of sodium ethoxide. Subsequent ring closure under acidic conditions (for mechanism see Scheme 2.16), chlorination and selective C-2 dehalogenation gave the required 4-chloropyrrolo[2,3-*d*]pyrimidine (**4.29**).[42]

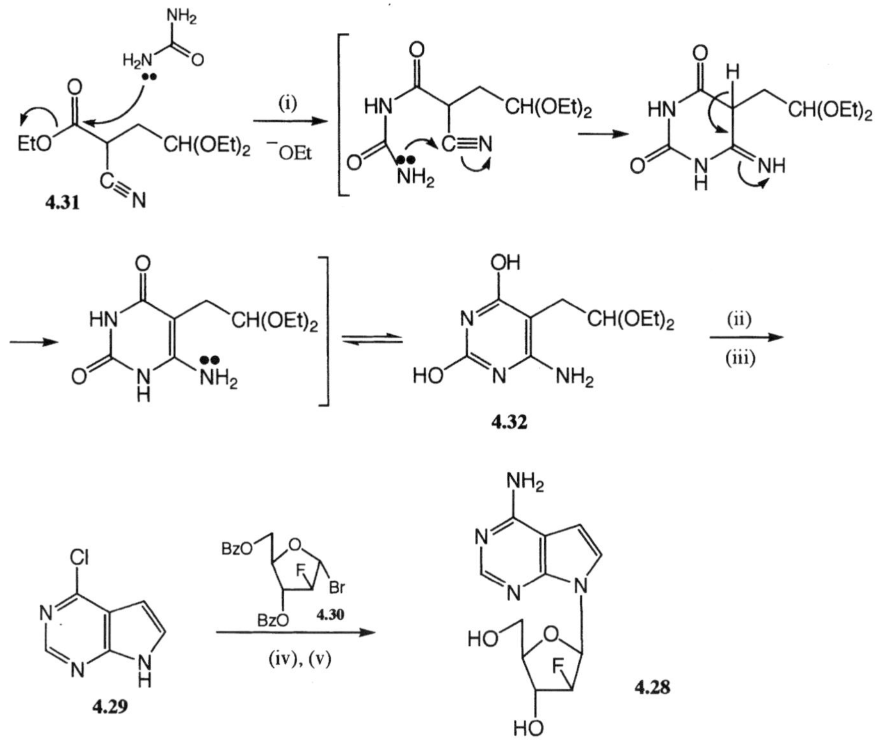

SCHEME 4.6 REAGENTS AND CONDITIONS: (I) EtOH, NaOEt, UREA, REFLUX, 3–4 H (II) aq. HCl (III) POCl₃ CHLORINATION THEN SELECTIVE C-2 DEHALOGENATION (IV) NaH, CH₃CN, 20H (V) NH₃, MeOH, 20H.

Several syntheses of toyocamycin and sangivamycin and their derivatives have been described,[43–45] a standard procedure (Scheme 4.7) starts from 4-amino-6-bromo-5-cyanopyrrolo[2,3-d]pyrimidine (**4.34**), which on reaction with triethylorthoformate gives the imino ether (**4.35**) required for coupling with the sugar moiety. Treatment with ethanolic ammonia results in ring closure and subsequent reductive debromination yields toyocamycin (**4.26**). Reaction of **4.26** with hydrogen peroxide and ammonium hydroxide gives sangivamycin (**4.27**), or reaction with dihydrogen sulphide produces thiosangivamycin (**4.38**).

4.3.3 1-Deazapurine Nucleosides

In Watson–Crick base pairing, the N-1 of purines is an important site for hydrogen bonding between the base pairs. Incorporation of 1-deazapurines into DNA may well disrupt the delicate balance of the double helix and impair the synthesis of proteins. This results in the 1-deazanucleosides, or pyrazolo[4,5-b]pyridine nucleosides displaying a wide spectrum of biological properties. 1-Deazaadenosine (**4.39**) acts as an inhibitor of blood platelet aggregation[46] and adenosine deaminase[47] and displays potent antitumour activity.[48]

Several syntheses of 1-deazaadenosine and its analogues have been described,[48–50] an efficient method starts from 3,4-diaminopyridine (**4.40**) which on treatment with

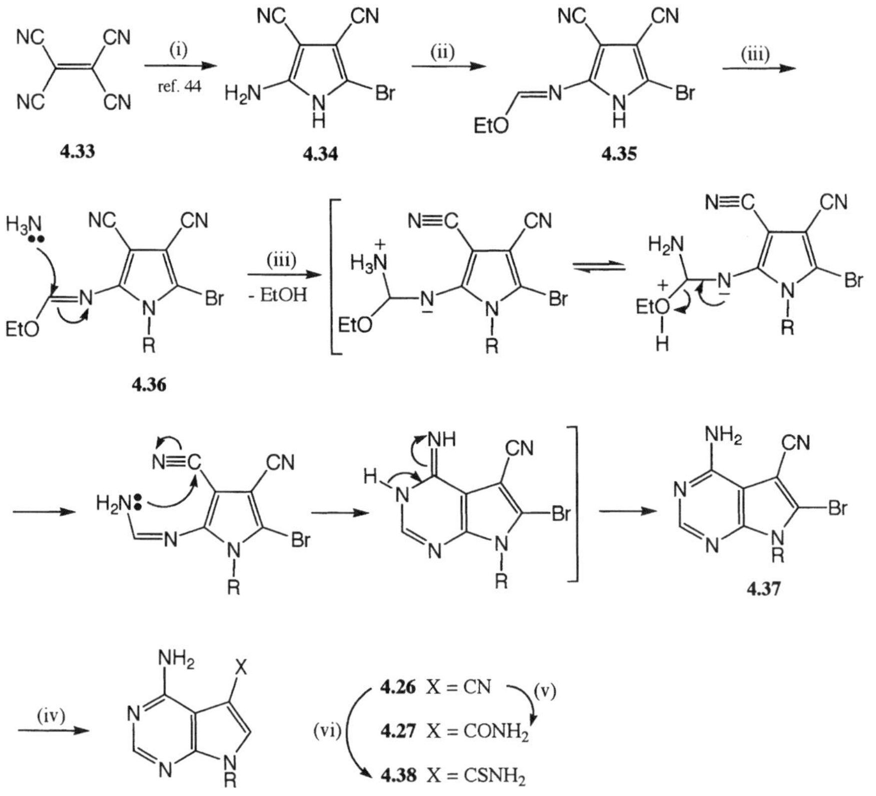

SCHEME 4.7 *REAGENTS AND CONDITIONS*: (I) HBr, ACETONE, −40°C (II) (EtO)$_3$CH, CH$_3$CN, REFLUX, 30 MIN. (III) PROTECTED SUGAR, COUPLING AGENT (IV) 10% Pd/C, NH$_4$CO$_2$H, EtOH, REFLUX (V) H$_2$O$_2$, NH$_4$OH, MeOH/H$_2$O (VI) H$_2$S, PYRIDINE, Et$_3$N, 2H. [R = FURANOSYL MOIETY].

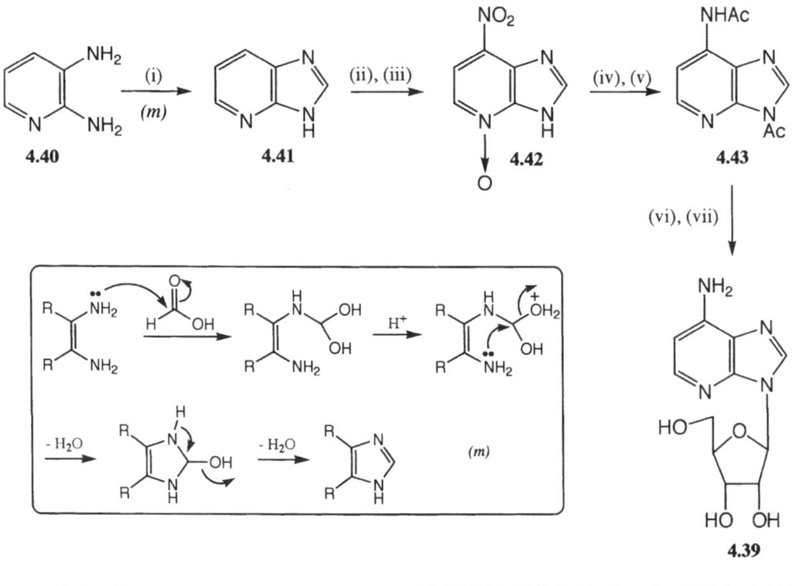

SCHEME 4.8 *REAGENTS AND CONDITIONS*: (I) HCO$_2$H (II) H$_2$O$_2$, AcOH (III) HNO$_3$, AcOH (IV) RANEY Ni, H$_2$ (V) Ac$_2$O, PYRIDINE (VI) **2.38**, SnCl$_4$, DCE (VII) NH$_3$, MeOH.

formic acid undergoes cyclization to pyrazolo[4,5-*b*]pyridine (**4.41**). Activation and electrophilic substitution yields the *N*-oxide (**4.42**), which is then reduced and the resulting amine (**4.43**) coupled under Vorbrüggen condition to give 1-deazaadenosine after deprotection (Scheme 4.8).[51] 1-Deazaguanosine (**4.44**) and its analogues can be prepared[52–54] from the intermediate 7-chloro-1*H*-imidazo[4,5-*b*]pyridin-5-acetylamine (**4.47**) (Scheme 4.9).[55,56]

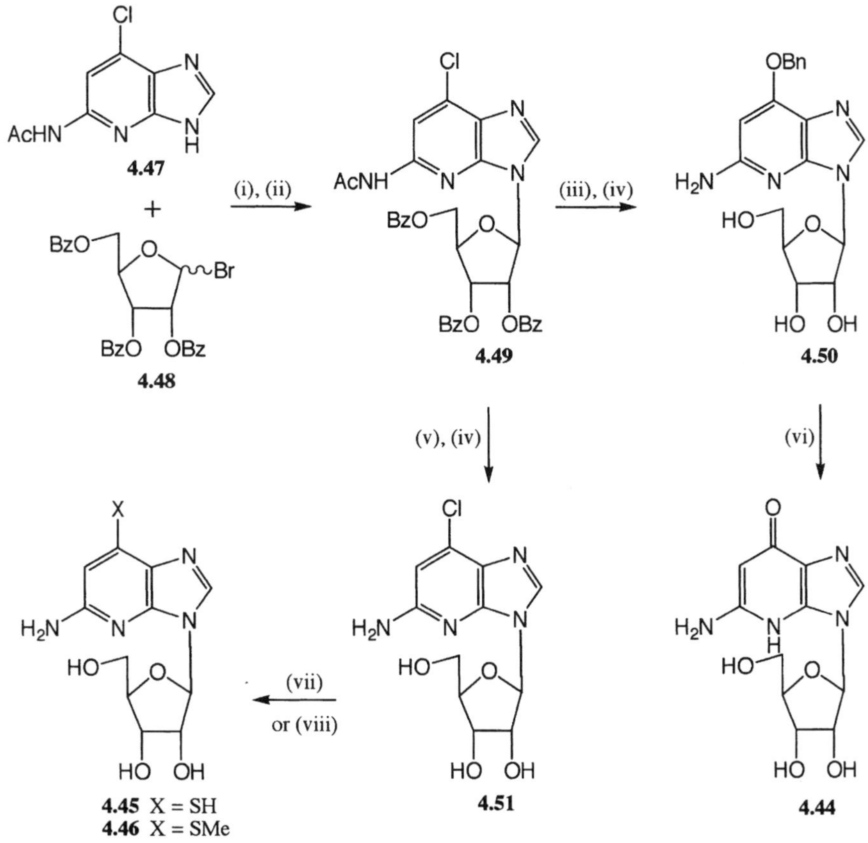

SCHEME 4.9 REAGENTS AND CONDITIONS: (I) **4.47**, HMDS, $(NH_4)_2SO_4$, REFLUX, 12H THEN (II) **4.48**, HgCN, C_6H_6, REFLUX, 1H (III) NaOBn, BnOH, 110°C, 48H (IV) EtOH, KOH, REFLUX, 3H (V) NaOMe, MeOH, 35°C, 15 MIN. (VI) EtOH, H_2O, 5% Pd/C, H_2 (42 PSI), 3H (VII) lq. H_2S, NaOMe, EtOH, SEALED TUBE, 114°C, 45H (VIII) lq. MeSH, NaOMe, EtOH, SEALED TUBE, 115°C, 42H.

The pyrazolo[3,4-*b*]pyridine nucleosides can also be described as 1-deazanucleosides; several syntheses have been described with the inosine analogue displaying inhibitory activity against Rhino 1-A virus and *de novo* purine nucleotide biosynthesis.[57,58]

4.3.4 3-Deazapurine Nucleosides

3-Deazapurine (imidazo[4,5-*c*]pyridine) nucleosides, act as inhibitors of *de novo* purine nucleotide biosynthesis and exhibit antiviral[59] and antibacterial activity.[60] 3-Deazaadenosine (**4.52**) acts as an alternate substrate competing with adenosine and is converted to 3-deazaadenosylhomocysteine, which competitively inhibits most of the methyltransferases which use adenosylmethionine (SAM). Interference with these biochemical transmethylations is believed to be responsible for the biological activity of 3-deazaadenosine.[61,62]

A convenient method for the synthesis of 3-deazaadenosine[63] involves formation of the 3-deazaadenosine precursor (**4.54**) by cyclization of 3,4-diamino-2,6-dichloropyridine (**4.53**)[64] (Scheme 4.10).

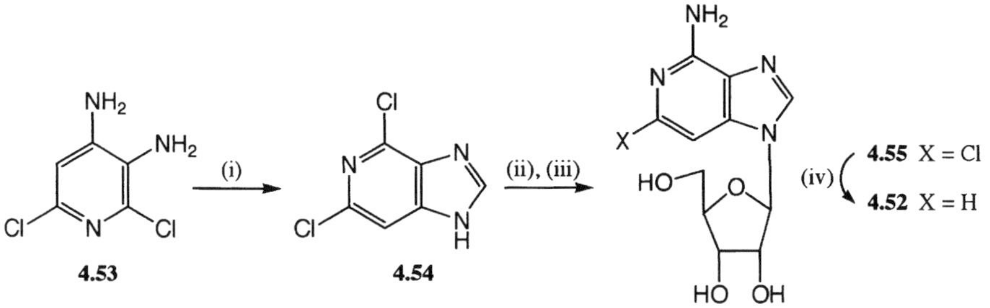

SCHEME 4.10 REAGENTS AND CONDITIONS: (I) (EtO)$_3$CH, Ac$_2$O, REFLUX, 30 MIN. (II) BSA, CH$_3$CN, **2.38**, LEWIS ACID (III) lq. NH$_3$, STEEL VESSEL, 100°C, 30H (IV) NaOH, H$_2$O, 20% Pd/C, H$_2$, 15H.

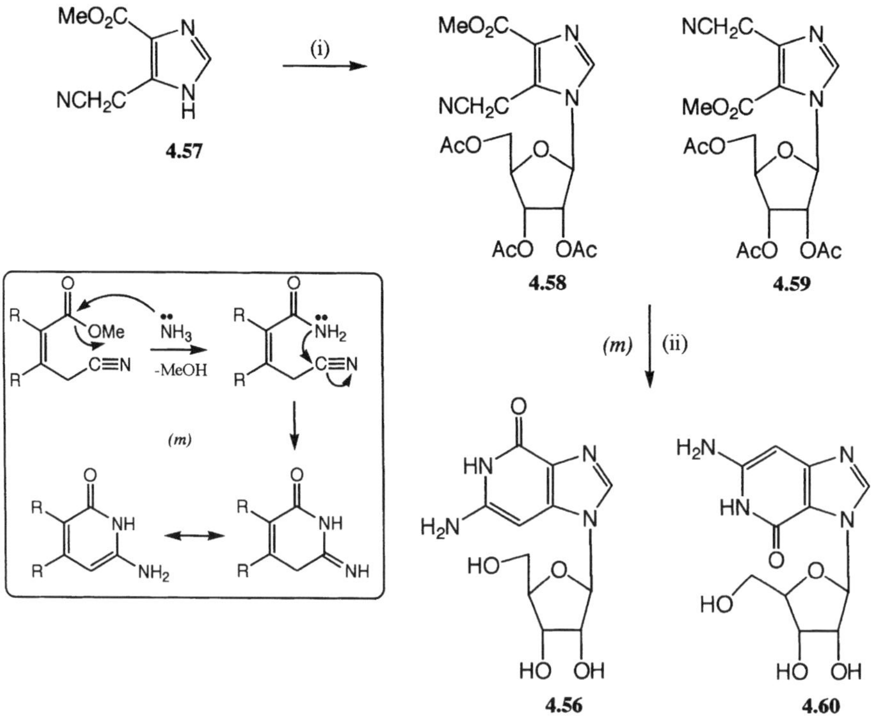

SCHEME 4.11 REAGENTS AND CONDITIONS: (I) **2.38**, CH$_3$CN, SnCl$_4$ (II) lq. NH$_3$, STEEL VESSEL, 110°C, 22H.

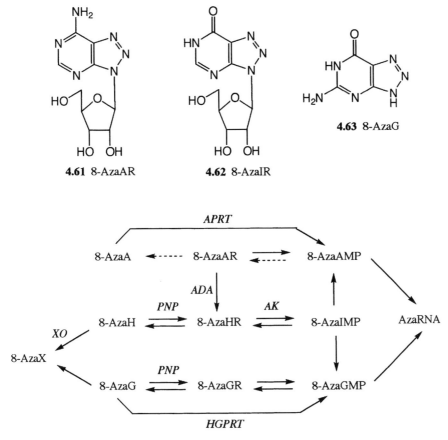

FIGURE 4.3 METABOLISM OF 8-AZANUCLEOSIDES.
[H, HYPOXANTHINE; X, XANTHINE; APRT, ADENINE PHOSPHORYLTRANSFERASE; HGPRT, HYPOXANTHINE-GUANINE PHOSPHORYLTRANSFERASE; ADA, ADENOSINE DEAMINASE; PNP, PURINE NUCLEOSIDE PHOSPHORYLASE; XO, XANTHINE OXIDASE.]

3-Deazaguanosine (**4.56**) can be prepared from an imidazole nucleoside precursor (**4.58**). Reaction of 5-cyanomethyl-4-carbomethoxyimidazole (**4.57**) with tetra-acetylated ribose (**2.38**) resulted in a mixture of products (**4.58** and **4.59**). Treatment of the imidazole nucleosides with liquid ammonia then led to the formation of 3-deazaguanosine (**4.56**) and the deazanucleoside (**4.60**) (Scheme 4.11).[65]

4.3.5 8-Azapurine Nucleosides

The interest in the 8-aza or 1,2,3-triazolo[4,5-*d*]pyrimidine nucleosides, stems from the *in vivo* anticancer activity displayed by 8-azaadenosine (**4.61**), 8-azainosine (**4.62**) and the purine 8-azaguanine (**4.63**).[66-68] The metabolism of the 8-azanucleosides has been extensively studied (Fig. 4.3), and has shown that these nucleosides are activated by conversion to their monophosphates followed by incorporation into RNA, where they disrupt protein synthesis.[62]

The synthesis of 8-azapurine nucleosides is hampered by additional problems of regioselectivity, owing to the additional N-8 atom, *i.e.* glycosylation can occur at

N-7, N-8 and N-9. 8-Azapurine ribonucleosides,[69,70] 2'-deoxyribonucleosides[69,71–73] and 2',3'-dideoxyribonucleosides[74,75] have been prepared.

Improved regioselectivity was observed for the synthesis of 8-azaadenosine (**4.61**) using the SnCl$_4$-catalyzed glycosylation of 8-azaadenine[76] (**4.66**) with **2.38** and acetonitrile as the solvent (Scheme 4.12). This led to 34% of the N^9-isomer (**4.61**) and 47% of the N^8-isomer (**4.67**, the kinetic product).[69]

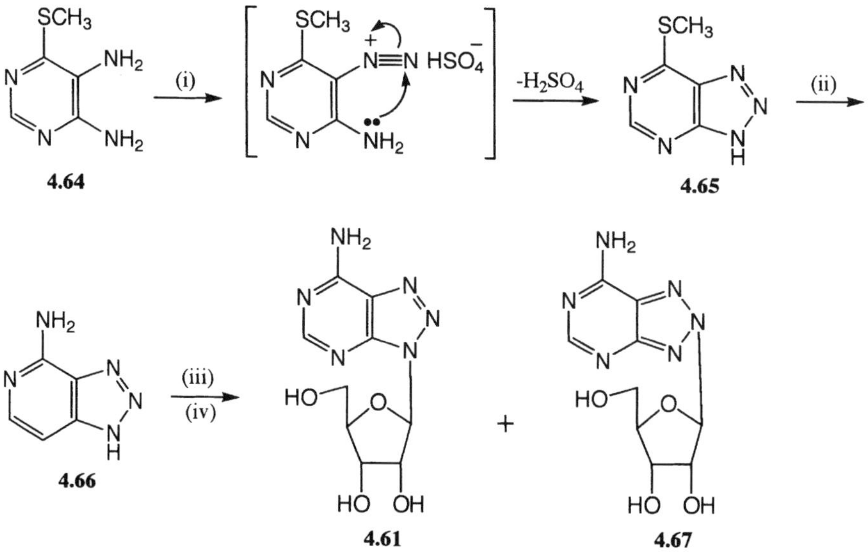

SCHEME 4.12   REAGENTS AND CONDITIONS: (I) NaNO$_2$, H$_2$SO$_4$, H$_2$O, 0°C (II) NH$_3$ (III) **2.38**, CH$_3$CN, SnCl$_4$ (IV) MeOH, aq. NH$_3$, 2 H.

4.3.6 Other Purine Modified Nucleosides

There are an increasing number of purine modified nucleosides of interest, either for their biological properties or for their novelty (Fig. 4.4). 2-Azaadenosine (**4.68**, 2-AzaAR), shows modest *in vivo* activity against L1210.[66] Combining both aza and deaza components in a heterocyclic moiety has resulted in the synthesis of 8-aza-1-deazapurine nucleosides (*e.g.* **4.69**),[77] which show inhibitory activity against adenosine deaminase and polio (Sb-1) and coxsackie B1 viruses, and the 2-aza-3-deaza- and 2,8-diaza-3-deaza-purine nucleosides (*e.g.* **4.70** and **4.71**).[78] Other more unusual purine mimics have been described, such as the cytotoxic imidazo[4,5-*d*]isothiazole nucleosides (*e.g.* **4.72**),[79] the naphtho[2,3-*d*]imidazole nucleosides (*e.g.* **4.73**) which show moderate activity against CMV,[80] and imidazo[4,5-*b*]quinolin-2-one nucleosides (*e.g.* **4.74**).[81]

4.4 PYRIMIDINE-MODIFIED NUCLEOSIDES

This area has been less extensively studied than purine modifications, and has concentrated primarily on aza/deaza-pyrimidine nucleosides and novel fused-ring systems. Azapyrimidine and 3-deazapyrimidine nucleosides have found use as anticancer, antiviral and antibacterial agents, the fused pyrimidine nucleosides, which are of more recent interest, have shown potential as inhibitors of DNA viruses.

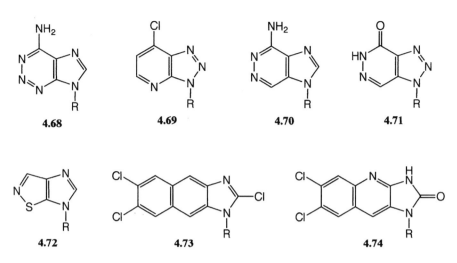

FIGURE 4.4 UNUSUAL PURINE-MODIFIED NUCLEOSIDES (R = RIBOSE).

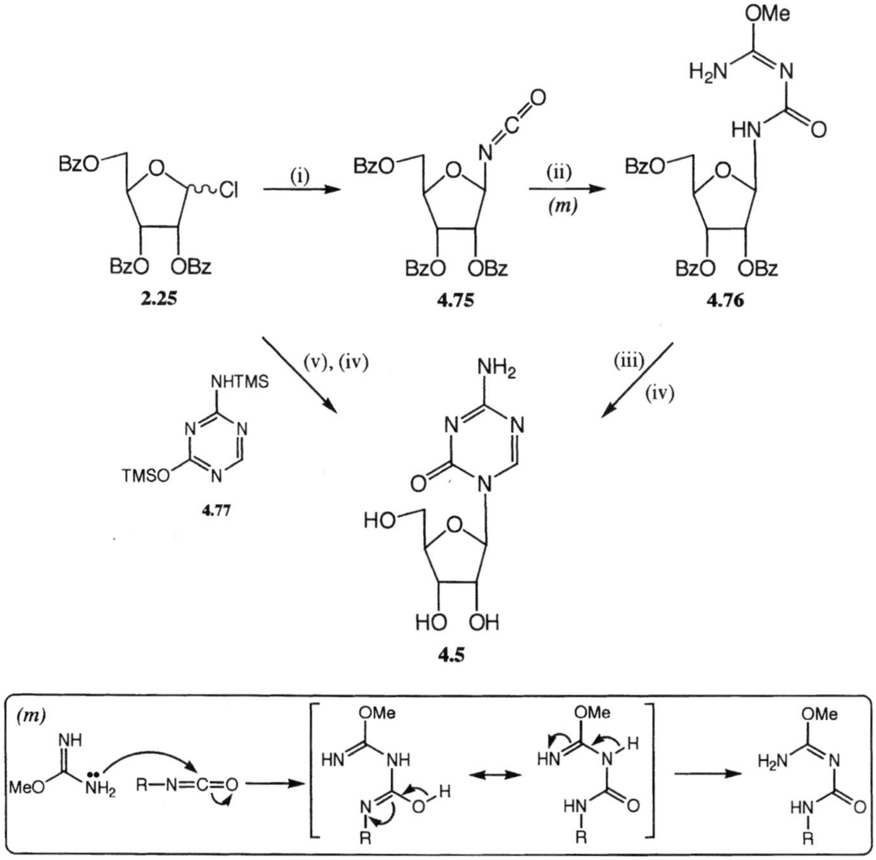

SCHEME 4.13 *REAGENTS AND CONDITIONS*: (I) AgNCO (II) METHYLISOUREA (III) HC(OEt)$_3$ (IV) MeOH, NH$_3$ (V) **2.25**, CH$_3$CN.

4.4.1 Azapyrimidine and 3-Deazapyrimidine Nucleosides

5-Azacytidine (**4.5**), (4-amino-1-β-D-ribofuranosyl-1,3,5-*s*-triazin-2-one), inhibits Gram-negative bacteria and is active against T-4 lymphomas and L-1210 leukaemia, both 5-azacytidine and 2'-deoxy-5-azacytidine are used clinically in the treatment of acute leukemia.[82,83] Either the isocyanate method, which involves building the triazine ring from the 1-isocyanate riboside (**4.75**),[84,85] or the direct coupling method have been employed for the synthesis of 5-azacytidine and its derivatives (Scheme 4.13).[86,87]

5,6-Dihydro-5-azathymidine (DHAdT, **4.78**), 1-(2-deoxy-β-D-ribofuranosyl)-5,6-dihydro-5-methyl-*s*-triazine-2,4(1*H*, 3*H*)-dione, is a nucleoside antibiotic isolated from *Streptomyces platensis* var. *clarensis*, which exhibits *in vitro* activity against DNA viruses, including HSV-1 and HSV-2, VZV and vaccinia viruses, and Gram-negative bacteria.[88] DHAdT can be prepared from the direct coupling of 5,6-dihydro-5-azathymine (**4.81**), which is prepared from *N*-methylbiuret (**4.79**), with the chlorosugar (**2.29**) (Scheme 4.14).[88]

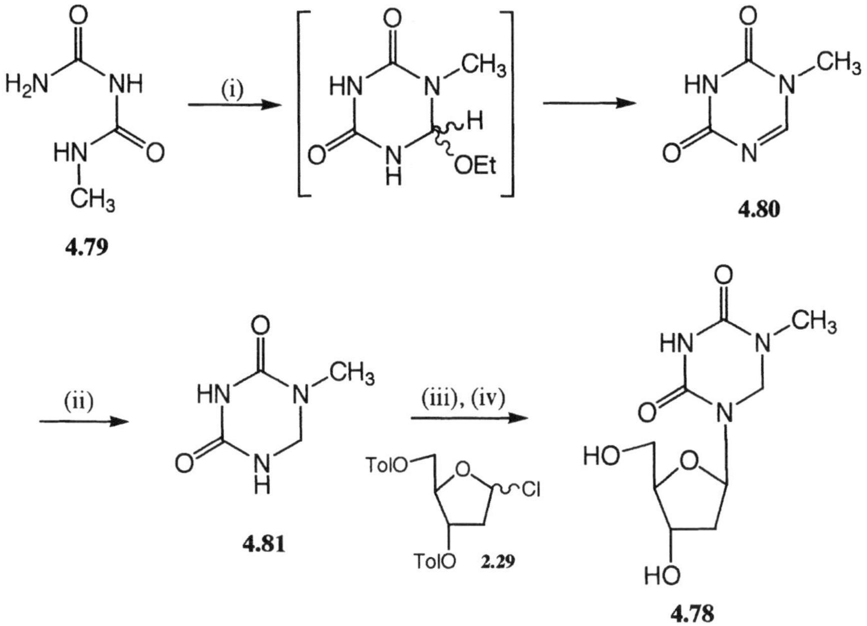

SCHEME 4.14 *REAGENTS AND CONDITIONS*: (I) (EtO)$_3$CH (II) RANEY Ni, H$_2$ (III) HMDS, CH$_3$CN, **2.29**, SnCl$_4$ (IV) NaOMe, MeOH.

Of the 6-azapyrimidine nucleosides (Fig. 4.5), 6-azauridine (**4.83**), in its monophosphate form,[89] is of most interest owing to its antiviral activity which results from its ability to act as a potent inhibitor of orotidine 5'-monophosphate decarboxylase (OCDase).[90] OCDase mediates the conversion of orotidine 5'-monophosphate (orotidylic acid) to UMP, with inhibition of this enzyme resulting in the disruption of *de novo* pyrimidine biosynthesis. 6-Azauridine (**4.83**) and its

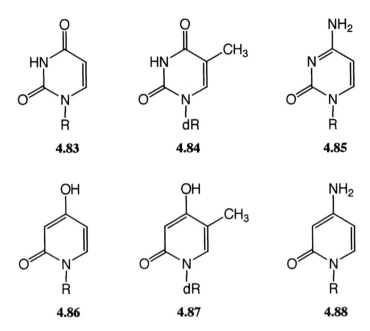

FIGURE 4.5 6-AZAPYRIMIDINE AND 3-DEAZAPYRIMIDINE NUCLEOSIDES
(R = RIBOSE, dR = 2-DEOXYRIBOSE).

derivatives are preferably prepared by Vorbrüggen coupling,[91-93] as is the case for 6-azathymidine (**4.84**),[94] 6-azacytidine (**4.85**) can be prepared directly from 6-azauridine.[51]

3-Deazauridine (**4.86**) is a potent antileukaemia agent[95,96] which, in its triphosphate form, acts as an inhibitor of CTP synthetase,[97] this enzyme effects the sole *de novo* pathway into cytosine nucleotide metabolism. As well as 3-deazauridine,[98] 3-deazathymidine[99,100] (**4.87**) and 3-deazacytidine (**4.88**)[101] have been prepared (Fig. 4.5) and a number of derivatives,[102-105] some of which display anticancer and antiviral activity.

4.4.2 Fused Pyrimidine Nucleosides

The 2,3-dihydrofuran[2,3-*d*]pyrimidin-2-one nucleoside (**4.89**, n = 4) structure was first observed as a by-product of the reaction of 5-hexynyl-2'-deoxyuridine with either copper iodide or mercury sulphate.[106] The unsubstituted 3-(2'-deoxy-β-D-ribofuranosyl)-2,3-dihydrofuran[2,3-*d*]pyrimidin-2-one (**4.89**, n = 0) has been prepared,[107-109] though was inactive against HSV, HCMV and VZV.[109]

A preliminary SAR study and synthesis of a series of 2,3-dihydrofuran[2,3-*d*]pyrimidin-2-one nucleosides has been described.[5] Reaction of 5-iodo-2'-deoxyuridine (**4.89**) with alkynes under Sonogashiri conditions (Scheme 2.38, *m2*) produced the 5-alkyne-2'-deoxyuridines (**4.90**), which on treatment with additional copper iodide produced the bicyclic nucleosides (**4.91**) (Scheme 4.15). The bicyclic nucleosides with C_8 (**4.6**) and C_9–C_{10} (**4.91**, n = 9–10) side chains were found to be the most active, with EC_{50} values against VZV of 3–9 nM and with no appreciable toxicity (SI > 5000).

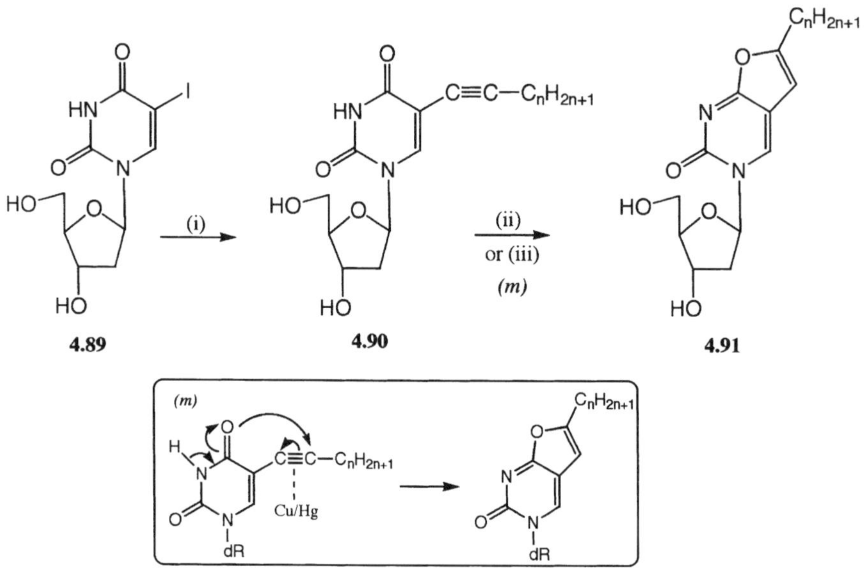

SCHEME 4.15 *REAGENTS AND CONDITIONS*: (I) HC≡C-C$_N$H$_{2N+1}$, ^{i}Pr$_2$EtNH, (Ph$_3$P)$_4$Pd(0), CuI, DMF, 19H (II) CuI, MeOH, Et$_3$N, REFLUX, 3–8H (III) HgSO$_4$, DIOXANE, H$_2$O, 48H.

Imidazo[1,2-*c*]pyrimidin-5(6*H*)-one dioxolane and oxathiolane nucleosides have been prepared,[110] with the *para*-nitrophenyl-substituted dioxolane (**4.93**) exhibiting potent activity against HBV. The imidazo[1,2-*c*]pyrimidine nucleosides were obtained by reaction of the cytidine nucleoside (**4.92**) with an electrophilic α-haloketone with subsequent deprotection (Scheme 4.16).

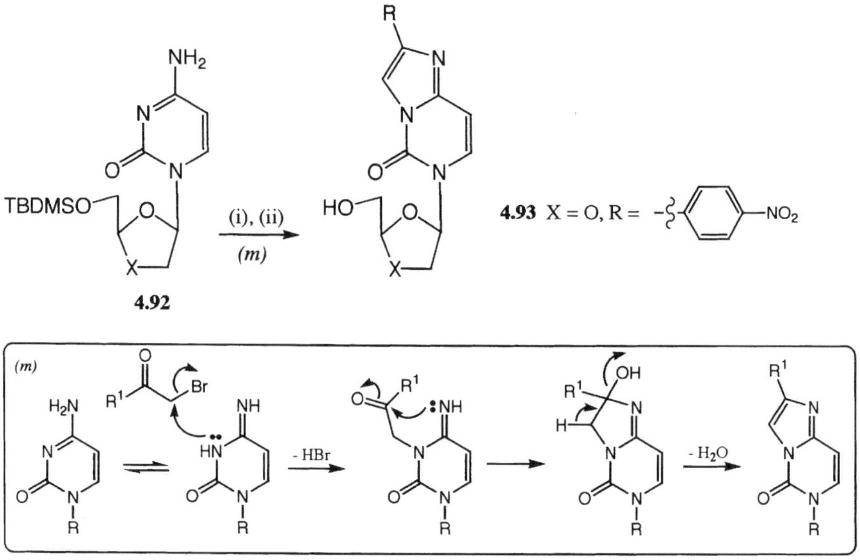

SCHEME 4.16 *REAGENTS AND CONDITIONS*: (I) α-HALOKETONE, MeOH, REFLUX, 20–48H (II) TBAF, AcOH, THF.

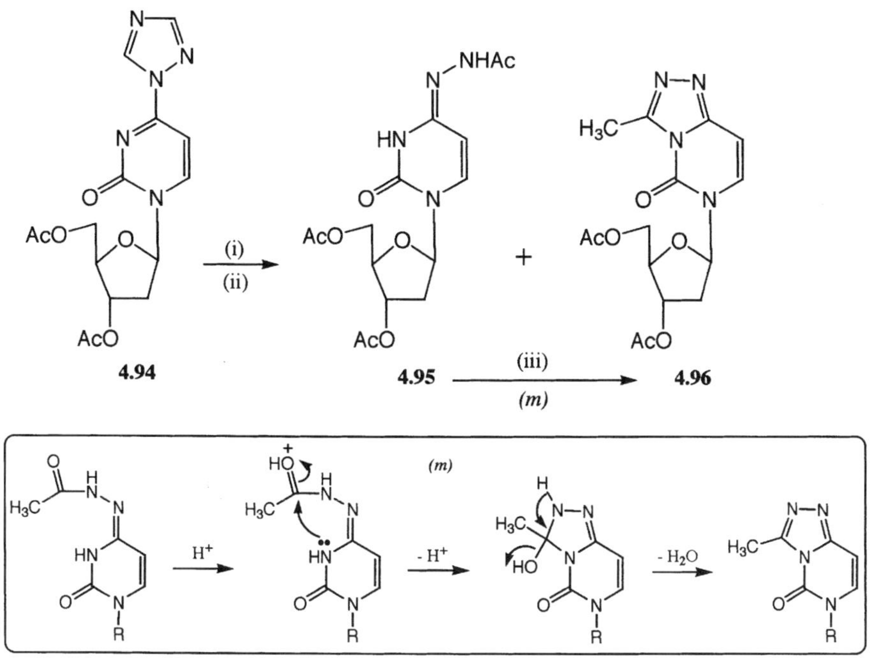

SCHEME 4.17 *Reagents and conditions* (i) NH_2NH_2 (ii) Ac_2O, PYRIDINE (iii) PYRIDINIUM HYDROCHLORIDE, PYRIDINE.

1,2,4-Triazolo[4,3-c]pyrimidone nucleosides have been described although no biological data was given,[111,112] compound **4.96** was noted as a by-product of the reaction of the triazolo-nucleoside (**4.94**) with hydrazine and acetic anhydride (Scheme 4.17), the product (**4.96**) is formed *via* acid-catalysed rearrangement of **4.95**.

4.5 REFERENCES

1. Y. Kosugi, Y. Saito, S. Mori, J. Watanabe, M. Baba, S. Shigeta, *Antiviral Chem. Chemother.*, **1994**, *5*, 366–371.
2. R.K. Robins, G.R. Revankar, *Med. Res. Rev.*, **1985**, *5*, 273–296.
3. L.B. Townsend, R.V. Devivar, S.R. Turk, M.R. Nassiri, J.C. Drach, *J. Med. Chem.*, **1995**, *38*, 4098–4105.
4. D.C. Case, *Oncology*, **1982**, *39*, 218–221.
5. C. McGuigan, C.J. Yarnold, G. Jones, S. Velázquez, H. Barucki, A. Brancale, G. Andrei, R. Snoeck, E. De Clercq, and J. Balzarini, *J. Med. Chem.*, **1999**, *42*, 4479–4484.
6. N. Minakawa, T. Takeda, T. Sasaki, A. Matsuda, T. Ueda, *J. Med. Chem.*, **1991**, *34*, 778–786 and references cited therein.
7. R. Dixon, J. Gourzis, D. McDermott, J. Fujitaki, P. Dewland, H. Gruber, *J. Clin. Pharmacol.*, **1991**, *31*, 342–347.
8. T. Inou, R. Kusaba, I. Takahashi, H. Sugimoto, K. Kuzuhara, Y. Yamada, J. Yamauchi, O. Otsubo, *Transplantation Proc.*, **1981**, *8*, 315–318.
9. K. Mizuno, M. Tsujino, M. Takada, M. Hayashi, K. Atsmi, K. Asano, T. Matsuda, *J. Antibiot.*, **1974**, *27*, 775–782.
10. M. Hayashi, T. Hirano, M. Yaso, K. Mizuno, T. Ueda, *Chem. Pharm. Bull.*, **1975**, *23*, 245–246.
11. K. Fukukawa, S. Shuto, T. Hirano, T. Ueda, *Chem. Pharm. Bull.*, **1984**, *32*, 1644–1646.
12. D.F. Ewing, R.W. Humble, G. Mackenzie, G. Shaw, *Nucleosides & Nucleotides*, **1995**, *14*, 369–372.
13. N. Minakawa, T. Takeda, T. Sasaki, A. Matsuda, T. Ueda, *J. Med. Chem.*, **1991**, *34*, 778–786.
14. N. Minakawa, N. Kojima, T. Sasaki, A. Matsuda, *Nucleosides & Nucleotides*, **1996**, *15*, 251–263.

15. P.C. Srivastava, D.G. Streeter, T.R. Matthews, L.B. Allen, R.W. Sidwell, R.K. Robins, *J. Med. Chem.*, **1976**, *19*, 1020–1026.
16. R.W. Sidwell, J.H. Huffman, G.P. Khare, L.B. Allen, J.T. Witkowski, R.K. Robins, *Science*, **1972**, *177*, 705–706.
17. R.W. Sidwell, G.R. Revankar, R.K. Robins, in "*Viral Chemotherapy*", D. Hugar (ed), Pergamon Press, Oxford, **1985**, *2*, pp 49–108.
18. J.T. Witkowski, R.K. Robins, R.W. Sidwell, L.N. Simon, *J. Med. Chem.*, **1972**, *15*, 1150–1154.
19. M.J. Slater, C. Gowrie, G.A. Freeman, S.A. Short, *Bioorg. Med. Chem. Lett.*, **1996**, *6*, 2787–2790.
20. J. Michael, S.B. Larson, M.M. Vaghefi, R.K. Robins, *J. Het. Chem.*, **1990**, *27*, 1063–1071.
21. J.T. Witkowski, R.K. Robins, R.W. Sidwell, *J. Med. Chem.*, **1973**, *16*, 935–937.
22. Y.S. Sanghvi, N.B. Hanna, S.B. Larson, J.M. Fujitaki, R.C. Willis, R.A. Smith, R.K. Robins, G.R. Revankar, *J. Med. Chem.*, **1988**, *31*, 330–335.
23. R.C. Willis, R.K. Robins, J.E. Seegmiller, *Mol. Pharmacol.*, **1980**, *18*, 287–295.
24. R.L. Walter, J. Symersky, A.F. Poirot, J.D. Stoekler, M.D. Erion, S.E. Ealick, *Nucleosides & Nucleotides*, **1994**, *13*, 689–706.
25. R. Alvarez, S. Vélazquez, A. San–Félix, S. Aquaro, E. De Clercq, C–F. Perno, A. Karlsson, J. Balzarini, M.J. Camarasa, *J. Med. Chem.*, **1994**, *37*, 4185–4194.
26. P. Norris, D. Horton, B.R. Levine, *Heterocycles*, **1996**, *43*, 2643–2655.
27. G. Inguaggiato, M. Jasamai, J.E. Smith, M.J. Slater, C. Simons, *Nucleosides & Nucleotides*, **1999**, *18*, 457–467.
28. S. Manfredini, R. Bazzanini, P.G. Baraldi, M. Guarneri, D. Simoni, M.E. Marongiu, A. Pani, E. Tramontano, P. La Colla, *J. Med. Chem.*, **1992**, *35*, 917–924.
29. S. Manfredini, R. Bazzanini, P.G. Baraldi, D. Simoni, S. Vertuani, A. Pani, E. Pinna, F. Scintu, D. Lichino, P. La Colla, *Bioog. Med. Chem. Lett.*, **1996**, *6*, 1279–1284.
30. R. Storer, C.J. Ashton, A.D. Baxter, M.M. Hann, C.L.P. Marr, A.M. Mason, C–L. Mo, P.L. Myers, S.A. Noble, C.R. Penn, N.G. Weir, J.M. Woods, P.L. Coe, *Nucleosides & Nucleotides*, **1999**, *18*, 203–216.
31. L.B. Townsend, G.R. Revankar, *Chem. Rev.*, **1970**, *70*, 389–438.
32. M.T. Migawa, J-L. Girardet, J.A. Walker, G.W. Koszalka, S.D. Chamberlain, J.C. Drach, L.B. Townsend, *J. Med. Chem.*, **1998**, *41*, 1242–1251.
33. R. Zou, J.C. Drach, L.B. Townsend, *J. Med. Chem.*, **1997**, *40*, 802–810 and 811–818.
34. K.S. Gudmundsson, J.C. Drach, L.L. Wotring, L.B. Townsend, *J. Med. Chem.*, **1997**, *40*, 785–793.
35. R.V. Devivar, E. Kawashima, G.R. Revankar, J.M. Breitenbach, E.D. Kreske, J.C. Drach, L.B. Townsend, *J. Med. Chem.*, **1994**, *37*, 2942–2949.
36. P.M. Krosky, M.R. Underwood, S.R. Turk, K.W.H. Feng, R.K. Jain, R.G. Ptak, A.C. Westerman, K.K. Biron, L.B. Townsend, J.C. Drach, *J. Virology*, **1998**, *72*, 4721–4728.
37. N.J. Leonard, D.Y. Curtin, K.M. Beck, *J. Am. Chem. Soc.*, **1947**, *69*, 2459–2461.
38. R.K. Robins, G.R. Revankar, in "*Advances in Antiviral Drug Design*", E. De Clercq (ed.), JAI Press, Greenwich, Conn., **1993**, pp 161–398.
39. R.H. Adamson, D.W. Zaharewitz, D.G. Johns, *Pharmacology*, **1977**, *15*, 84.
40. A. Bloch, R.J. Leonard, C.A. Nichol, *Biochim. Biophys. Acta*, **1967**, *138*, 10–25.
41. B.K. Bhattacharya, T.S. Rao, G.R. Revankar, *J. Chem. Soc. Perkin Trans. 1*, **1995**, 1543–1550.
42. J. Davoll, *J. Chem. Soc.*, **1960**, 131–138.
42. A.R. Porcari, L.B. Townsend, *Nucleosides & Nucleotides*, **1999**, *18*, 153–159.
44. S.H. Krawczyk, M.R. Nassiri, L.S. Kucera, E.R. Kern, R.G. Ptak, L.L. Wotring, J.C. Drach, L.B. Townsend, *J. Med. Chem.*, **1995**, *38*, 4106–4114.
45. K. Ramasamy, R.K. Robins, G.R. Revankar, *Tetrahedron*, **1986**, *42*, 5869–5878.
46. I. Antonini, G. Cristalli, P. Franchetti, M. Grifantini, S. Martelli, F. Petrelli, *J. Pharm. Sci.*, **1984**, *73*, 366–369.
47. G. Cristalli, M. Grifantini, S. Vittori, W. Balduini, F. Cattabeni, *Nucleosides & Nucleotides*, **1985**, *4*, 237.
48. G. Cristalli, P. Franchetti, M. Grifantini, S. Vittori, T. Bordoni, C. Geroni, *J. Med. Chem.*, **1987**, *30*, 1686–1688.
49. T. Itoh, S. Kitano, Y. Mizuno, *Heterocycles*, **1982**, *17*, 305–309.
50. T. Itoh, J. Inaba, Y. Mizuno, *Heterocycles*, **1977**, *8*, 433.
51. Y. Mizuno, T. Itoh, A. Nomura, *Heterocycles*, **1982**, *8*, 615–644.
52. R.D. Elliott, J.A. Montgomery, *J. Med. Chem.*, **1978**, *21*, 112–114.
53. B.L. Cline, R.P. Panzica, L.B. Townsend, *J. Het. Chem.*, **1978**, *15*, 839–847.
54. B.L. Cline, R.P. Panzica, L.B. Townsend, *J. Het. Chem.*, **1975**, *12*, 603–604.
55. C. Temple, B.H. Smith, J.A. Montgomery, *J. Org. Chem.*, **1973**, *38*, 613–615.
56. C. Temple, B.H. Smith, C.L. Kussner, J.A. Montgomery, *J. Org. Chem.*, **1976**, *41*, 3784–3788.
57. J.A. Montgomery, S.J. Clayton, W.E. Fitzgibbon, *J. Het. Chem.*, **1964**, *1*, 215–216.

58. Y.S. Sanghvi, S.B. Larson, R.K. Robins, G.R. Revankar, *Nucleosides & Nucleotides*, **1989**, *8*, 887–890.
59. P.D. Cook, L.B. Allen, D.G. Streeter, J.H. Huffman, R.W. Sidwell, R.K. Robins, *J. Med. Chem.*, **1978**, *21*, 1212–1218.
60. L.B. Allen, J.H. Huffman, P.D. Cook, R.B. Meyer Jr., R.K. Robins, R.W. Sidwell, *Antimicrob. Agents Chemother.*, **1977**, *12*, 144–119.
61. P.K. Chiang, H.H. Richards, G.L. Cantoni, *Mol. Pharmacol.*, **1977**, *13*, 939–947.
62. J.A. Montgomery, *Med. Res. Rev.*, **1982**, *2*, 271–308.
63. J.A. Montgomery, A.T. Shortnacy, S.D. Clayton, *J. Het. Chem.*, **1977**, *14*, 195–197.
64. R.J. Rousseau, R.K. Robins, *J. Het. Chem.*, **1965**, *2*, 196–201.
65. P.D. Cook, R.J. Rousseau, A.M. Mian, P. Dea, R.B. Meyer, R.K. Robins, *J. Am. Chem. Soc.*, **1976**, *98*, 1492–1498.
66. J.A. Montgomery, R.D. Elliott, H.J. Thomas, *Ann. N.Y. Acad. Sci.*, **1975**, *255*, 292.
67. R.D. Elliott, J.A. Montgomery, *J. Med. Chem.*, **1977**, *20*, 116–120.
68. G.W. Kidder, V.C. Dewey, R.E. Parks, G.L. Woodside, *Science*, **1949**, *109*, 511–514.
69. F. Seela, I. Münster, U. Löchner, H. Rosemeyer, *Helv. Chim. Acta*, **1998**, *81*, 1139–1155.
70. J.A. Montgomery, A.T. Shortnacy, G. Arnett, W.M. Shannon, *J. Med. Chem.*, **1977**, *20*, 401–404.
71. F. Seela, S. Lampe, *Helv. Chim. Acta*, **1993**, *76*, 2388–2397.
72. Z. Kazimierczuk, U. Bindig, F. Seela, *Helv. Chim. Acta*, **1989**, *72*, 1527–1536.
73. J.A. Montgomery, A.T. Shortnacy, J.A. Secrist III, *J. Med. Chem.*, **1983**, *26*, 1483–1489.
74. F. Seela, K. Mersmann, *Helv. Chim. Acta*, **1993**, *76*, 2184–2193.
75. F. Seela, K. Mersmann, *Helv. Chim. Acta*, **1992**, *75*, 1885–1896.
76. R. Weiss, R.K. Robins, C.W. Noell, *J. Org. Chem.*, **1960**, *25*, 765–770.
77. P. Franchetti, L. Messini, L. Cappellacci, G. Abu Sheikha, M. Grifantini, P. Guarracino, A. De Montis, A.G. Loi, M.E. Marongiu, P. La Colla, *Nucleosides & Nucleotides*, **1994**, *13*, 1739–1755.
78. J.C. Bussolari, K. Ramesh, J.D. Stoekler, S–F. Chen, R.P. Panzica, *J. Med. Chem.*, **1993**, *36*, 4113–4120.
79. E.E. Swayze, J.C. Drach, L.L. Wotring, L.B. Townsend, *J. Med. Chem.*, **1997**, *40*, 771–784.
80. Z. Zhu, J.C. Drach, L.B. Townsend, *J. Org. Chem.*, **1998**, *63*, 977–983.
81. Z. Zhu, B. Lippa, L.B. Townsend, *J. Org. Chem.*, **1999**, *64*, 4159–4168.
82. G.E. Rivard, R.L. Momparler, J. Demers, P. Benoit, R. Raymond, K–T. Lin, L.F. Momparler, *Leukemia Res.*, **1981**, *5*, 453–462.
83. A.D. Riggs, P.A. Jones, *Adv. Cancer Res.*, **1983**, *40*, 1–30.
84. A. Piskala, F. Sorm, *Coll. Czech. Chem. Commun.*, **1964**, *29*, 2060.
85. N.B. Hanna, J. Zajícek, A. Piskala, *Nucleosides & Nucleotides*, **1997**, *16*, 129–144.
86. L. Cappellacci, K.N. Tiwari, J.A. Montgomery, J.A. Secrist III, *Nucleosides & Nucleotides*, **1999**, *18*, 613–614.
87. J. Ben–Hattar, J. Jiricny, *J. Org. Chem.*, **1986**, *51*, 3211–3213.
88. H. Skulnick, *J. Org. Chem.*, **1978**, *43*, 3188–3194.
89. B. Gabrielsen, J.J. Kirsi, C.D. Kwong, D.A. Carter, C.A. Krauth, L.K. Hanna, J.W. Huggins, T.P. Monath, D.F. Kefauver, H.A. Blough, J.T. Rankin, C.M. Bartz, J.H. Huffman, D.F. Smee, R.W. Sidwell, W.M. Shannon, J.A. Secrist III, *Antiviral Chem. & Chemother.*, **1994**, *5*, 209–220.
90. C.A. Pasternak, R.E. Handschumaker, *J. Biol. Chem.*, **1959**, *234*, 2992–2997.
91. U. Niedballa, H. Vorbrüggen, *J. Org. Chem.*, **1974**, *39*, 3654–3660.
92. G. Inguaggiato, D. Hughes, E. De Clercq, J. Balzarini, C. Simons, *Antiviral Chem. & Chemother.*, **1999**, *10*, 241–249.
93. I. Basnak, P.L. Coe, R.T. Walker, *Nucleosides & Nucleotides*, **1994**, *13*, 163–175.
94. G.T. Shiau, W.H. Prusoff, *Carbohydr. Res.*, **1978**, *62*, 175–177.
95. B.S. Yap, K.B. McCredie, M.J. Keating, G.P. Bodey, E.J. Freireich, *Cancer Treatment Reps.*, **1981**, *65*, 521–524.
96. R.L. Momparler, L.F. Momparler, *Cancer Chemother. Pharmacol.*, **1989**, *25*, 51–54.
97. W.J. Moriconi, M. Slavik, S. Taylor, *Investigational New Drugs*, **1986**, *4*, 67–84.
98. M.J. Robins, C. Kaneko, M. Kaneko, *Can. J. Chem.*, **1981**, *59*, 3356–3359.
99. S. Nesnow, T. Miyazaki, T. Khwaja, R.B. Meyer, C. Heidelberger, *J. Med. Chem.*, **1973**, *16*, 524–528.
100. H. Eschenhof, P. Strazewski, C. Tamm, *Tetrahedron*, **1992**, *48*, 6225–6230.
101. M. Legraverend, C.H. Nguyen, A. Zerial, E. Bisagne, *Nucleosides & Nucleotides*, **1985**, *5*, 125–134.
102. L.B. Allen, A.G. Teepe, M.J. Kehoe, C.S. Holland, D.J. McNamara, P.D. Cook, *Antiviral Res.*, **1989**, *12*, 259–268.
103. K. Miyai, R.L. Tolman, R.K. Robins, *J. Med. Chem.*, **1978**, *21*, 427–430.
104. B. Devadas, T.E. Rogers, S.H. Gray, *Synth. Commun.*, **1995**, *20*, 3199–3210.
105. G.E.H. Elgemeie, A.M.E. Attia, D.S. Forag, S.M. Sherif, *J. Chem. Soc., Perkin Trans. 1*, **1994**, 1285–1288.
106. M.J. Morris, P.J. Barr, *J. Org. Chem.*, **1983**, *48*, 1854–1862.

107. R. Kumar, E.E. Knaus, L.I. Wiebe, *J. Het. Chem.*, **1991**, *28*, 1917–1925.
108. K. Eger, M. Jalalian, M. Schmidt, *J. Het. Chem.*, **1995**, *32*, 211–218.
109. R. Kumar, L.I. Wiebe, E.E. Knaus, *Can. J. Chem.*, **1996**, *74*, 1609–1615.
110. T.S. Mansour, C.A. Evans, M. Charron, B.E. Korba, *Bioorg. Med. Chem. Lett.*, **1997**, *7*, 303–308.
111. D. Loakes, D.M. Brown, S.A. Salisbury, *Tetrahedron Lett.*, **1998**, *39*, 3865–3868.
112. H. Hayatsu, A. Kitajo, K. Sugihara, N. Nitta, K. Negishi, *Nucleic Acids Res.*, **1978**, *5*, 315–318.

Chapter 5

L-NUCLEOSIDES

5.1 INTRODUCTION

L-Nucleosides, the enantiomers of the natural D-nucleosides, are not generally recognised by normal mammalian enzymes but are recognised by virus-encoded or bacterial enzymes, this results in minimal host toxicity but good antiviral/antibacterial activity. L-Nucleosides have emerged as potent antiviral agents against HIV, HBV and other DNA and RNA viruses, with 3TC™ (5.1, Lamivudine®) being the first L-nucleoside approved by the FDA for use in combination therapy against HIV and HBV.[1,2]

Since this discovery, many other active L-nucleosides have been identified[3] (Fig. 5.1) such as L-FMAU (5.3),[4] the 2',3'-dideoxy-2',3'-didehydro-L-nucleoside

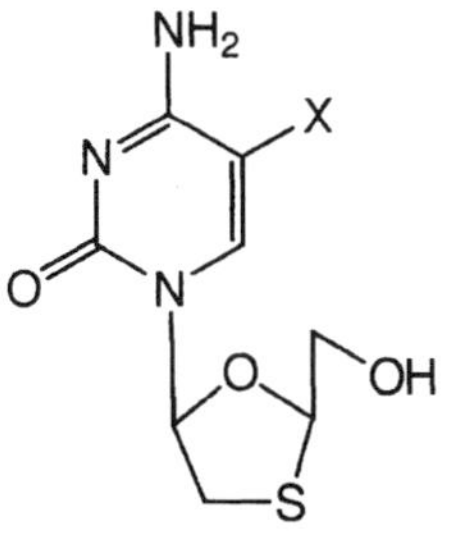

5.1 X = H, 3TC
5.2 X = F, β-L-FTC
5.3 L-FMAU 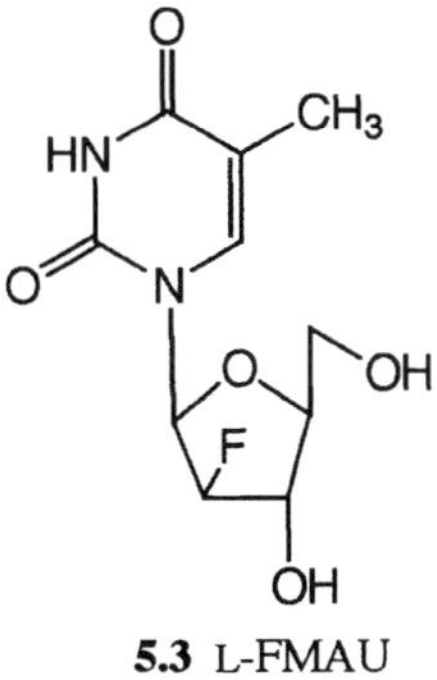
5.4 β-L-FD$_4$C 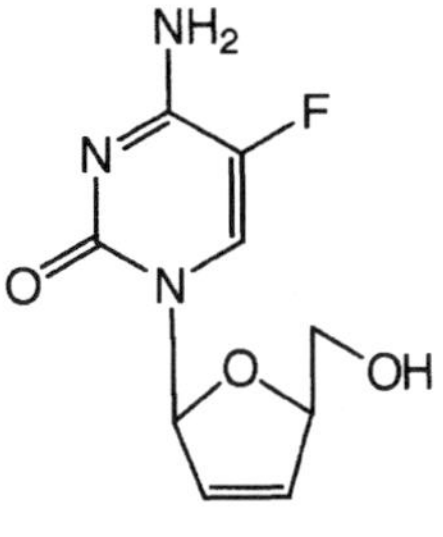

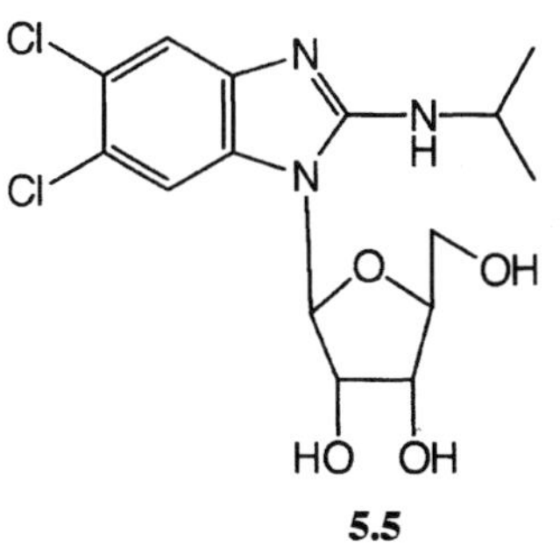

5.5
5.6 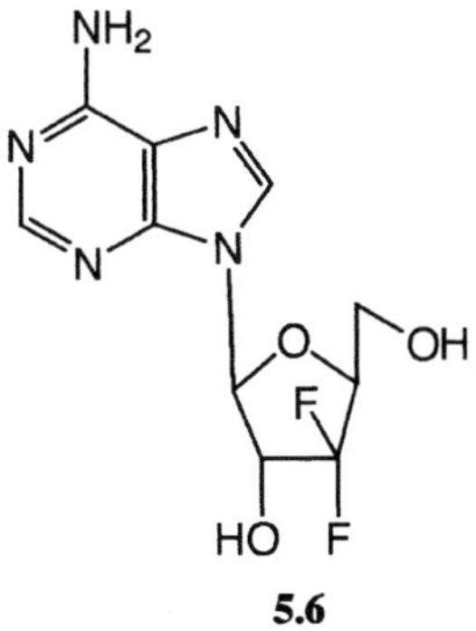

FIGURE 5.1 ANTIVIRAL L-NUCLEOSIDES.

β-L-FD$_4$C (**5.4**),[5] the benzimidazole-β-L-riboside (**5.5**),[6] and the 2',2'-difluoro-L-nucleoside (**5.6**).[7]

5.2 L-OXATHIOLANE AND L-DIOXOLANE NUCLEOSIDES

Of the four possible stereoisomers of 2',3'-dideoxy-3'-thiacytidine, (−)-β-L-(2R, 5S)-1,3-oxathiolanylcytosine (3TC, **5.1**) was found to be the most potent against HIV and HBV with the least toxic effects.[8] The asymmetric synthesis of the key L-oxathiolane (**5.13**) was achieved from L-gulose (**5.7**) *via* the 1,6-thioanhydro-L-gulopyranose intermediate (**5.10**) (Scheme 5.1).[9]

The 1,6-thioanhydro-L-gulopyranose intermediate (**5.10**) was obtained by reaction of the tosyl sugar (**5.9**) with the potassium salt of O-ethylxanthic acid (see *m*, Scheme 5.1) and deprotection of the acetyl protecting groups, with selective oxidative cleavage of the vicinal *cis*-diol of **5.10** to give the oxathiolane structure. Coupling

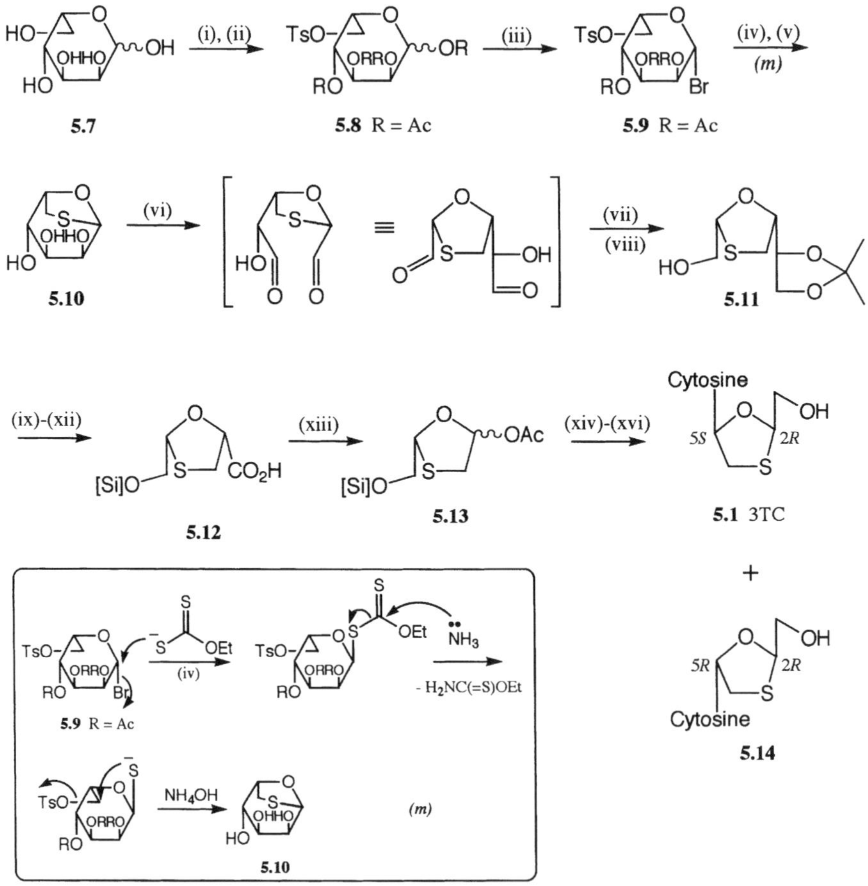

SCHEME 5.1 REAGENTS AND CONDITIONS: (I) TsCl, PYRIDINE, 4 H (II) Ac$_2$O, PYRIDINE, 4 H (III) HBr, AcOH, 15 H (IV) EtOC(=S)S$^-$K$^+$, ACETONE, REFLUX, 3 H (V) NH$_4$OH, MeOH, O/N (VI) NaIO$_4$, MeOH-H$_2$O, 0°C (VII) NaBH$_4$, MeOH-H$_2$O, 0°C (VIII), *p*-TsOH, ACETONE (IX) TBDPSCl, IMIDAZOLE, DMF, 1 H (X) *p*-TsOH, MeOH, 30 MIN (XI) Pb(OAc)$_4$, EtOAc, 10 MIN. (XII) PDC, DMF, 8 H (XIII) Pb(OAc)$_4$, THF, PYRIDINE, 30 MIN. (XIV) N^4-Ac-CYTOSINE, HMDS, (NH$_4$)$_2$SO$_4$, REFLUX, 3 H, TMSOTf, DCE, 1.5 H (XV) NH$_3$, MeOH, 3 H (XVI) TBAF, THF, 30 MIN. [Si] = TBDMS.

of (5.13) with N^4-acetyl-cytosine gave a mixture of anomers (5.14 : 5.1 ≈ 1:2), for the cytosine derivatives antiviral potency was found to be in the following decreasing order: 5-iodocytosine (β-anomer) > 5-fluorocytosine (α-anomer) > 5-methyl-cytosine (α-anomer) > 5-methylcytosine (β-anomer) > 5-bromocytosine (β-anomer) > 5-chlorocytosine (β-anomer). In the purine series, antiviral potency was found to be in the following decreasing order: adenine (β-anomer) > 6-chloropurine (β-anomer) > 6-chloropurine (α-anomer) > 2-NH_2-6-Cl-purine (β-anomer) > guanine (β-anomer) > N^6-methyladenine (β-anomer) > N^6-methyladenine (α-anomer).[10]

As with the oxathiolane nucleosides, the L-enantiomers of the dioxolane nucleosides were more active than their D-enantiomers against HIV-1 and Epstein Barr virus (EBV).[11,12] The asymmetric synthesis of L-β-(2S,4S)- and L-α-(2S,4R)-dioxolanyl nucleosides (5.15 and 5.16 respectively) proceeded *via* the 1,6-anhydrogulopyranose (5.18), prepared by acidic treatment of 2,3:5,6-di-*O*-isopropylidene-L-gulofuranose (5.17) (Scheme 5.2). The β-L-5-fluorocytosine dioxolane nucleoside (5.15, X = F) showed the most potent anti-HIV activity among the L-(2S)-dioxolanylpyrimidines and -purines.[11]

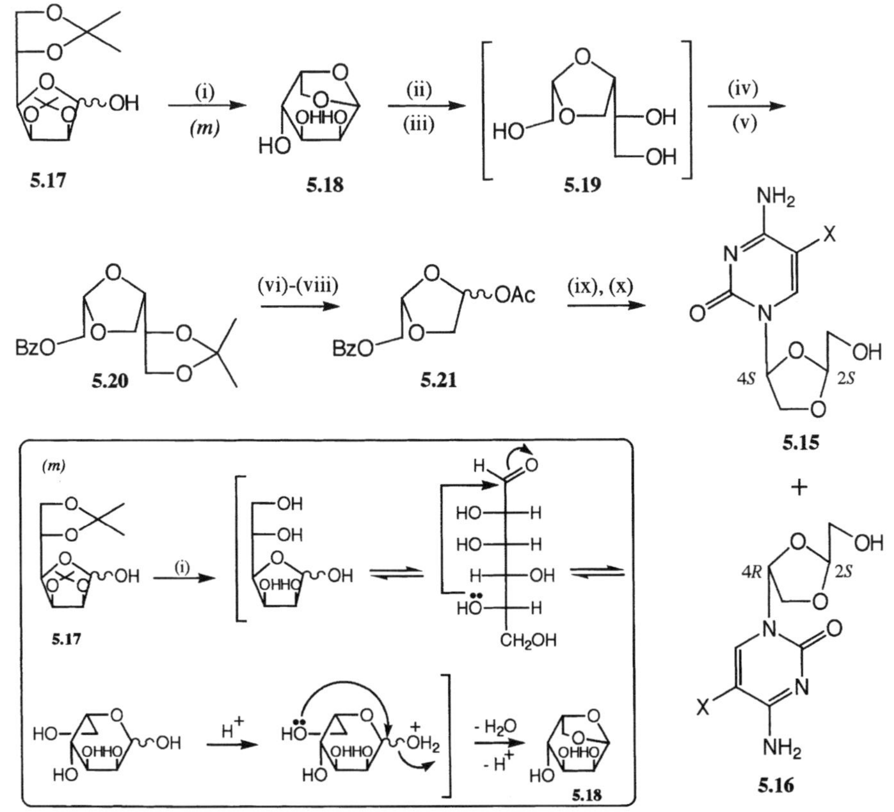

SCHEME 5.2 REAGENTS AND CONDITIONS: (I) 0.5N HCl, REFLUX, 20 H (II) $NaIO_4$, MeOH-H_2O, 0°C (III) $NaBH_4$, MeOH-H_2O, 0°C (IV) *p*-TsOH, ACETONE, 6 H (V) BzCl, PYRIDINE, CH_2Cl_2, 2 H (VI) *p*-TsOH, MeOH, 2 H (VII) $NaIO_4$, RuO_2, CH_3CN, CCl_4, H_2O, 5 H (VIII) $Pb(OAc)_4$, THF, PYRIDINE, 45 MIN. (IX) N^4-Bz-5-X-CYTOSINE, HMDS, $(NH_4)_2SO_4$, REFLUX, 2.5 H THEN TMSOTf, DCE, 2 H (X) NH_3, MeOH, 2 DAYS.

5.2.1 Mechanism of Action

The heterosubstituted L-nucleosides 3TC (**5.1**) and β-L-FTC (**5.2**) are readily taken up into cells where they are phosphorylated to their active triphosphate forms by deoxycytidine kinase, deoxycytidylate kinase and possibly nucleoside diphosphate kinases.[13,14] The anti-HIV activity of 3TC and β-L-FTC is owing to efficient stereoselective phosphorylation to their respective triphosphates and their stability towards deamination by cytidine deaminase, the triphosphates have also been shown to inhibit incorporation of the natural deoxyribonucleoside triphosphates into HBV DNA.[15,16] After incorporation into viral RNA or DNA the mechanism of action follows that described in either section 1.7.1.1 for anti-HIV activity or section 1.7.2.1 for anti-HBV activity.

5.3 2′,3′-DIDEOXY-L-NUCLEOSIDES

The dose-limiting toxicity of the antiviral drug DDC (**1.27**, Zalcitabine®) is severe neuropathy caused by inhibition of the synthesis of mitochondrial DNA. The 2′,3′-dideoxy-L-nucleosides, β-L-DDC (**5.22**) and β-L-FDDC (**5.23**) displayed low toxicity and a negligible inhibitory effect on host mitochondrial DNA synthesis, and were also shown to be 280 and 1000 times more potent respectively against HBV than their D-enantiomeric counterparts.[17]

2′,3′-Dideoxy-L-nucleosides have been prepared from D-glutamic acid,[17–19] D-glyceraldehyde[20] and L-arabinose.[21,22] Preparation of the required 2,3-dideoxy-L-ribofuranose (**5.28**) from D-glutamic acid (**5.24**) was achieved according to the methodology of Okabe et al.[23] Nitrous acid deamination of **5.24** and subsequent

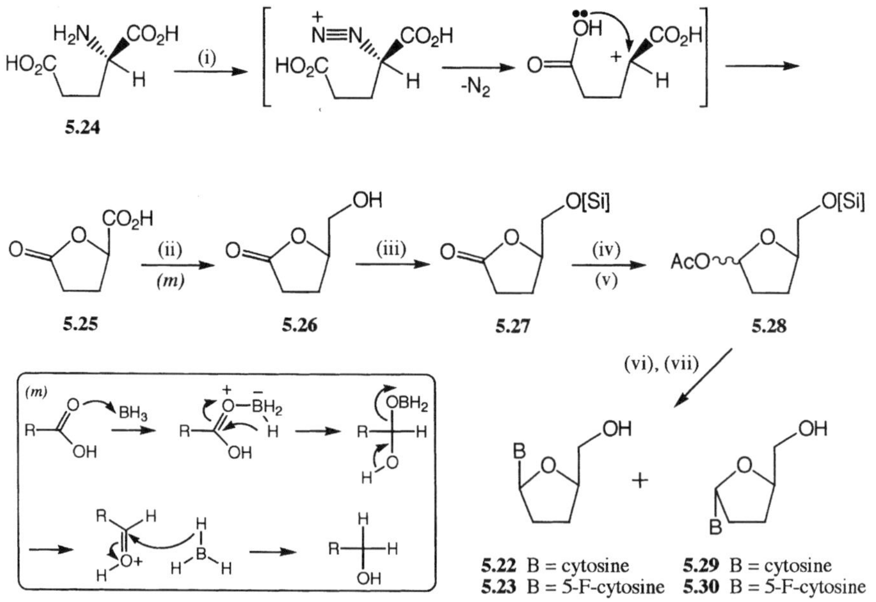

SCHEME 5.3 REAGENTS AND CONDITIONS: (I) NaNO$_2$, H$_2$O, 5.6 N HCl, O/N (II) BH$_3$-SMe$_2$, THF, 2 H (III) TBDMSCl, IMIDAZOLE, CH$_2$Cl$_2$, 2 H (IV) DIBAL, TOLUENE, −78°C, 1 H (V) Ac$_2$O, Et$_3$N, O/N (VI) PYRIDIMIDINE/PURINE BASE (B), HMDS, (NH$_4$)$_2$SO$_4$, REFLUX, 1 H THEN EtAlCl$_2$, CH$_2$Cl$_2$, 40 MIN. (VII) p-TsOH, MeOH, H$_2$O, 7 H. [Si] = TBDMS.

reduction with borane-methyl sulphide complex produces (R)-(−)-γ-(hydroxymethyl)-γ-butyrolactone (5.26) which is converted to 5.28 in three steps. Optimal yields of the β-L-nucleosides (5.22 and 5.23) were achieved using either ethylaluminium dichloride or potassium nonafluoro-1-butane sulphonate as the coupling Lewis acids (Scheme 5.3).[17]

Synthesis from L-arabinose (5.31) avoids anomeric mixtures on coupling and also allows access to 2',3'-dideoxy-2',3'-didehydro-L-nucleosides. The synthesis described by T-S. Lin et al.[21] proceeds via 2,2'-anhydro-β-L-uridine (5.33),[24] with subsequent deoxygenation/ elimination to give protected β-L-D$_4$U (5.36), which was converted to the protected β-L-DDU (5.37) by hydrogenation (Scheme 5.4).

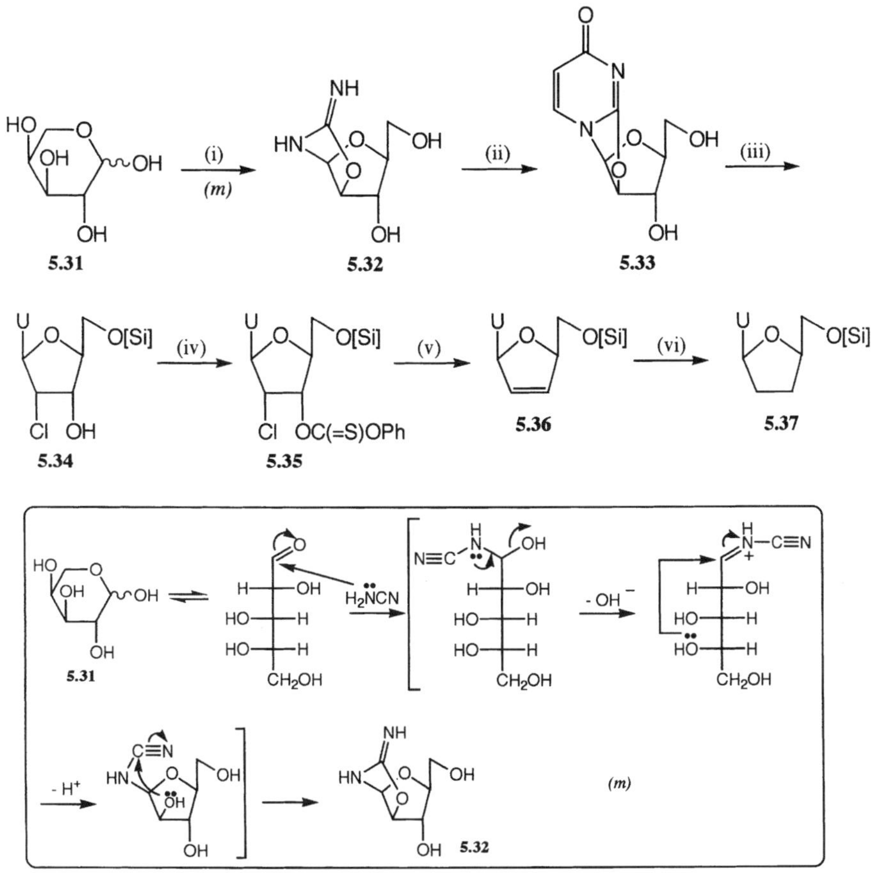

SCHEME 5.4 REAGENTS AND CONDITIONS: (I) H$_2$NCN, NH$_4$OH, MeOH (II) HC≡C-CO$_2$Me (III) t-BuMe$_2$SiCl, PYRIDINE, R.T. 3 DAYS THEN REFLUX 2 H (IV) PhOC(=S)Cl, CH$_3$CN, DMAP (V) AIBN, n-Bu$_3$SnH, TOLUENE, 110°C (VI) H$_2$, Pd/C, EtOH. [Si] = TBDMS.

5.4 D₄-L-NUCLEOSIDES

2',3'-Dideoxy-2',3'-didehydro-β-L-cytidine (**5.38**, β-L-D₄C) and its 5-fluoro-derivative (**5.4**, β-L-FD₄C) display potent activity against HBV and significant activity against HIV.[5] More recently 2'-fluoro-2',3'-unsaturated L-nucleosides, such as 2',3'-dideoxy-2',3'-didehydro-β-L-2'-fluoroadenine (**5.39**, β-L-2'-FD₄A), have been described and exhibit activity against HIV and HBV.[25,26]

The D₄-L-nucleoside β-L-Fd₄C (**5.4**) was prepared by a transglycosylation reaction[27] of silylated 5-fluorouracil with the β-L-uridine nucleoside (**5.40**).[24] Dimesylation and base treatment of **5.41** gave the cyclic ether (**5.42**), the 5-fluorouracil base was then converted to 5-flourocytosine (**5.43**) following previously described methodology (Scheme 2.36) and finally reaction with *tert*-butoxide furnished the target compound β-L-FD₄C (**5.4**) (Scheme 5.5).

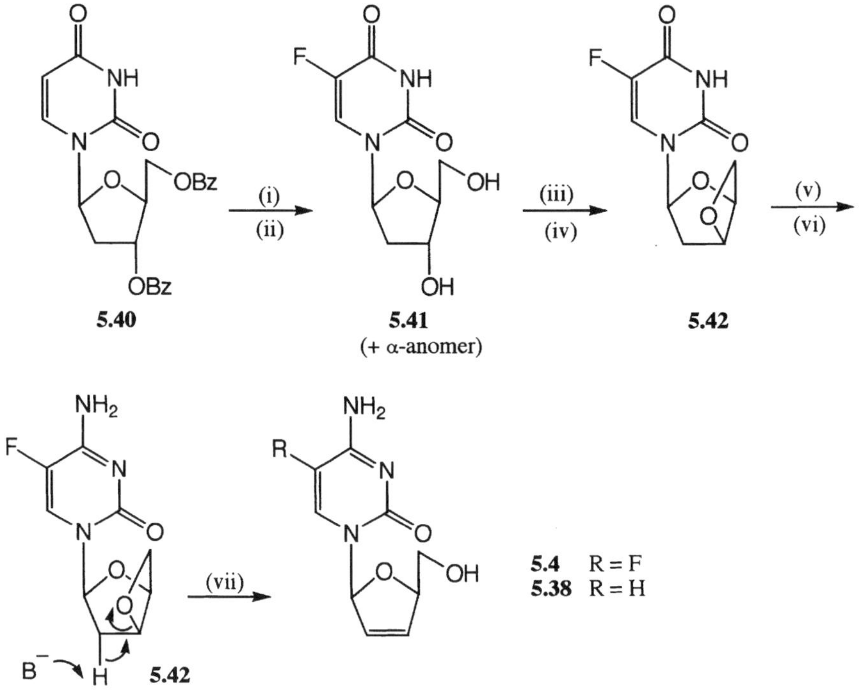

SCHEME 5.5 *REAGENTS AND CONDITIONS*: (I) (TMS)₂-5-F-U, TMSOTf, CH₃CN (II) NH₃, MeOH (III) MsCl, PYRIDINE (IV) 1N NaOH, EtOH, H₂O (V) 1,2,4-TRIAZOLE, *p*-ClC₆H₄OPOCl₂, PYRIDINE (VI) NH₄OH, DIOXANE (VII) ᵗBuOK, DMSO.

The synthesis of (*R*)-(+)-4-[(tert-butyldimethylsilyloxy)methyl]-2-fluoro-2-buten-4-olide (**5.43**), required for the preparation of the 2'-fluoro-2',3'-unsaturated L-nucleosides, can be approached from two routes (Scheme 5.6). Route (a) involved introduction of the 2-phenylseleno moiety to the lactone (**5.44**), treatment of (**5.45α/β**) with lithium hexamethyldisilizane (LiHMDS) and FN(SO₂Ph)₂ gave the 2-α-fluoro-lactone (**5.46**) with oxidative elimination of the 2-phenylseleno moiety yielding the required lactone (**5.43**).[25] Route (b) commenced from L-glyceraldehyde acetonide[28] (5.47) which was subjected to a Horner-Emmons reaction in the presence of triethyl α-fluorophosphonoacetate and NaHMDS to give (5.48) as a mixture of (E)/(Z)-isomers. Cyclization under acidic conditions and silyl protection gave the required lactone (5.43) and the protected diol (5.49) which were readily separated.[26]

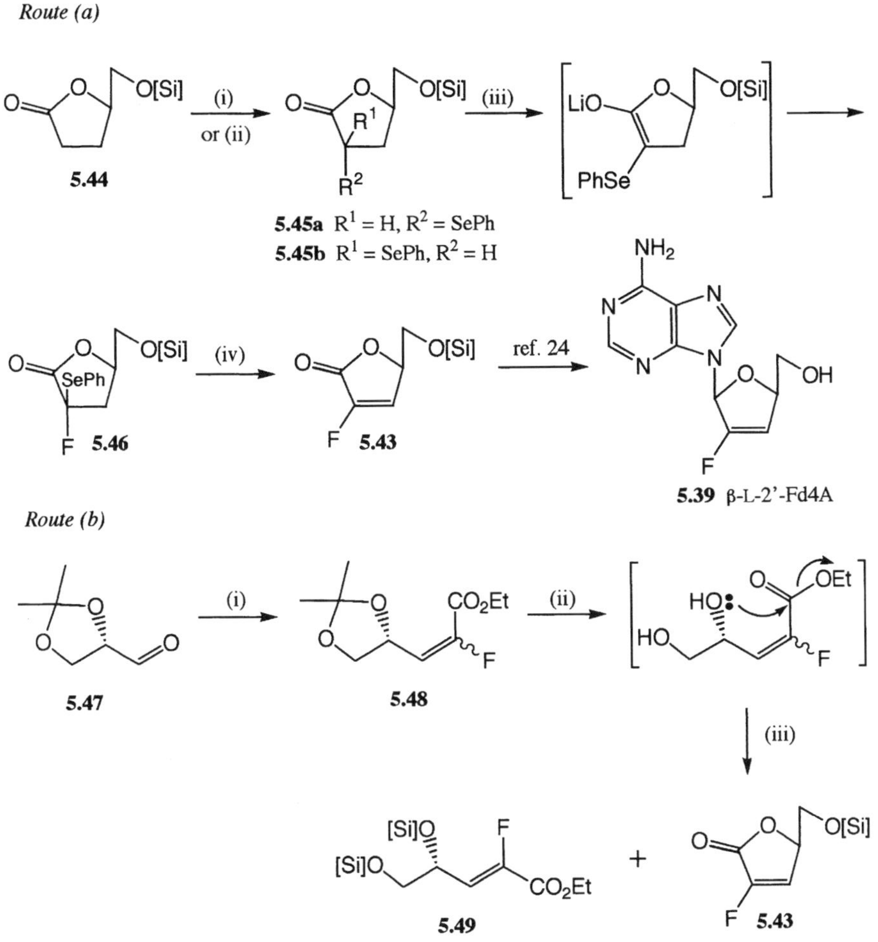

SCHEME 5.6 REAGENTS AND CONDITIONS: ROUTE (A) (I) LiHMDS, THF, PhSeBr, −78°C (II) LiHMDS, THF, N-(PhSe)PHTHALIMIDE, −78°C (III) LiHMDS, THF, FN(SO$_2$Ph)$_2$, −78°C (IV) 30% H$_2$O$_2$, PYRIDINE. ROUTE (B) (I) NaHMDS, (EtO)$_2$P(O)CHFCO$_2$Et, THF, −78°C (II) c-HCl, EtOH (III) TBDMSCl, IMIDAZOLE, CH$_2$Cl$_2$. [Si] = TBDMS.

5.5 OTHER L-NUCLEOSIDES

L-Nucleosides with varying sugar conformations and modified heterocyclic bases have been prepared and evaluated for biological activity, including L-*ribo*-, L-*arabino*-, L-*xylo*, L-*lyxo*-furanosyl-nucleosides, 2' and 3'-substituted L-nucleosides and L-hexopyranosyl-nucleosides,[3] however only a few of these modified L-nucleosides have displayed significant biological activity.

5.5.1 L-*Ribo*-furanosyl-nucleosides

L-*Ribo*-furanosyl-nucleosides can be prepared by modification of a L-nucleoside precursor[29] or by condensation of a suitable L-ribofuranosyl-donor, which can be prepared from L-xylose,[30–32] L-arabinose[33,34] and D-ribose,[35,36] and a heterocyclic base. L-Ribose (**5.54**) is available from D-ribose (**1.1**) in 5 steps (Scheme 5.7),[36] with the protected L-*ribo*-furanoside (**5.55**) suitable for coupling with a heterocyclic base. The antiviral agent 5,6-dichloro-2-(isopropylamino)-1-(β-L-ribofuranosyl) benzimidazole (**5.5**, 1263W94),[6] which is currently in clinical trials to evaluate its antiviral activity and tolerability in HIV-infected patients, has been prepared from this L-*ribo*-furanosyl precursor[37] and is the most promising L-ribose nucleoside prepared.

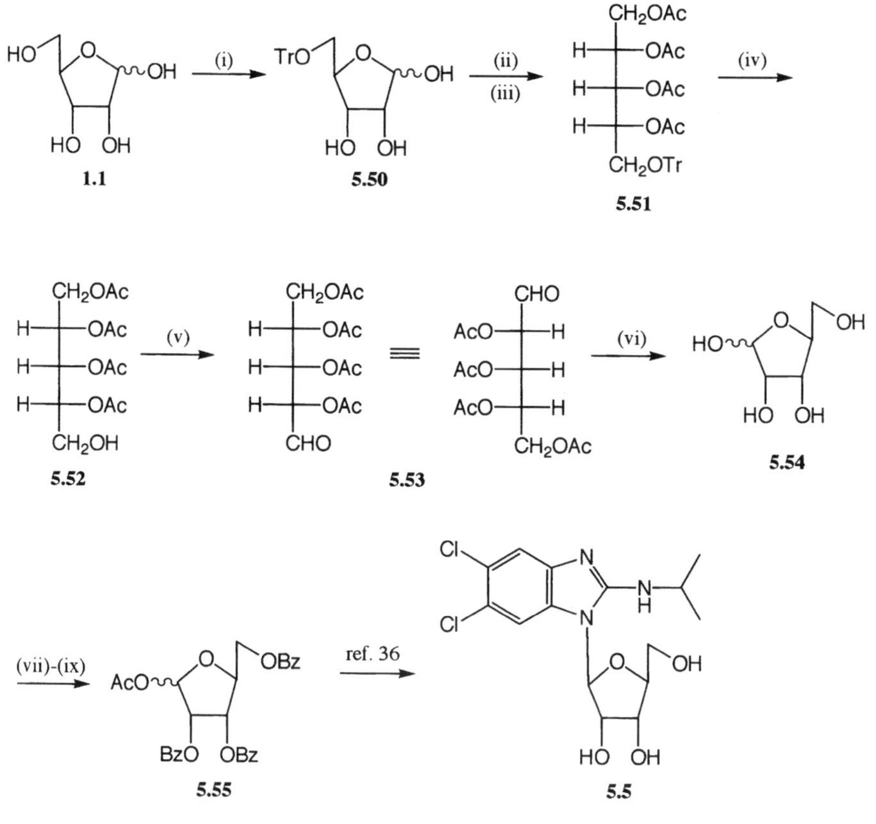

SCHEME 5.7 REAGENTS AND CONDITIONS: (I) TrCl, PYRIDINE (II) NaBH$_4$ (III) Ac$_2$O, PYRIDINE (IV) HCO$_2$H, Et$_2$O (V) DMSO, TFAA, Et$_3$N, −78 °C (VI) K$_2$CO$_3$, EtOH (VII) MeOH, H$_2$SO$_4$ (VIII) BzCl, PYRIDINE (IX) Ac$_2$O, AcOH, H$_2$SO$_4$.

5.5.1.1 L-2′-Deoxyribo-furanosyl-nucleosides

The five naturally occurring 2′-deoxy-L-nucleosides, L-dU, L-T, L-dC, L-dA and L-dG, have been prepared[38] from the common intermediate 3′,5′-di-*O*-benzoyl-2′-deoxy-β-L-uridine (**5.40**)[24] using the acid-catalysed glycosylation procedure, as well as from L-ribose or other sugar precursors (see references 29-36). L-Thymidine is not recognised by human TK but acts as a specific substrate for HSV-1 TK and reduces HSV-1 multiplication in HeLa cells.[38] L-2'-Deoxy-4'-thionucleosides have also been prepared[39,40] but do not possess any significant biological activity.

The most active series in this category are the 2'-deoxy-2',2'-difluoro-L-*erythro*-pentofuranosyl nucleosides.[41,42] The key sugar moiety (**5.60**) was prepared from the protected L-gulono-γ-lactone (**5.56**) which was subjected to oxidative cleavage to give 2,3-*O*-isopropylidene-L-glyceraldehyde (**5.57**). Coupling of **5.57** with ethyl bromodifluoroacetate under Reformatsky conditions gave the propionates (**5.58a/5.58b**) which were cyclized on acid treatment and the resulting lactone protected (**5.59**) (Scheme 5.8).[42] The lactol (**5.60**) was then coupled to the pyrimidine or purine heterocycle using a Mitsunobu procedure to give the 2'-deoxy-2',2'-difluoro-L-*erythro*-pentofuranosyl pyrimidine or purine nucleosides, the most active of these nucleosides being the adenine derivative (**5.61**). L-3'-Deoxyribo-furanosyl-nucleosides have also been prepared from L-xylose however antiviral activity was not observed.[43]

5.5.2 L-*Arabino*-furanosyl-nucleosides

Pyrimidine and purine L-nucleosides with an *arabino*-sugar conformation have been synthesized,[44-49] with β-L-2'-fluoro-5-methyl-arabinofuranosyluracil (**5.3**, β-L-

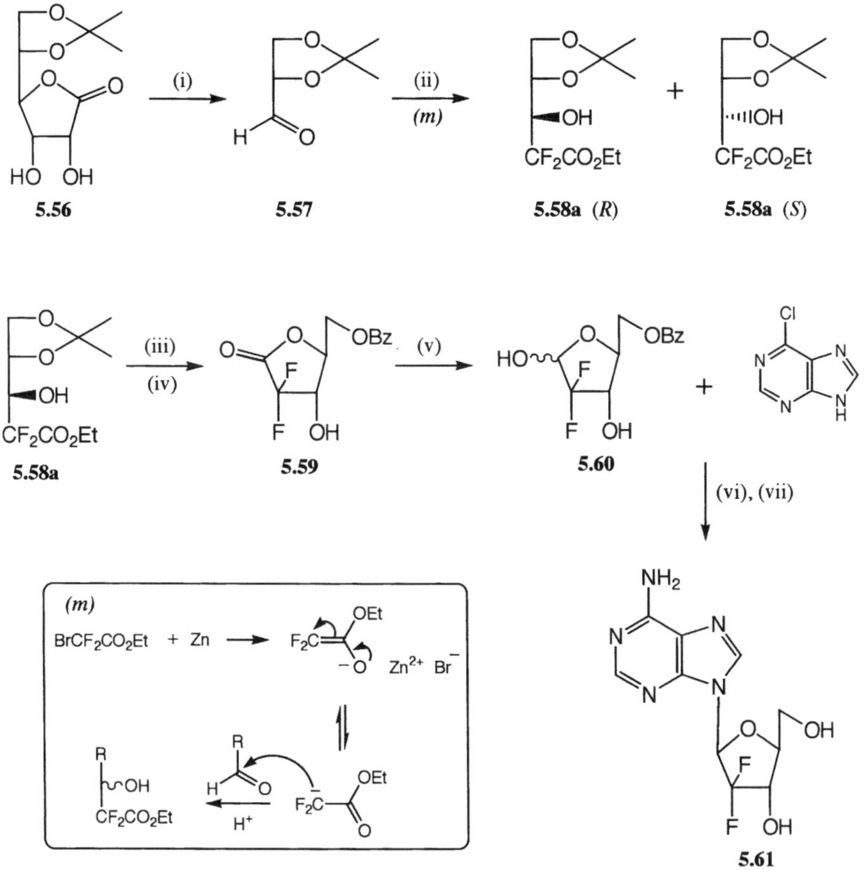

SCHEME 5.8 REAGENTS AND CONDITIONS: (I) NaIO₄, H₂O, pH 5–6, 2 H (II) BrCF₂CO₂Et, Zn, THF, 60°C (III) DOWEX-50 (H⁺), MeOH, H₂O, 4 DAYS (IV) BzCl, 2,6-LUTIDINE, CH₂Cl₂, (V) Li(*t*-BuO)₃AlH, THF, −78°C, 2 H (VI) DEAD, PH₃P, THF, 8 H (VII) NH₃, MeOH, STEEL BOMB, 100°C, 20 H.

FMAU),[4,32] a potent anti-HBV and anti-EBV L-nucleoside with a favourable toxicity profile, the most promising of this class.

The required 2-fluoro-L-*arabino*-furanoside (**5.64**) was prepared in four steps from the L-ribose sugar (**5.55**).[32] 1,3,5-Tri-*O*-benzoyl-α-L-ribofuranose (**5.62**) was obtained in a one-step process involving initial halogenation at C-1 and subsequent hydrolysis.[50] The C2-hydroxy group was then converted into the imidazolyl sulphonate which was readily displaced with fluoride anion to give the *arabino*-fluoro-sugar (**5.63**). The brominated precursor (**5.64**) was prepared *in situ* and coupled with the silylated thymine base under reflux to give β-L-FMAU (Scheme 5.9).

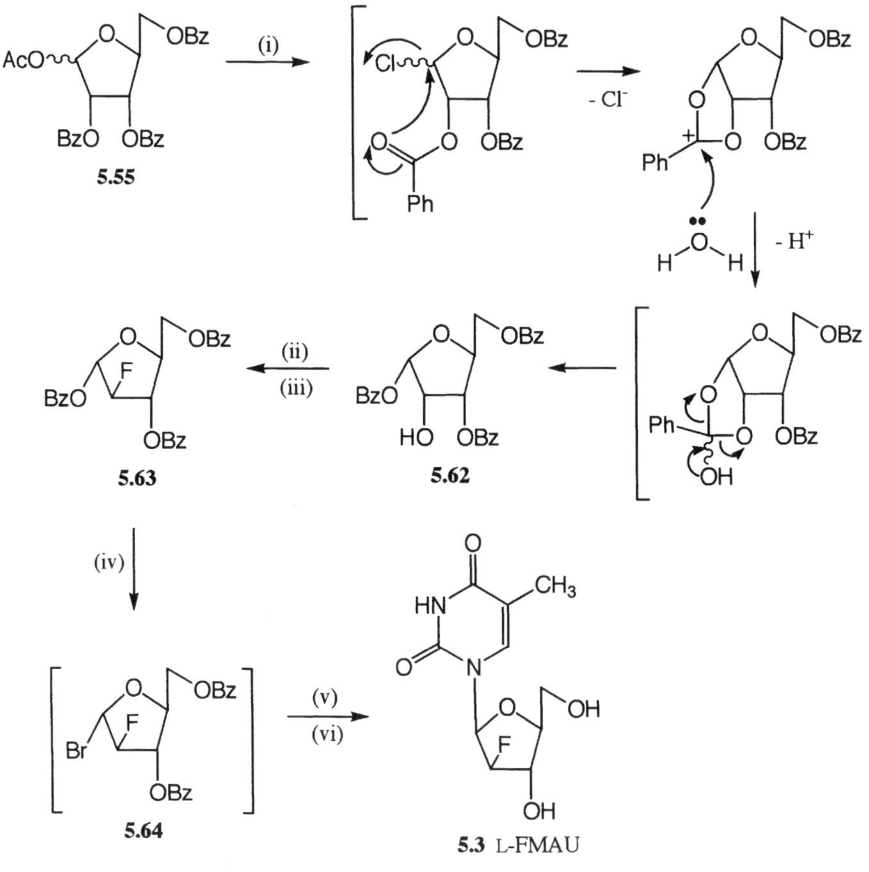

SCHEME 5.9 REAGENTS AND CONDITIONS: (I) HCl (G), AcCl, CH$_2$Cl$_2$, 0°C THEN H$_2$O, CH$_3$CN (II) SO$_2$Cl$_2$, DMF, CH$_2$Cl$_2$, −15°C, IMIDAZOLE (III) KHF$_2$, 2,3-BUTANEDIOL, 48% HF/H$_2$O, REFLUX (IV) HBr/AcOH, CH$_2$Cl$_2$ (V) T(TMS)$_2$, 1,2-DICHLOROETHANE, REFLUX (VI) NH$_3$, MeOH.

5.5.3 L-*Xylo*-furanosyl-nucleosides

The β-L-*xylo*-furanosyl-nucleosides of the five naturally occurring nucleic acid bases have been prepared by direct condensation of a suitably protected L-*xylo*-furanose (**5.68**) and a purine or pyrimidine base with either SnCl$_4$ or TMSOTf as the Lewis acid (Scheme 5.10).[51] Compounds **5.69**–**5.73** lacked any significant activity against a range of DNA and RNA viruses.

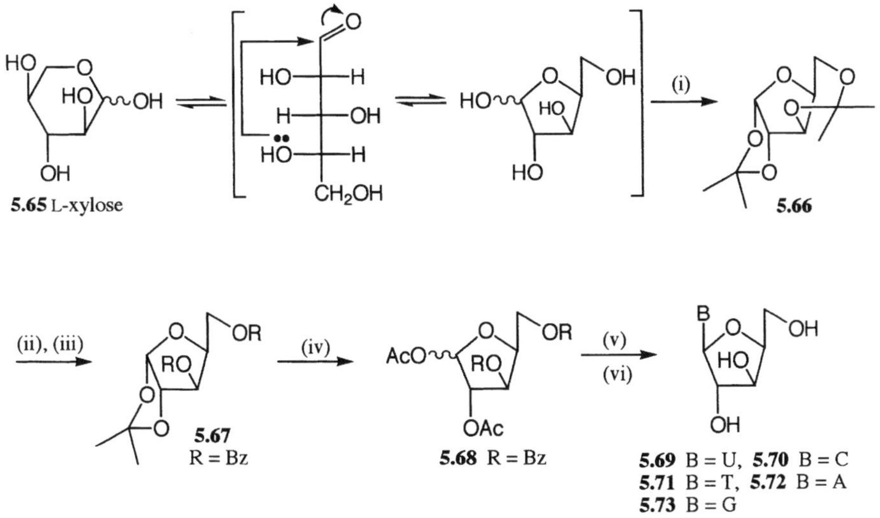

L-NUCLEOSIDES

SCHEME 5.10 *REAGENTS AND CONDITIONS*: (I) ACETONE, H_2SO_4, $CuSO_4$ (II) HCl, H_2O (III) BzCl, PYRIDINE, $CHCl_3$ (IV) Ac_2O, AcOH, H_2SO_4 (V) PYRIMIDINE/PURINE (B), HMDS, TMSCl THEN $SnCl_4$ OR TMSOTf, CH_3CN (VI) NH_3, MeOH.

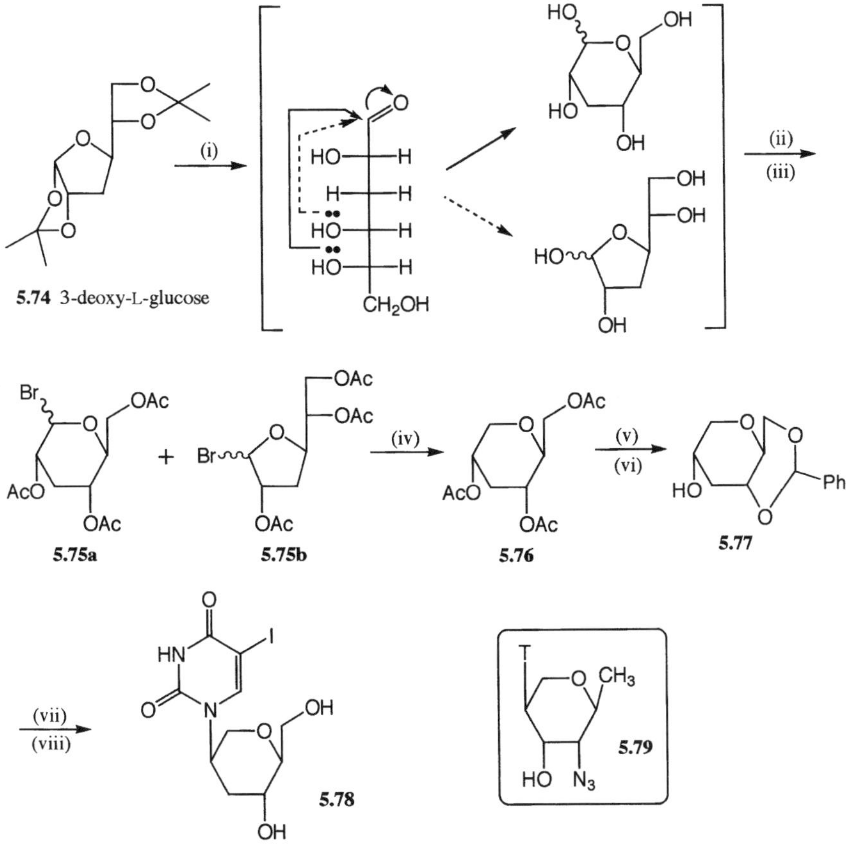

5.65 L-xylose
5.66
5.67 R = Bz
5.68 R = Bz
5.69 B = U, 5.70 B = C
5.71 B = T, 5.72 B = A
5.73 B = G

5.74 3-deoxy-L-glucose

5.75a 5.75b 5.76 5.77

5.78 5.79

SCHEME 5.11 *REAGENTS AND CONDITIONS*: (I) IRA-120 (H^+) RESIN, EtOH, H_2O, REFLUX. (II) Ac_2O, PYRIDINE (III) HBr, AcOH (IV) Bu_3SnH, Et_2O (V) NaOMe, MeOH (VI) $PhCH(OMe)_2$, DIOXANE (VII) NaH, TsCl THEN Na SALT OF 5-IODO-URACIL, DMF, 90°C, 1 H (VIII) 80% AcOH, REFLUX, 1 H.

5.5.4 L-Hexopyranosyl-nucleosides

L-Anhydrohexitol nucleosides have been prepared from 3-deoxy-L-glucose (5.74).[52] Acid deprotection of (5.74), peracetylation and conversion to the bromo-sugars gave a 5:1 ratio of L-pyranose (5.75a) and L-furanose (5.75b) sugars. Dehalogenation with Bu$_3$SnH gave the pyranose (5.76), which was converted to the desired L-anhydrohexitol precursor (5.77) for reaction with a pyrimidine or purine base (Scheme 5.11).

The L-anhydrohexitol nucleosides (*e.g.* 5.78) were devoid of antiviral activity, however the 3-azido-L-hexopyranosyl nucleoside (5.79) has shown significant dose-dependent activity against HIV *in vitro*.[53]

5.6 REFERENCES

1. S.L. Doong, C.H. Tsai, R.F. Schinazi, D.C. Liotta, Y.C. Cheng, *Proc. Natl. Acad. Sci. USA*, **1991**, *88*, 8495–8499.
2. J.L. Dienstag, R.P. Perillo, E.R. Schiff, M. Barholomew, C. Vicary, M. Rubin, *New Engl. J. Med.*, **1995**, *333*, 1657–1661.
3. P.Y. Wang, J.H. Hong, J.S. Cooperwood, C.K. Chu, *Antiviral Res.*, **1998**, *40*, 19–44.
4. C.K. Chu, T.W. Ma, K. Shanmuganathan, C.G. Wang, Y.J. Xiang, S.B. Pai, G.Q. Yao, J-P. Sommadossi, Y-C. Cheng, *Antimicrob. Ag. Chemother.*, **1995**, *39*, 979–981.
5. T-S. Lin, M-Z. Luo, M-C. Liu, Y-L. Zhu, E. Gullen, G.E. Dutschman, Y–C. Cheng, *J. Med. Chem.*, **1996**, *39*, 1757–1759.
6. G.W. Koszalka, S.D. Chamberlain, R.J. Harvey, L.W. Frick, S.S. Good, M.L. Davis, A. Smith, J.C. Drach, L.B. Townsend, K.K. Biron, *Antiviral Res.*, **1996**, *30*, A43.
7. Y. Xiang, L.P. Kotra, C.K. Chu, *Bioorg. Med. Chem. Lett.*, **1995**, *5*, 743–748.
8. C-N. Chang, S-L. Doong, J.H. Zhou, J.W. Beach, L.S. Jeong, C.K. Chu, C–H. Tsa, Y–C. Cheng, *J. Biol. Chem.*, **1992**, *33*, 6899–6902.
9. J.W. Beach, L.S. Jeong, A.J. Alves, D. Pohl, H.O. Kim, C-N. Chang, S-L. Doong, R.F. Schinazi, Y–C. Cheng, C.K. Chu, *J. Org. Chem.*, **1992**, *57*, 2217–2219.
10. L.S. Jeong, R.F. Schinazi, J.W. Beach, H.O. Kim, S. Nampalli, K. Shanmuganathan, A.J. Alves, A. McMillan, C.K. Chu, R. Mathis, *J. Med. Chem.*, **1993**, *36*, 181–195.
11. H.O. Kim, R.F. Schinazi, K. Shanmuganathan, L.S. Jeong, J.W. Beach, S. Nampalli, D.L. Cannon, C.K. Chu, *J. Med. Chem.*, **1993**, *36*, 519–528.
12. J-S. Lin, T. Kira, E. Gullen, Y. Choi, F. Qu, C.K. Chu, Y-C. Cheng, *J. Med. Chem.*, **1999**, *42*, 2212–2217.
13. M.T. Paff, D.R. Averett, K.L. Prus, W.H. Miller, J.D. Nelson, *Antimicrob. Ag. Chemother.*, **1994**, *38*, 1230–1238.
14. D.S. Shewach, D.C. Liotta, R.F. Schinazi, *Biochem. Pharmacol.*, **1993**, *45*, 1540–1543.
15. N.M, Gray, C.L.P. Marr, C.R. Penn, J.M. Cameron, R.C. Bethell, *Biochem. Pharmacol.*, **1995**, *50*, 1043–1051.
16. M.G. Davis, J.E. Wilson, N.A. VanDraanen, W.H. Miller, G.A. Freeman, S.M. Daluge, F.L. Boyd, A.E. Painter, L.R. Boone, *Antiviral Res.*, **1996**, *30*, 133–145.
17. T-S. Lin, M-Z. Luo, M-C. Liu, B. Pai, G.E. Dutschman, Y–C. Cheng, *J. Med. Chem.*, **1994**, *37*, 798–803.
18. A. Tse, T.S. Mansour, *Tetrahedron Lett.*, **1995**, *36*, 7807–7810.
19. P.J. Bolon, P. Wang, C.K. Chu, G. Gosselin, V. Boudou, C. Pierra, C. Mathé, J-L. Imbach, A. Faraj, M.A. el Alaoui, J-P. Sommadossi, S.B. Pai, Y-L. Zhu, J-S. Lin, Y–C. Cheng, R.F. Schinazi, *Bioorg. Med. Chem. Lett.*, **1996**, *6*, 1657–1662.
20. G. Rassu, F. Zanardi, L. Battistini, E. Gaetani, G. Casiraghi, *J. Med. Chem.*, **1997**, *40*, 168–180.
21. T-S. Lin, M-Z. Luo, M-C. Liu, *Tetrahedron Lett.*, **1994**, *35*, 3477–3480.
22. C. Génu–Dellac, G. Gosselin, J-L. Imbach, *Tetrahedron Lett.*, **1991**, *32*, 79–82.
23. M. Okabe, R.C. Sun, S.Y.K. Tam, L.J. Todaro, D.L. Coffen, *J. Org. Chem.*, **1988**, *53*, 4780–4786.
24. A. Holy, *Coll. Czech. Chem. Commun.*, **1972**, *37*, 4072–4086.
25. S-H. Chen, Q. Wang, J. Mao, I. King, G.E. Dutschman, E.A. Gullen, Y–C. Cheng, T.W. Doyle, *Bioorg. Med. Chem. Lett.*, **1998**, *8*, 1589–1594.
26. K. Lee, Y. Choi, E. Gullen, S. Schlueter-Wirtz, R.F. Schinazi, Y–C. Cheng, C.K. Chu, *J. Med. Chem.*, **1999**, *42*, 1320–1328.
27. M. Imazawa, F. Eckstein, *J. Org. Chem.*, **1978**, *43*, 3044–3048.
28. C.A. Hubschwerlen, *Synthesis*, **1986**, 962–964.
29. V. Boudou, G. Gosselin, J-L. Imbach, *Nucleosides & Nucleotides*, **1999**, *18*, 607–609.
30. E. Moyroud, O. Botta, P. Strazewski, *Tetrahedron*, **1999**, *55*, 1277–1284.

31. E. Moyroud, O. Botta, C. Lobato, P. Strazewski, *Tetrahedron*, **1998**, *54*, 13529–13546.
32. T. Ma, S.B. Pai, Y.L. Zhu, J.S. Lin, K. Shanmuganathan, J. Du, C. Wang, H. Kim, M.G. Newton, Y.C. Cheng, C.K. Chu, *J. Med. Chem.*, **1996**, *39*, 2835–2843.
33. G.M. Visser, J. van Westrenen, C.A.A. van Boeckel, J.H. van Boom, *Recl. Trav. Chim. Pays–Bas*, **1986**, *105*, 528–537.
34. Y. Abe, T. Takizawa, T. Kunieda, *Chem. Pharm. Bull*, **1980**, *28*, 1324–1326.
35. M.E. Jung, Y. Xu, *Tetrahedron Lett.*, **1998**, *38*, 4199–4202.
36. M.E. Jung, C.J. Nichols, O. Kretschik, Y. Xu, *Nucleosides & Nucleotides*, **1999**, *18*, 541–546.
37. J.H. Tidwell, S.D. Chamberlain, R. Hammitt, C.L. Burns, G.W. Koszalka, *Abs. Papers Am. Chem. Soc.*, **1998**, *216*, 172–orgn.
38. S. Spadari, G. Maga, F. Focher, G. Ciarrocchi, R. Manservigi, F. Arcamone, M. Capobianco, A. Carcuro, F. Colonna, S. Iotti, A. Garbesi, *J. Med. Chem.*, **1992**, *35*, 4214–4220.
39. J. Uenishi, K. Takahashi, M. Motoyama, H. Akashi, T. Sasaki, *Nucleosides & Nucleotides*, **1994**, *13*, 1347–1361.
40. F. De Valette, J-L. Barascut, J-L. Imbach, *Nucleosides & Nucleotides*, **1998**, *17*, 2289–2310.
41. Y.J. Xiang, L.P. Kotra, C.K. Chu, *Bioorg. Med. Chem. Lett.*, **1995**, *5*, 743–748.
42. L.P. Kotra, Y. Xiang, M.G. Newton, R.F. Schinazi, Y-C. Cheng, C.K. Chu, *J. Med. Chem.*, **1997**, *40*, 3635–3644.
43. C. Mathé, G. Gosselin, M-C. Bergogne, A-M. Aubertin, G. Obert, A. Kirn, J–L. Imbach, *Nucleosides & Nucleotides*, **1995**, *14*, 549–550.
44. J. Jansons, Y. Maurinsch, M. Lidaks, *Nucleosides & Nucleotides*, **1995**, *14*, 1709–1724.
45. T-S. Lin, M-Z. Luo, M-C. Lin, *Nucleosides & Nucleotides*, **1994**, *13*, 1861–1870.
46. T.S. Mansour, A.R. Cimpoia, H. Lin, P.J. Hunter, C.A. Evans, H.L.A. Tse, J.W. Gillard, A.D. Borthwick, D.J. Knight, J.A.V. Coates, *Antiviral Chem. Chemother.*, **1995**, *6*, 138–142.
48. J–L. Girardet, J.C. Drach, S.D. Chamberlain, G.W. Koszalka, L.B. Townsend, *Nucleosides & Nucleotides*, **1998**, *17*, 2389–2401.
49. T. Ma, J.S. Lin, M.G. Newton, Y-C. Cheng, C.K. Chu, *J. Med. Chem.*, **1997**, *40*, 2750–2754.
50. P.R. Brodfuehrer, C. Sapino, H.G. Howell, *J. Org. Chem.*, **1985**, *50*, 2598–2600.
51. G. Gosselin, M-C. Bergogne, J-L. Imbach, *J. Het. Chem.*, **1993**, *30*, 1229–1233.52. M.W. Andersen, S.M. Daluge, L. Kerremans, P. Herdewijn, *Tetrahedron Lett.*, **1996**, *37*, 8147–8150.
53. F. Sztaricskai, Z. Dinya, G. Batta, L. Gergely, B. Szabó, *Nucleosides & Nucleotides*, **1992**, *11*, 11–21.

CHAPTER 6

C-NUCLEOSIDES

6.1 INTRODUCTION

C-Nucleosides have C1' of their sugar moieties linked to different nitrogen-containing heterocycles through a carbon-carbon bond. Although N-nucleosides which have a C1'-N glycosidic link predominate in nature, there are many naturally occurring C-nucleosides which have a C1'-C glycosidic link.[1–4] The first C-nucleoside isolated was pseudouridine (**6.1**),[5] many other members of this class of nucleoside have now been isolated from natural sources, such as showdomycin (**6.2**),[6] formycin (**6.3**),[7] pyrazofurin (**1.58**)[8] and oxazinomycin (**6.4**),[9] or prepared synthetically such as tiazofurin (**6.5**),[10] selenazofurin (**6.6**),[11] ethyl 3,5-dichloro-6-(β-D-ribofuranosyl)pyrazine-2-carboxylate (**6.7**)[12] and 7-(5'-deoxy-5'-iodo-β-D-ribofuranosyl)thieno[3,4-d]pyrimidin-4(3H)-one (**6.8**)[13] (Fig. 6.1).

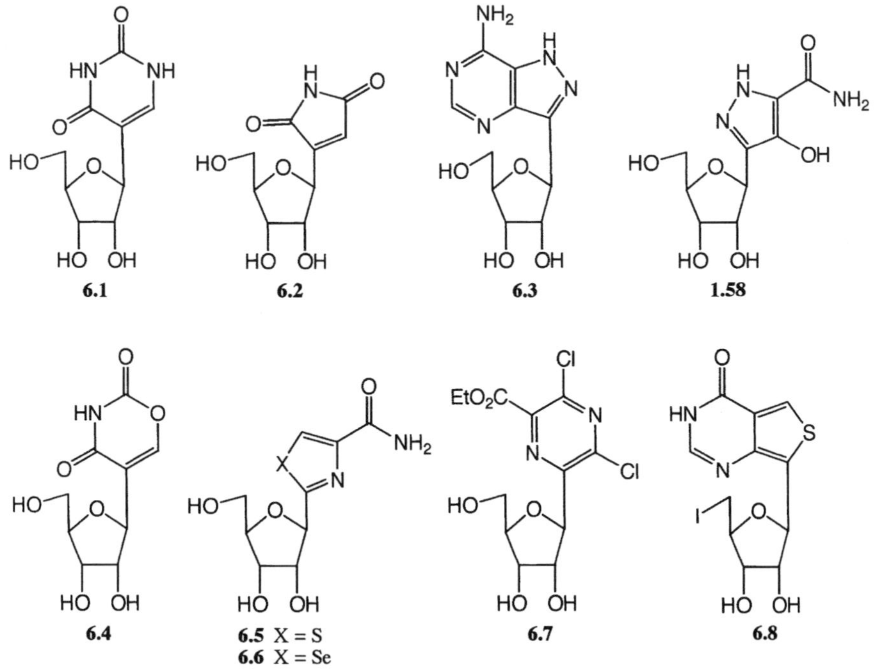

FIGURE 6.1 EXAMPLES OF NATURAL AND SYNTHETIC C-NUCLEOSIDES.

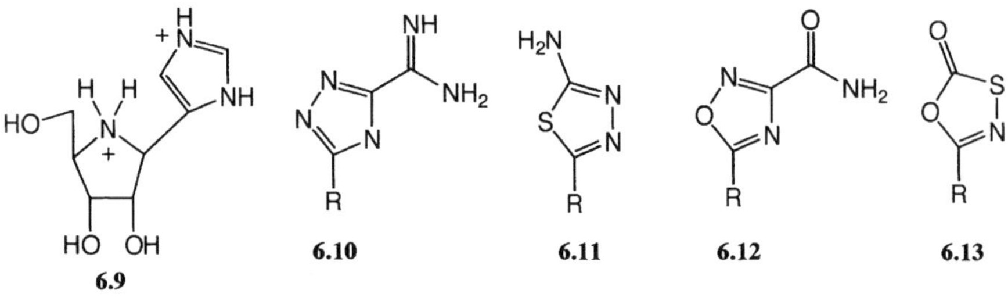

C-NUCLEOSIDES

FIGURE 6.2 EXAMPLES OF 5-RING HETEROCYCLIC BASE MOIETIES OF C-NUCLEOSIDES (R = RIBOSE).

C-Nucleosides act as isosteric mimics of the N-nucleoside metabolites and as such often display antibacterial, antiviral and antitumour properties.[1–4] Importantly, the C-C glycosidic link renders C-nucleosides stable to enzymatic hydrolysis.

6.2 C-NUCLEOSIDES WITH A 5-RING HETEROCYCLIC BASE MOIETY

The interest in C-nucleosides with a 5-ring heterocyclic base moiety results from the biological activity observed for the naturally occurring C-nucleosides showdomycin (**6.2**) and pyrazofurin (**1.58**). A large variety of this category of C-nucleosides have been synthesised with base moieties ranging from azole (*e.g.* **6.2**), imidazole (**6.9**),[14] 1,3-thiazole (*e.g.* **6.5**), 1,2,4-triazole (**6.10**),[15] 1,3,4-thiadiazole (**6.11**),[16] 1,2,4-oxadiazole (**6.12**)[17] and 1,3,4-oxathiazole (**6.13**)[18] (Fig. 6.2).

6.2.1 Showdomycin

Showdomycin, 2-(β-D-ribofuranosyl)maleimide (**6.2**), isolated from *Streptomyces showdoensis*,[6] exhibits antibacterial and antitumour activity, resulting from its ability to alkylate thiol groups in enzymes through addition to the maleimide double bond.[19,20] Numerous syntheses of showdomycin have been described,[1] many of which involve the stepwise synthesis of the maleimide moiety from a sugar precursor.[21–24] The phenyl thioglycoside (**6.14**) was shown to be a convenient starting material for the construction of the maleimide unit. Decarbomethoxylation[25] of (**6.14**) and subsequent oxidation gave the sulphoxide (**6.15**), which was subjected to a Pummerer rearrangement. Dethiophenylation gave the keto-ester (**6.16**), which on reaction with the Wittig reagent carbamoylmethylenetriphenylphosphorane and deprotection of the acetyl protecting groups provided showdomycin (**6.2**) (Scheme 6.1).[24]

Other methods for the synthesis of showdomycin involve direct coupling of the maleimide unit to a suitably protected ribose unit,[26–29] and transformation of a more readily available C-nucleoside into showdomycin.[30,31]

Generation of the C1'–C1 bond of showdomycin can be achieved using an anisyl tellurium sugar (**6.19**) which readily undergoes radical exchange *via* the radical mechanism described (*m*, Scheme 6.2).[28,32] The reaction is initiated by photolysis of 2-thiopyridone ethanoate (**6.20**) which produces methyl radicals which can then react with the tellurium sugar to give the carbohydrate radical (R•) which then reacts

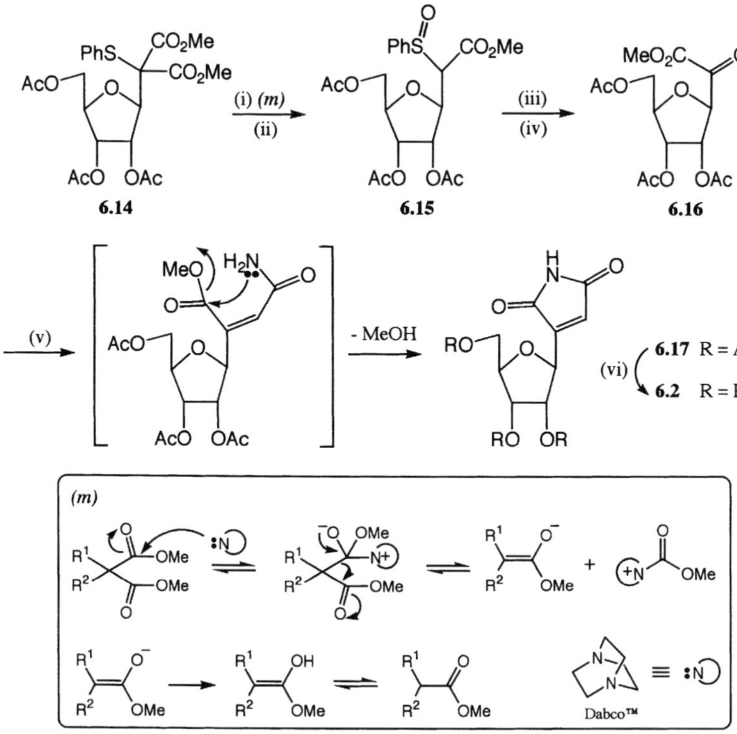

SCHEME 6.1 *REAGENTS AND CONDITIONS*: (I) DABCO, Me$_2$S, BENZENE, REFLUX, 24 H (II) *M*-CPBA, CH$_2$Cl$_2$, 2 H (III) (TfO)$_2$O, 2,6-LUTIDINE, CH$_3$CN, 3 H (IV) HgCl$_2$, H$_2$O, 1 H (V) Ph$_3$P=CHCONH$_2$, CHCl$_3$, 1 H (VI) ACETYL DEPROTECTION.

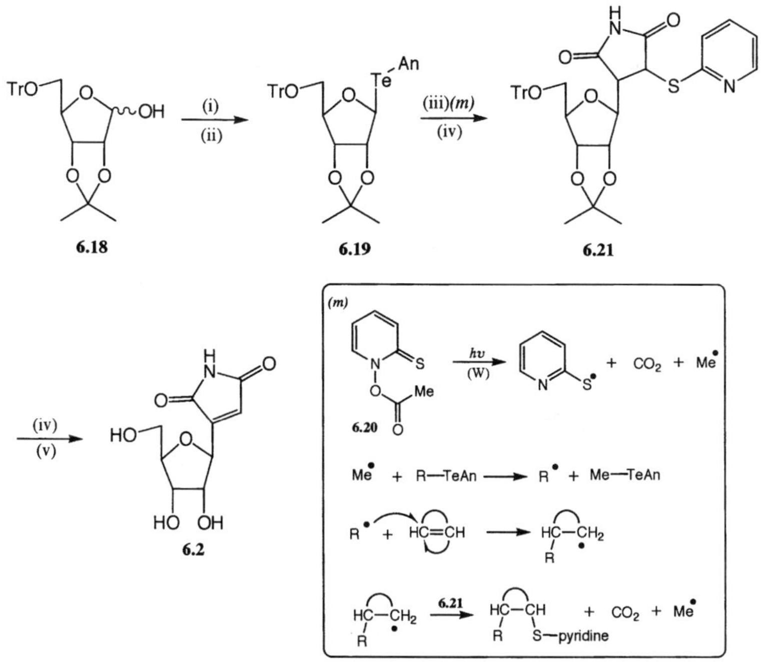

SCHEME 6.2 *REAGENTS AND CONDITIONS*: (I) MsCl, PYRIDINE (II) NaBH$_4$, An$_2$Te$_2$, EtOH, THF (III) **6.20**, CH$_2$Cl$_2$, 150W TUNGSTEN LAMP, MALEIMIDE, 5°C (IV) *M*-CPBA, CH$_2$Cl$_2$, 1 H THEN C$_6$H$_6$, REFLUX, 10 H (V) H$^+$ DEPROTECTION.

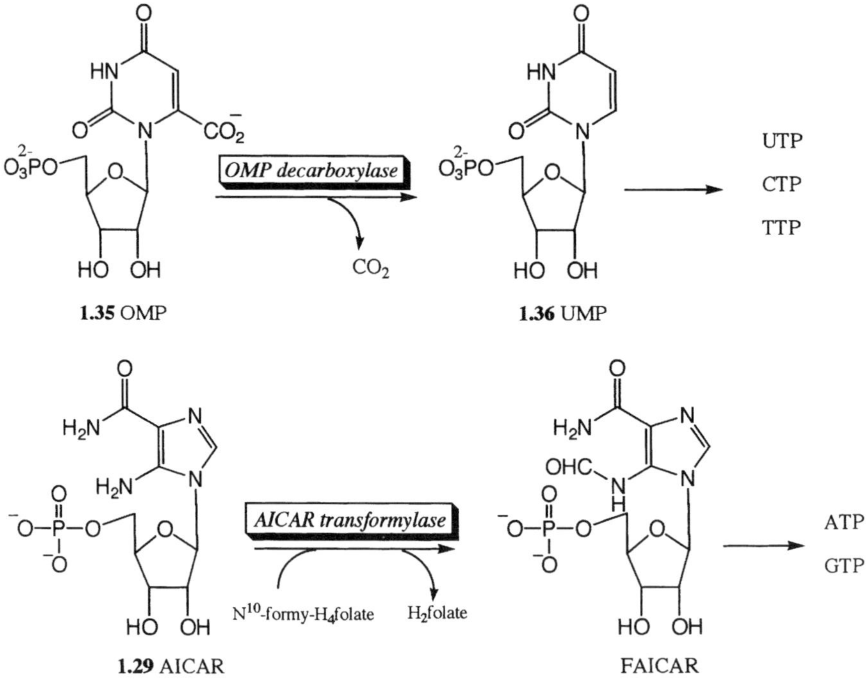

FIGURE 6.3 ENZYME TARGETS OF PYRAZOFURIN.

with maleimide to form (**6.21**). Subsequent oxidation/elimination and deprotection of the ribose protecting groups yielded showdomycin.

Several sugar modified analogues of showdomycin have been prepared such as the 2'-deoxy-ribofuranosyl,[33] arabinofuranosyl,[34] carbocyclic[35] and L-ribofuranosyl derivatives,[36] however in general the derivatives were either devoid of biological activity or no data was given.

6.2.2 Pyrazofurin

Pyrazofurin, 5-carboxy-amido-4-hydroxy-3-(β-D-ribofuranosyl)pyrazole (**1.58**), isolated from *Streptomyces candidus* NRL3601,[8] exhibits antifungal, antitumour and broad-spectrum antiviral activity. The 5'-monophosphate of pyrazofurin is the bioactive species, with the biological activities of pyrazofurin primarily owing to its ability to inhibit orotidine-5'-phosphate decarboxylase (OMP decarboxylase) and therefore *de novo* pyrimidine synthesis (Fig. 6.3).[1,3]

The 5'-monophosphate of pyrazofurin has also been shown to inhibit *de novo* purine synthesis by inhibition of 5-aminoimidazole-4-carboxamide-1-β-D-ribofuranosyl-5'-monophosphate formyltransferase (AICAR transformylase),[1,3] though this has less of an influence on biological activity. Toxicity, owing to poor drug specificity, resulted in pyrazofurin being withdrawn from clinical trials, however topical drug application has been suggested in order to provide a higher therapeutic ratio.[37]

As with showdomycin, the majority of synthetic procedures for the preparation of pyrazofurin involve construction of the 1,2-diazole heterocycle from a suitable

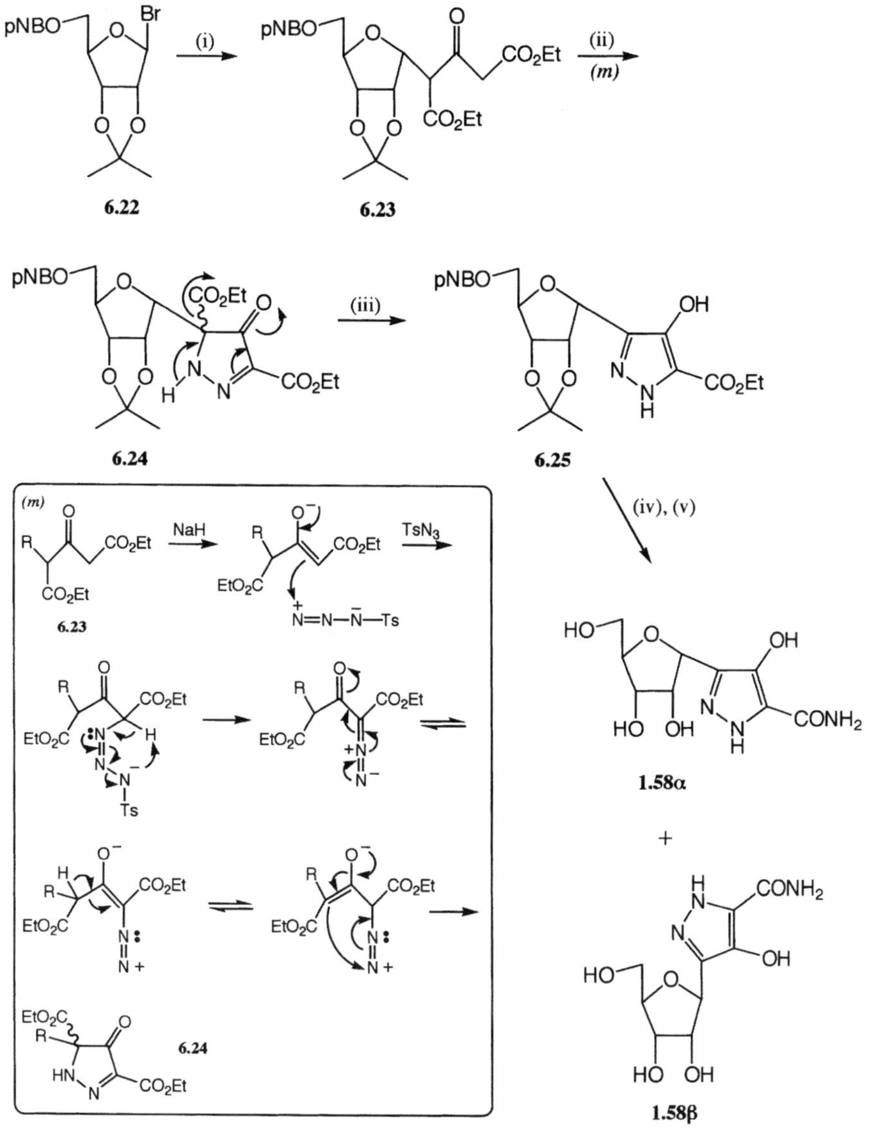

SCHEME 6.3 REAGENTS AND CONDITIONS: (I) KH, C_6H_6, DIETHYL 1,3-ACETONEDICARBOXYLATE, 18-CROWN-6, 16 H (II) NaH, DME, p-TsN₃, 3 H (III) NaOEt, EtOH, 45 MIN. (IV) NH₃, MeOH, 100°C, 3 H (V) 90% TFA, 45 MIN. pNB = PARA-NITROBENZOYL.

sugar precursor,[38–41] an example of which is shown in Scheme 6.3.[39] Reaction of the bromo sugar (**6.22**) with the potassium salt of diethyl 1,3-acetonedicarboxylate gave compound (**6.23**) which was subjected to base catalysed diazo transfer[42] with tosyl azide resulting in formation of the cyclic product (**6.24**). Treatment with sodium ethoxide resulted in removal of the quaternary ethoxycarbonyl group and aromatisation of the heterocycle (**6.25**). Reaction with methanolic ammonia converted the ester to the amide and resulted in epimerisation to give a mixture of

C-NUCLEOSIDES

α- and β-pyrazofurin (**1.58α** and **1.58β**) after acid deprotection of the isopropylidene group.

Modification of the base moiety of pyrazofurin at positions 1, 4 or 5 resulted in loss of activity,[43] of the sugar modified anlogues[1,3,4] the 5'-deoxy-pyrazofurin retained biological activity against a number of viruses including influenza A virus.[44]

6.2.3 Tiazofurin and Selenazofurin

Tiazofurin (**6.5**), 2-(β-D-ribofuranosyl)thiazole-4-carboxamide, displays significant antiviral[45] and antitumour activity.[46] Tiazofurin acts as an inhibitor of IMP dehydrogenase[47] with subsequent inhibition of guanine nucleotide biosynthesis, the biological activity results from its metabolism to tiazofurin-adenine dinucleotide (**6.26**), which binds to the NAD co-factor site, where the thiazole-4-carboxamide moiety occupies the pocket normally filled by the nicotinamide ring.[48]

It has been shown that the β-D-ribofuranosyl and 4-carboxamidothiazol-2-yl moieties are essential features for the biological activity of tiazofurin, with any modifications resulting in a significant or complete loss of activity.

The main exception is the isosteric substitution of sulphur for selenium, to give selenazofurin (**6.6**), 2-(β-D-ribofuranosyl)selenazole-4-carboxamide.[11] Selenazofurin operates with the same mechanism of action as tiazofurin and was found to be 5 times more potent than tiazofurin against L1210 and P388 cancer cell lines.[11]

An efficient synthesis[49] of both **6.5** and **6.6** which avoids the formation of unwanted elimination products observed in the earlier syntheses, involves condensation of either the methyl 2,5-anhydroallonothioate (**6.27**) or the corresponding selenoate (**6.28**) with 2-amino-2-cyanoacetate. Reductive dediazotization of the cyclic compounds (**6.29** and **6.30**) gave ethyl 2-(β-D-ribofuranosyl)thiazole-4-carboxylate (**6.31**) and the corresponding selenazole (**6.32**), which on treatment with methanolic ammonia gave the target compounds (**6.5**) and (**6.6**) (Scheme 6.4).

6.3 C-NUCLEOSIDES WITH A 6-RING HETEROCYCLIC BASE MOIETY

This category of C-nucleosides includes nucleosides with pyridine (*e.g.* **6.33**),[50] pyridazine (*e.g.* **6.34**),[51] pyrimidine (pseudouridine, **6.1**),[5] pyrazine (*e.g.* **6.7**),[12] 1,3-oxazine (oxazinomycin, **6.4**),[9] 1,3-thiazine (*e.g.* **6.35**)[52] and 1,2,4-triazine (*e.g.* **6.36**)[53] base moieties (Fig. 6.4).

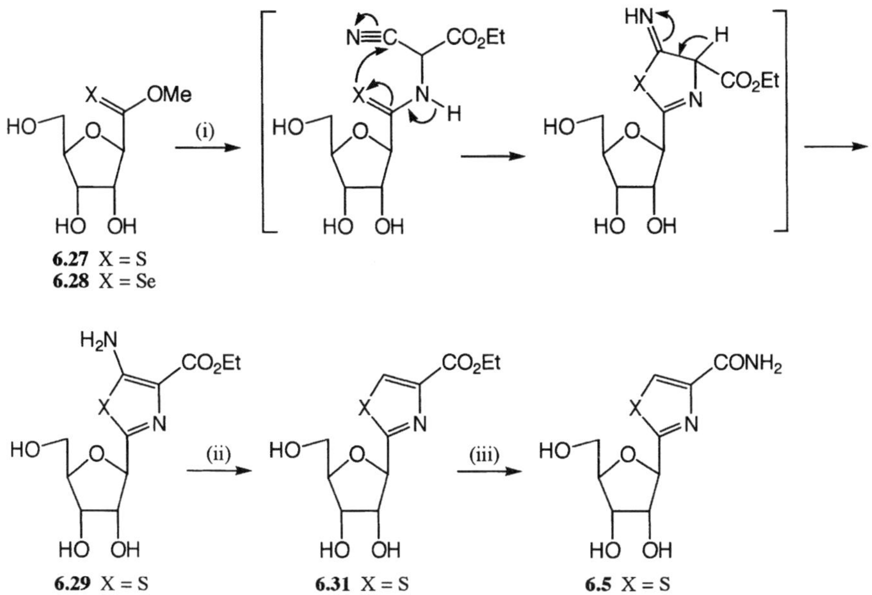

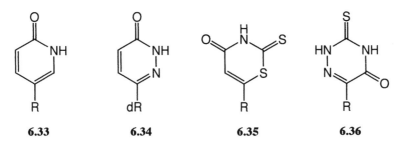

SCHEME 6.4 REAGENTS AND CONDITIONS: (I) $H_2NCH(CN)CO_2Et$, MeOH, PYRIDINE (II) $NaNO_2$, HCl, H_3PO_2 (III) NH_3, MeOH.

FIGURE 6.4 EXAMPLES OF 6-RING HETEROCYCLIC BASE MOIETIES OF C-NUCLEOSIDES (R = RIBOSE, dR = 2-DEOXYRIBOSE).

6.3.1 Pyrimidine C-nucleosides

Pseudouridine (**6.1**) is the most common RNA modification known in nature, occurring in tRNA (2–3%) and rRNA (~8%) in which it plays pivotal structural and functional roles.[54,55] Several syntheses of pseudouridine have been reported,[1] a practical synthesis involving a coupling reaction of 5-lithium-2,4-dimethoxypyrimidine (**6.37**) and the ribonolactone (**6.38**) has recently been reported (Scheme 6.5).[56] Concomitant reduction of the lactol (**6.39**) and removal of the isopropylidene group was achieved in one step using BF_3 as the Lewis acid and triethylsilane as the hydride donor.[57] Deprotection of (**6.40**) gave pseudouridine (**6.1**) in 20% overall yield.

Using standard methodology pseudouridine can be readily transformed to a number of other compounds of therapeutic interest including pseudoisocytidine,

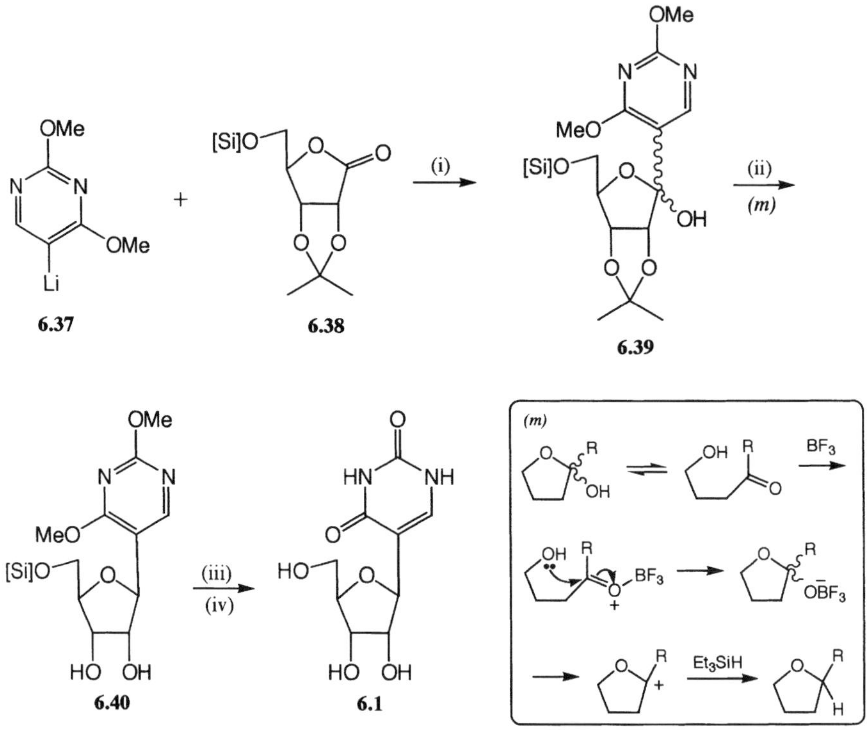

SCHEME 6.5 REAGENTS AND CONDITIONS: (I) THF, −78°C, 1.5 H (II) BF₃•Et₂O, Et₃SiH, CH₂Cl₂, 4°C, 1 H THEN R.T. O/N (III) NaI, AcOH, REFLUX, 25 MIN. (IV) 3% HCl, MeOH, 4 H. [Si] = TBDMS.

which displays antitumour properties,[58] and the antileukaemic 2,4-dimethoxypyrimidin-5-yl C-riboside.[59]

The acetonitrile riboside (**6.41**) is also a useful sugar for the synthesis of both pseudoisocytidine (**6.42**) and the 2-thio-derivative (**6.43**).[60] Formylation of **6.41** gave the crude sodium enolate, which was then treated with methyl iodide to give (**6.44**). After separation of the anomers by chromatography, pseudoisocytidine (**6.42**) and the 2-thio-derivative (**6.43**) were obtained after treatment with guanidine and thiourea respectively, followed by acidic deprotection (Scheme 6.6).

6.3.2 Pyrazine C-nucleosides

Pyrazine (1,4-diazine) C-nucleosides are a relatively new class of nucleoside mimetics which have shown activity against HCMV.[12] These nucleosides are prepared[61] in a similar manner to that described for pseudouridine (see Scheme 6.5), with subsequent modification of the pyrazine C-nucleoside base.[12] Reaction of 3,5-dichloro-2-(2,3,5-tri-O-benzyl-β-D-ribofuranosyl)pyrazine (**6.48**) with lithium 2,2,6,6-tetramethyl piperidine (LTMP) at low temperature followed by quenching with ethyl cyanoformate gave, after debenzylation, ethyl 3,5-dichloro-6-(β-D-ribofuranosyl)pyrazine-2-carboxylate (**6.48**) (Scheme 6.7).

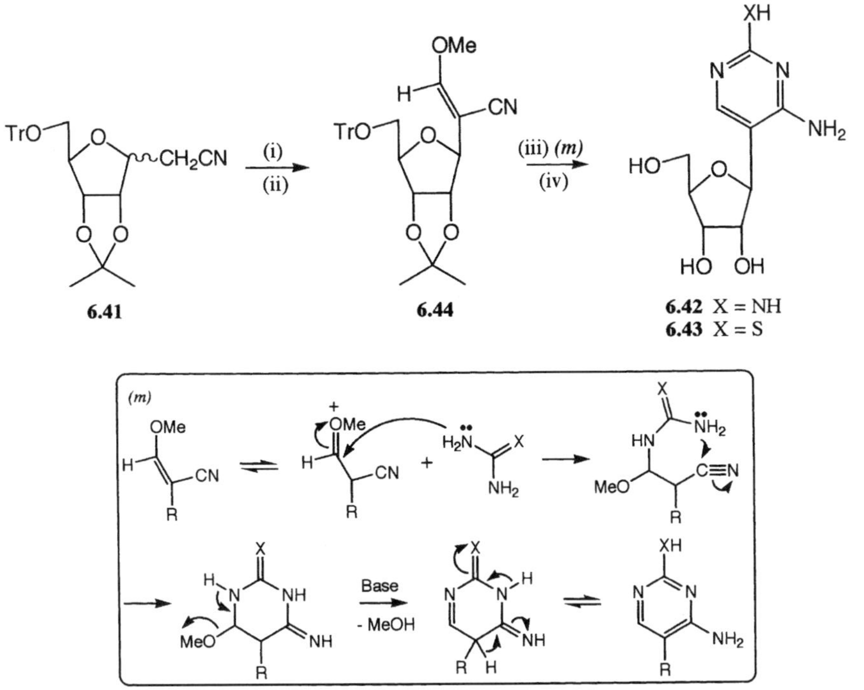

SCHEME 6.6 *REAGENTS AND CONDITIONS*: (I) NaH, Et$_2$O, EtOH, OVERNIGHT (II) MeI, DMF, 5 H (III) GUANIDINE HYDROCHLORIDE (X = NH) OR THIOUREA (X = S), NaOEt, EtOH, REFLUX, 15 H (IV) 10% HCl, MeOH, 1 H.

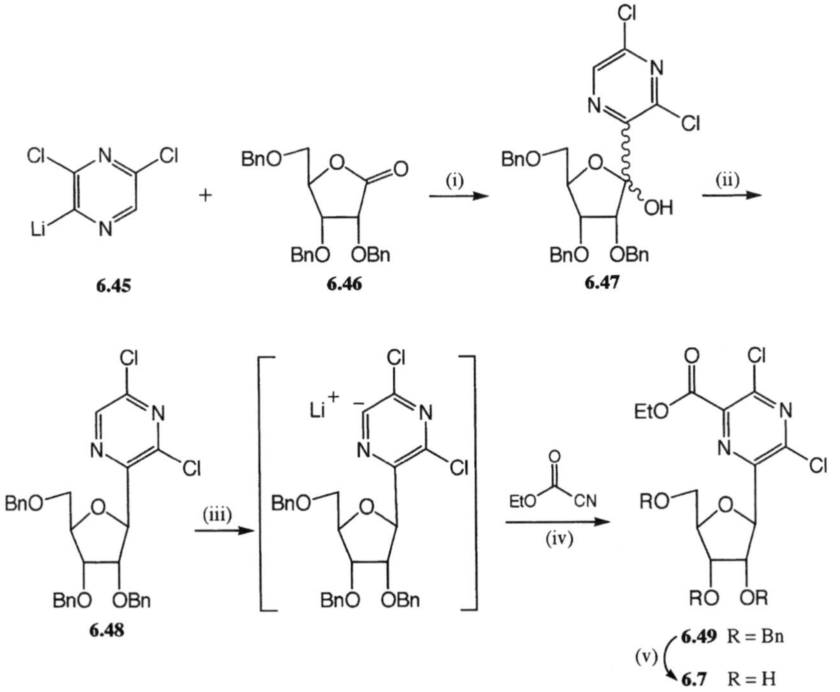

SCHEME 6.7 *REAGENTS AND CONDITIONS*: (I) THF, −78°C TO R.T. (II) BF$_3$•Et$_2$O, Et$_3$SiH, CH$_2$Cl$_2$, −78°C TO R.T. (III) LTMP, THF, −94°C, 2 H THEN (IV) ETHYL CYANOFORMATE, −94°C, 1 H (V) BCl$_3$, CH$_2$Cl$_2$, −78°C, 2.5 H.

6.3.3 Oxazine C-nucleosides

The naturally occurring 1,3-oxazine C-nucleoside, oxazinomycin or minimycin (**6.4**) has been isolated from several strains of *Streptomyces*, and displays anticancer and antibacterial activities.[9,62] The synthesis of oxazinomycin was reported in 1977 and commenced from the enamine (**6.50**) which when treated with hydroxylamine gave the amino-isoxazole (**6.51**).[63] Catalytic hydrogenation of the isoxazole gave the aminoacrylamide (**6.52**) which was hydrolysed to the formylacetamide (**6.53**). Reaction of (**6.53**) with *N,N'*-carbonyldiimidazole (CDI) resulted in formation of the 1,3-oxazine heterocycle, with subsequent deprotection giving oxazinomycin (**6.4**) (Scheme 6.8).

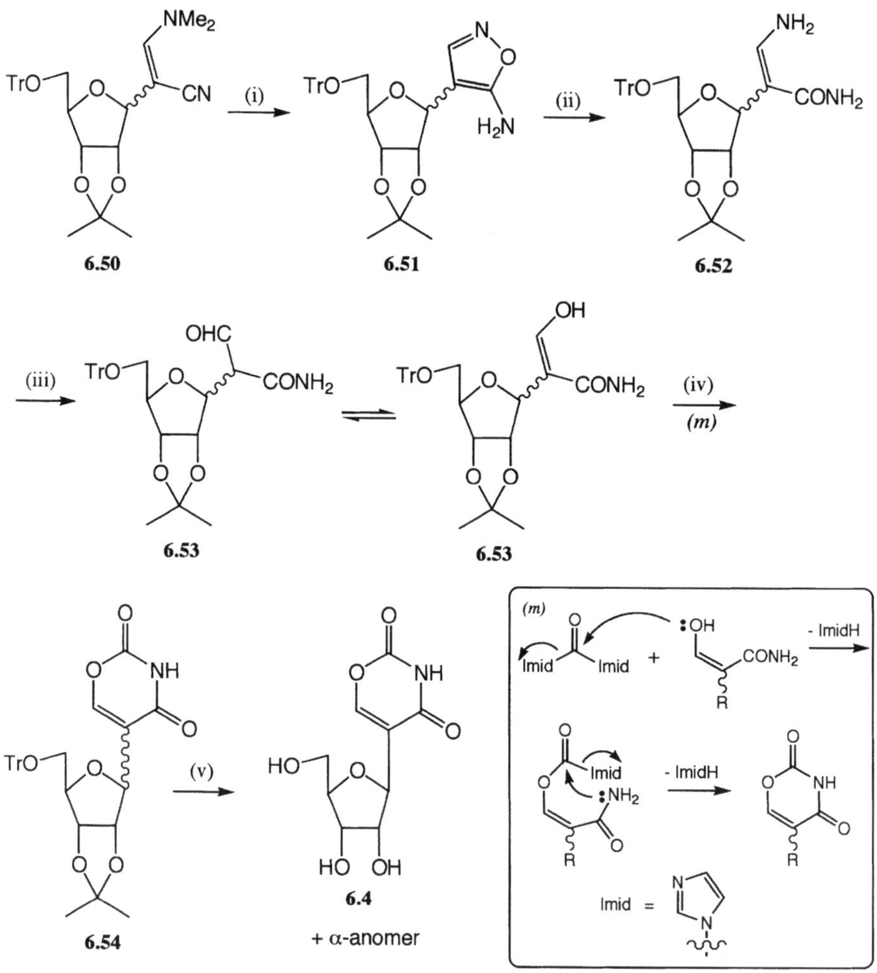

SCHEME 6.8 REAGENTS AND CONDITIONS: (I) $NH_2OH \cdot HCl$, DMF, PYRIDINE, 70°C, 6.5 H (II) H_2, PtO_2, DME, 25 MIN. (III) 0.05N HCl, $CHCl_3$, 7 H (IV) KH, CDI, DME, 6 H (V) 90% TFA, 3 H.

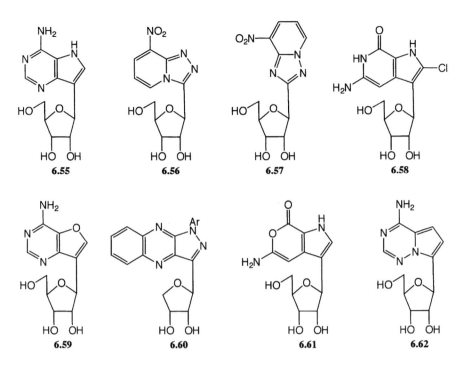

FIGURE 6.5 EXAMPLES OF BICYCLIC HETEROCYCLIC BASE MOIETIES OF C-NUCLEOSIDES.

6.4 C-NUCLEOSIDES WITH A BICYCLIC HETEROCYCLIC BASE MOIETY

A broad range of C-nucleosides with bicyclic heterocyclic base moieties 'purine-mimics' have been prepared and recently reviewed.[2] Many exhibit biological activity such as formycin (**6.3**),[7] the thieno[3,4-d]pyrimidine C-nucleosides (*e.g.* **6.8**)[13], the 9-deazapurine C-nucleosides (*e.g.* 9-deazaadenosine, **6.55**),[64] the 1,2,4-triazolo[1,5-a]pyridin-2-yl C-nucleosides[65] (*e.g.* **6.56**) which exhibit antiviral activity against Sindbis virus,[66] the cytotoxic 1,2,4-triazolo[4,3-a]pyridin-3-yl C-nucleosides (*e.g.* **6.57**),[67] pyrrolo[2,3-d]pyrimidin-6-yl C-nucleosides (*e.g.* **6.58**) which are active against Semliki Forest virus,[68] the furo[3,2-d]pyrimidine C-nucleoside pyrrolosine (**6.59**),[69] the anticancer flavazol-3-yl C-nucleosides (*e.g.* **6.60**),[70] the immunostimulant pyrrolo[3,2-d]oxazin-3-yl C-nucleoside (**6.61**)[71] and the 4-aminopyrrolo[2,1-f]1,2,4-triazin-7-yl C-nucleoside (**6.62**) which exhibits pronounced *in vitro* growth inhibitory activity against a number of cancer cell lines (Fig. 6.5).[72]

The isosteric relationship between formycin and pyrrolosine with adenosine, and formycin B with inosine, enables these C-nucleosides to replace adenosine and inosine respectively in many biochemical processes and results in a wide spectrum of biological activity.

6.4.1 Formycin and Formycin B

7-Amino-3-β-D-ribofuranosylpyrazolo[4,3-d]pyrimidine, formycin (**6.3**) and its metabolite 3-β-D-ribofuranosylpyrazolo[4,3-d]pyrimidin-7-one, formycin B (laurusin, **6.63**) were isolated from *Streptomyces lavendula*.[7] Formycin acts as the C-nucleoside

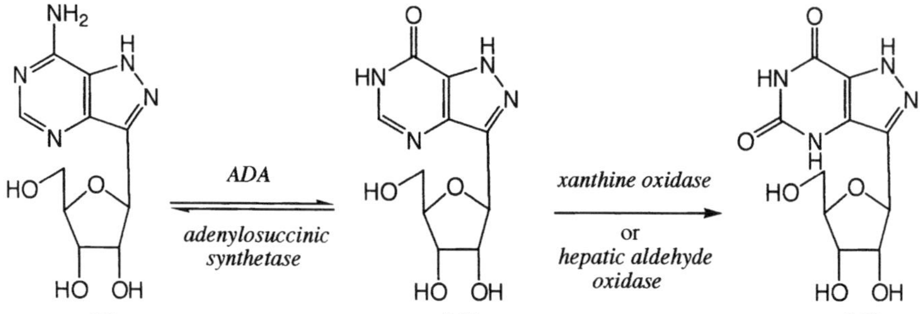

FIGURE 6.6 METABOLISM OF FORMYCIN.

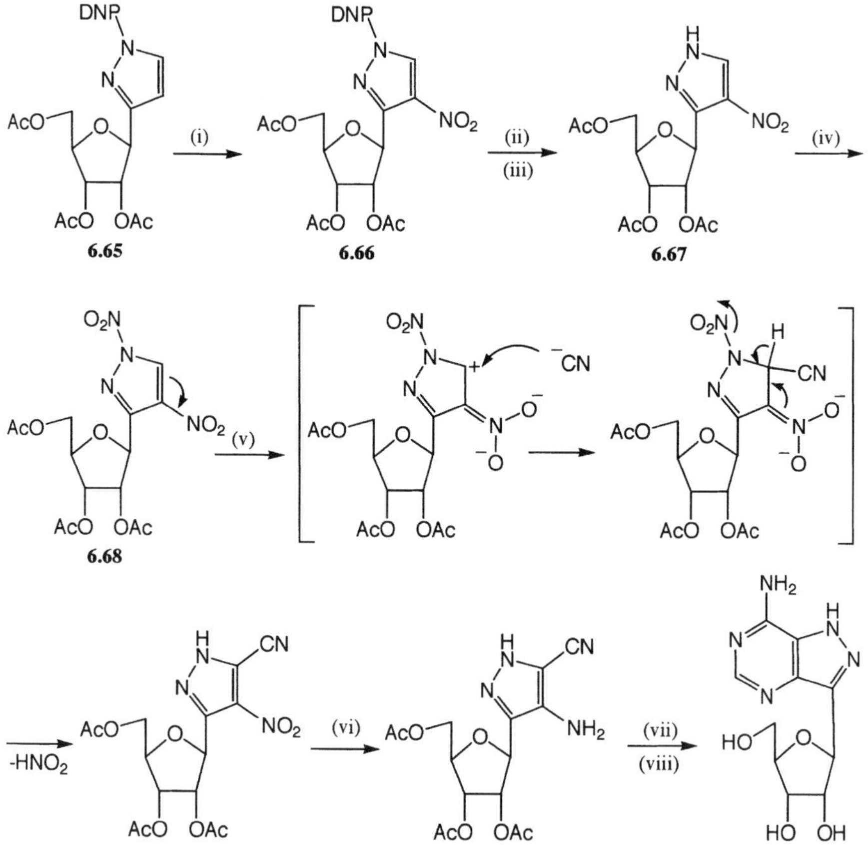

SCHEME 6.9 *REAGENTS AND CONDITIONS*: (I) Cu(NO$_3$)$_2$•3H$_2$O, Ac$_2$O (II) NaOMe, MeOH (III) Ac$_2$O, PYRIDINE (IV) *N*-NITRATION, ACETYL NITRATE OR Cu(NO$_3$)$_2$ (V) KCN, EtOH (VI) H$_2$, Pd/C, EtOH (VII) FORMAMIDINE ACETATE, 2-ETHOXYETHANOL, REFLUX (VIII) NaOMe, MeOH. DNP = 2,4-DINITROPHENYLHYDRAZINE.

isostere of adenosine resulting in anticancer, antiviral and antibacterial properties. However, the chemotherapeutic potential of formycin is reduced over time owing to the facile conversion of formycin to the less active formycin B (C-nucleoside isostere of inosine) by adenosine deaminase and subsequent oxidation to the xanthosine isoster (**6.64**) (Fig. 6.6).

Several syntheses of formycin and formycin B have been described,[2] elaboration of the pyrazole C-nucleoside (**6.65**) provided a novel route to formycin,[73] the key step involving thermal rearrangement of the 1,4-dinitro-3-pyrazol-3-yl C-nucleoside (**6.68**) and *cine*-substitution with cyanide ion to give the 3(5)cyano-4-nitro-5(3)-pyrazole C-nucleoside (**6.69**) (Scheme 6.9).

6.4.2 9-Deazapurine C-nucleosides

The anticancer agent 9-deazaadenosine, 4-amino-7-(β-D-ribofuranosyl)-5H-pyrrolo[3,4-d]pyrimidine (**6.55**), was isolated from blue green-alga[64] and its structure confirmed by comparison with previously synthesised 9-deazaadenosine.[74] 9-Deazainosine (**6.71**) exhibits both antitumour and antiprotozoan activity, of importance are the antibacterial properties against Pneumocystis carnii pneumonia (PCP), a disease common in immunosuppressed patients.[75] Both 9-deazaadenosine and 9-deazainosine can be prepared from the common 2-formylacetonitrile intermediate (**6.72**).[74,76]

Initial formation of the respective pyrrole C-nucleosides (**6.75** and **6.76**) was achieved on prolonged reaction with 1,5-diazabicyclo[4.3.0]nonene-5 (DBN), with subsequent reaction with formamidine acetate resulting in formation of the required pyrrolopyrimidine C-nucleosides (**6.77** and **6.78**), which were deprotected to give 9-deazaadenosine (**6.55**) and 9-deazainosine (**6.72**) (Scheme 6.10).

6.4.3 Pyrrolosine

Pyrrolosine, 4-amino-7-β-D-ribofuranosyl[3,2-d]pyrimidine, 7-oxa-7,9-dideaza adenosine (**6.59**),[69] inhibits RNA synthesis and displays significant anticancer activity while the furo[3,2-d]inosine C-nucleoside displays antiprotozoan activity.[77] Pyrrolosine was prepared from the 2-formylacetonitrile intermediate (**6.72**) using the methodology described for 9-deazaadenosine and 9-deazainosine (section 6.4.2) (Scheme 6.11).[78]

6.4.4 Thieno[3,4-d]pyrimidine C-nucleosides

The thieno[3,4-d]inosine C-nucleoside (**6.81**)[79] is an analogue of inosine and formycin B and displays inhibitory activity against PNP. Synthesis involves condensation of the thiophene (**6.82**) with the acylated ribose sugar (**2.33**) to give the thiophene C-nucleoside (**6.83**). Deformylation and annulation resulted in cyclization to the thieno[3,4-d]pyrimidine C-nucleoside (**6.84**) which was then deprotected to give the inosine derivative (**6.81**). The 5'-iodo derivative (**6.8**)[13] was prepared by the reaction of (**6.81**) with methyl triphenoxyphosphonium iodide (Rydon's reagent) and found to be a good inhibitor of PNP (Scheme 6.12).

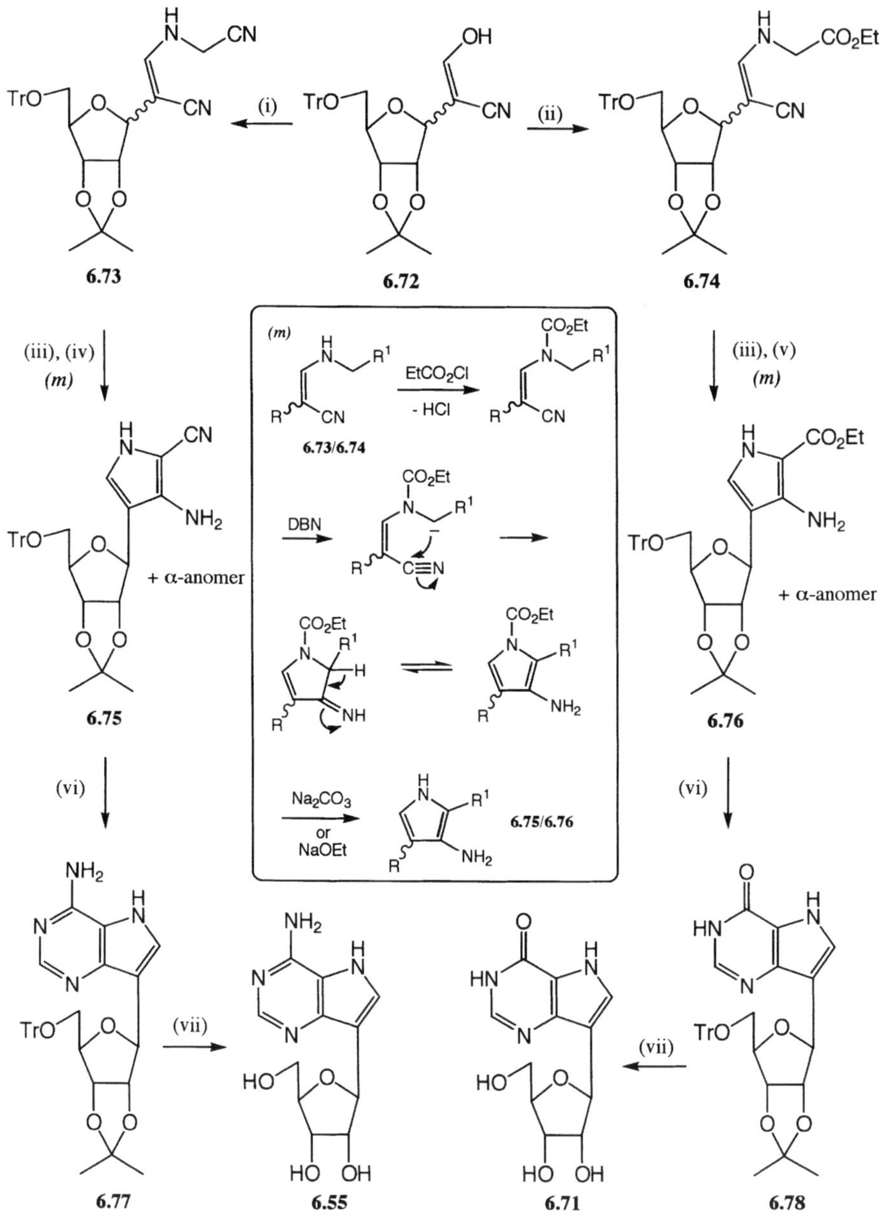

SCHEME 6.10 *REAGENTS AND CONDITIONS*: (I) NCCH$_2$NH$_2$•HCl, NaOAc, MeOH, 20 H (II) EtO$_2$CCH$_2$NH$_2$•HCl, NaOAc, MeOH, H$_2$O, 16 H (III) EtCO$_2$Cl, DBN, CH$_2$Cl$_2$, 16–20 H (IV) Na$_2$CO$_3$, MeOH, 50 MIN. (V) NaOEt, EtOH, 40 MIN. (VI) FORMAMIDINE ACETATE, EtOH, REFLUX, 5 H FOR **6.77**, 4 DAYS FOR **6.78** (VII) 6–12% HCl, MeOH.

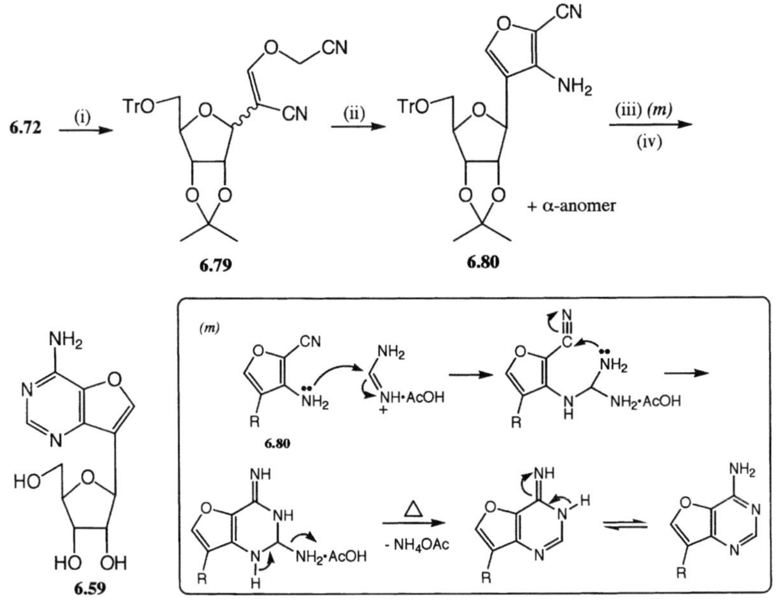

SCHEME 6.11 REAGENTS AND CONDITIONS: (I) ClCH$_2$CN, KF, 18-CROWN-6, DMF, 20 H (II) LDA, THF, −70°C (III) FORMAMIDINE ACETATE, EtOH, REFLUX, 48 H (IV) 6% HCl, MeOH, 1 H.

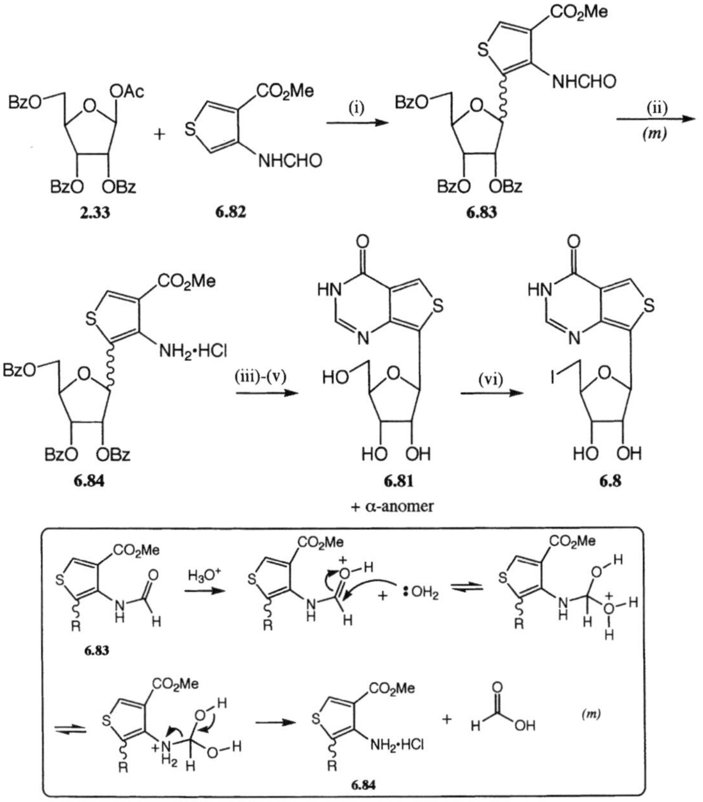

SCHEME 6.12 REAGENTS AND CONDITIONS: (I) SnCl$_4$, MeNO$_2$, 60–65°C (II) 15% HCl, MeOH (III) Et$_3$N (IV) FORMAMIDINE ACETATE, EtOH, REFLUX (V) NaOMe, MeOH (VI) CH$_3$P$^+$(OPh)$_3$I$^-$, DMF, 1 H.

Natural Nucleosides

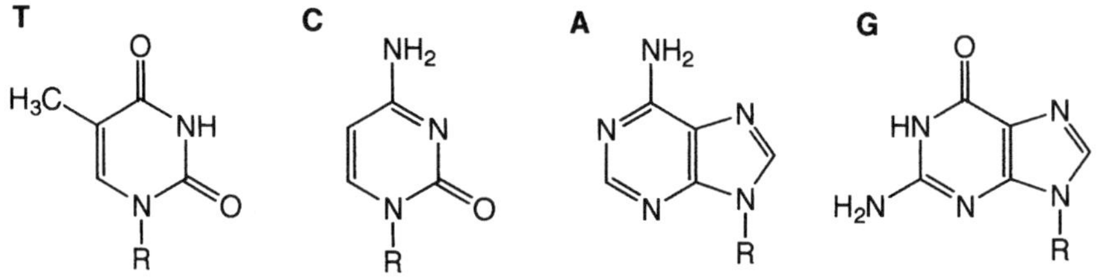

T 1.22, **C** 1.23, **A** 1.24, **G** 1.25

Nonpolar Isosteres

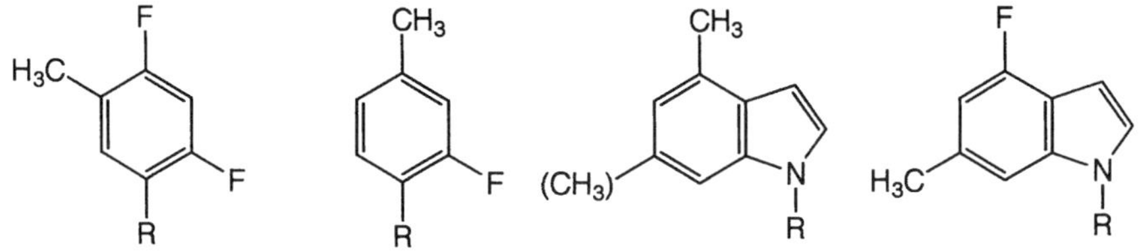

6.85, 6.86, 6.87, 6.88

FIGURE 6.7 NON-POLAR ISOSTERES OF THE NATURAL NUCLEOSIDES (R = 2-DEOXYRIBOSE).

6.5 C-ARYLGLYCOSIDES AS NUCLEOSIDE MIMETICS

The C-arylglycosides (or non-nitrogen heterocyclic C-nucleosides) are widespread in nature and extremely diverse with respect to the aglycone moiety. The chemistry,[80,81] biochemistry and occurrence of the C-arylglycosides have been the subject of several reviews.[3,82]

More recently the use of C-arylglycosides as nonpolar isosteres of the natural nucleosides has been reported by E.T. Kool *et al.* with surprising results.[83,84] The interest in nonpolar isosteres of nucleosides was to study non-covalent interactions involving DNA and RNA, such as molecular recognition and catalytic events, these isosteres would be required to closely mimic the size and shape of the natural nucleosides but lack any significant hydrogen bonding capability. Therefore substituted benzenes were prepared as mimetics of pyrimidine nucleosides (**6.85** and **6.86**) and indoles as mimetics of the purine nucleosides (**6.87** and **6.88**) (Fig. 6.7).[85]

Initial synthesis of the thymidine isostere involved reaction of the Grignard reagent (**6.89**) with the glycosyl halide (**2.29**) however, this reaction resulted in low yields of the β-arylglycoside (**6.90**) owing to α,β-elimination of the chloride under the basic reaction conditions resulting in formation of the furan (**6.91**) (Scheme 6.13).[85]

The yields of the coupling reaction between the aryl moiety and the glycosyl halide were considerably improved by using diarylcadmium reagents, prepared by reaction of the Grignard reagents with cadmium chloride.[86] Reaction of the diarylcadmium reagent (**6.92**) with the glycosyl halide (**2.29**) gave yields of approximately 60% of

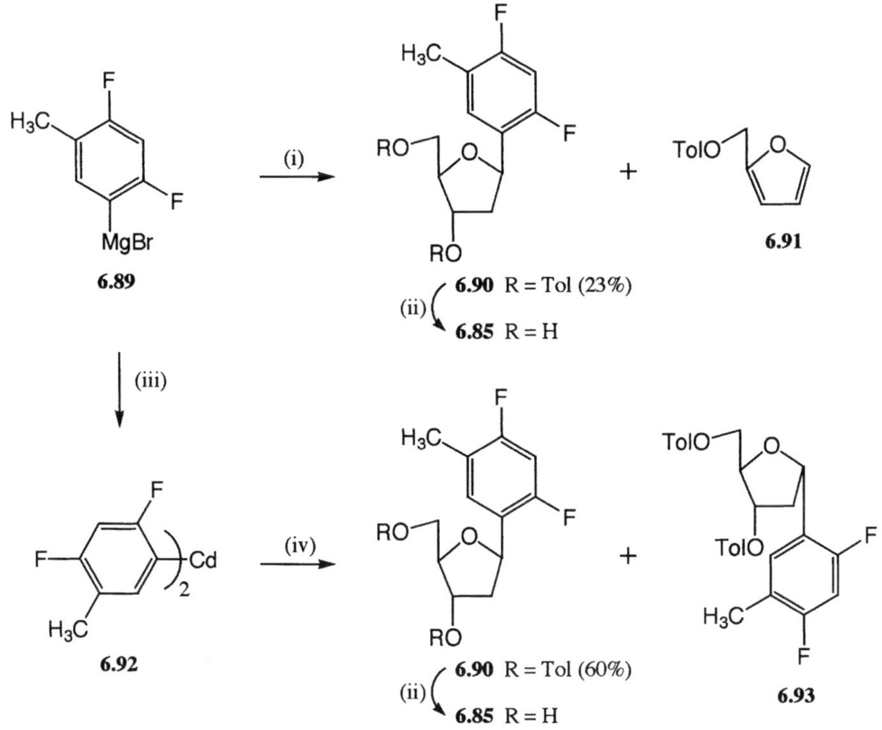

SCHEME 6.13 *REAGENTS AND CONDITIONS*: (I) **2.29**, THF, O/N (II) NaOMe, MeOH, 3 H (III) CdCl$_2$, THF, REFLUX, 2–3 H (IV) **2.29**, THF, REFLUX, 2 H.

6.90 with a minor amount of the α-anomer (**6.93**) which was easily separated by column chromatography. Deprotection of the β-arylglycoside (**6.90**) with sodium methoxide gave the required thymidine mimetic **6.85** (Scheme 6.13).

The adenosine isostere required formation of the 4,6-dimethylindole (**6.96**) from N-(trifluoroacetyl)-2-anilino acetal (**6.94**) *via* an acid-catalysed cyclisation-elimination reaction (Scheme 6.14 (*m*)).[87] The coupling of the indole (**6.96**) with the glycosyl halide (**2.29**) was achieved using the direct sodium-salt-glycosylation procedure (see section 2.2.1.2) to give the β-arylglycoside (**6.97**) in 27% yield, with subsequent deprotection providing the adenosine mimetic (**6.87**) (Scheme 6.14).[85]

Unexpectedly studies involving DNA template strands containing the thymidine mimetic (**6.85**) showed that it was an excellent and highly specific template for replication,[83] *i.e.* despite its low hydrogen bonding capability the enzyme (*E. coli* DNA polymerase I) recognised the isostere as thymidine. The results obtained from this research suggested that steric factors rather than hydrogen bonding may be the more important factor in replication.

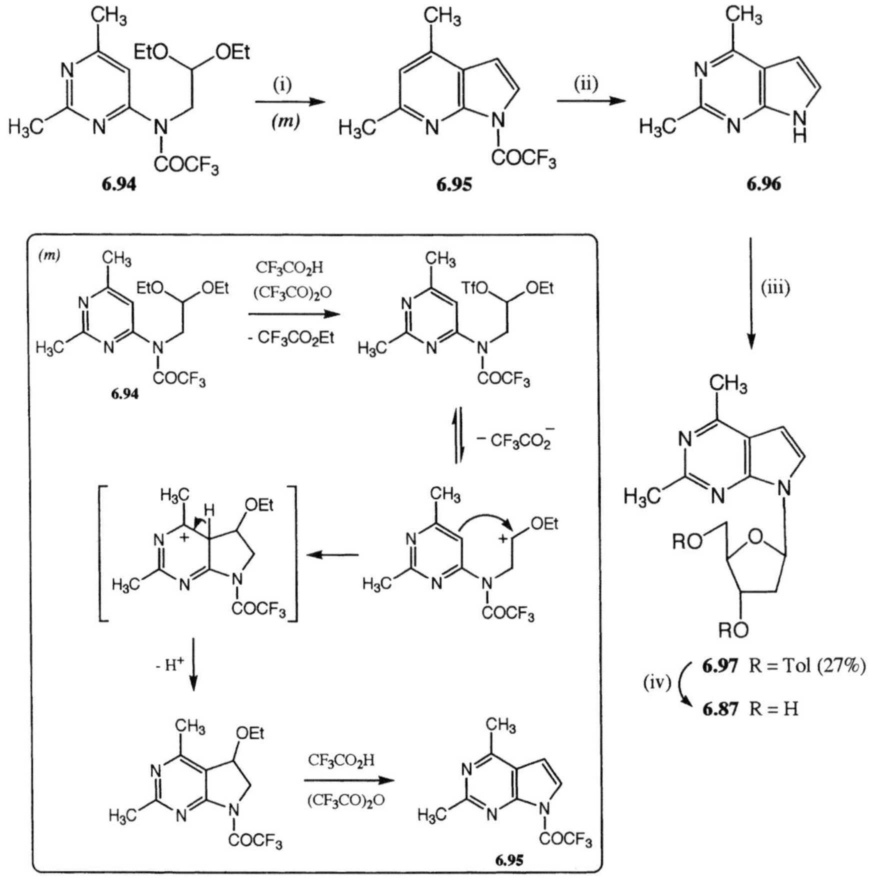

SCHEME 6.14 *REAGENTS AND CONDITIONS*: (I) $(CF_3CO)_2O$, CF_3CO_2H, 56°C, 72 H (II) 5% METHANOLIC KOH, O/N (III) NaH, CH_3CN, 0°C, 30 MIN., THEN **2.29**, O/N (IV) NaOMe, MeOH, 3 H.

6.6 REFERENCES

1. M.A.E. Shaban, A.Z. Nasr, *Adv. Het. Chem.*, **1997**, *68*, 223–432.
2. M.A.E. Shaban, *Adv. Het. Chem.*, **1998**, *70*, 163–337.
3. U. Hacksell, G.D. Daves, *Prog. Med. Chem.*, **1985**, *22*, 1–65.
4. L.J.S. Knutsen, *Nucleosides & Nucleotides*, **1992**, *11*, 961–983.
5. W.E. Cohn, E. Volkin, *Nature*, **1951**, *167*, 483–484.
6. G.D. Daves, C.C. Cheng, *Prog. Med. Chem.*, **1976**, *13*, 303–349.
7. M. Hori, E. Ito, T. Takita, G. Koyama, T. Takeuchi, H. Umezawa, *J. Antibiot.*, **1964**, *17A*, 96–99.
8. K. Gerzon, D.C. DeLong, J.C. Cline, *Pure Appl. Chem.*, **1971**, *28*, 489–497.
9. T. Haneishi, T. Okazaki, T. Hata, C. Tamura, M. Nomura, A. Naito, I. Seki, M. Arai, *J. Antibiot.*, **1971**, *24*, 797–799.
10. M. Fuertes, T. Garcia Lopez, G. Garcia Munoz, M. Stud, *J. Org. Chem.*, **1976**, *41*, 4074–4077.
11. P.C. Srivastava, R.K. Robins, *J. Med. Chem.*, **1983**, *26*, 445–448.
12. J.A. Walker II, W. Liu, D.S. Wise, J.C. Drach, L.B. Townsend, *J. Med. Chem.*, **1998**, *41*, 1236–1241.
13. S.A. Patil, B.A. Otter, R.S. Klein, *J. Het. Chem.*, **1993**, *30*, 509–515.
14. B.A. Horenstein, R.F. Zabinski, V.L. Schramm, *Tetrahedron Lett.*, **1993**, *34*, 7213–7216.
15. Y.S. Sanghvi, N.B. Hanna, S.B. Larson, J.M. Fujitaki, R.C. Willis, R.A. Smith, R.K. Robins, G.R. Revankar, *J. Med. Chem.*, **1988**, *31*, 330–335.

16. H. Hrebabecky, *Collect. Czech. Chem. Commun.*, **1986**, *51*, 1311–1315.
17. W.J. Hennen, R.K. Robins, *J. Het. Chem.*, **1985**, *22*, 1747–1748.
18. D.K. Buffel, B.P. Simons, J.A. Deceuninck, G.J. Hoornaert, *J. Org. Chem.*, **1984**, *49*, 2165–2168.
19. M.S. Kang, J.P. Spencer, A.D. Elbein, *J. Biol. Chem.*, **1979**, *254*, 10037–10043.
20. Y. Uehara, M. Rabinovitz, *Biochem. Pharmacol.*, **1981**, *30*, 3165–3169.
21. G. Trummlitz, J.G. Moffatt, *J. Org. Chem.*, **1973**, *38*, 1841–1845.
22. N. Katagiri, T. Haneda, N. Takahashi, *Heterocycles*, **1984**, *22*, 2195–2198.
23. A. Kaye, S. Neidle, C.B. Reese, *Tetrahedron Lett.*, **1988**, *29*, 2711–2714.
24. T. Kametani, K. Kawamura, T. Honda, *J. Am. Chem. Soc.*, **1987**, *109*, 3010–3017.
25. D.H. Miles, B-S. Huang, *J. Org. Chem.*, **1977**, *41*, 208–214.
26. A.G. Barrett, H.B. Broughton, S.V. Attwood, A.A.L. Gunatilaka, *J. Org. Chem.*, **1986**, *51*, 495–503.
27. Y. Araki, T. Endo, M. Tanji, J. Nagasawa, Y. Ishido, *Tetrahedron Lett.*, **1988**, *29*, 351–354.
28. D.H.R. Barton, M. Ramesh, *J. Am. Chem. Soc.*, **1990**, *112*, 891–892.
29. M.W. Mumper, C. Aurenge, R.S. Hosmane, *Bioorg. Med. Chem. Lett.*, **1993**, *3*, 2847–2850.
30. L. Kalvoda, J. Farkas, F. Sorm, *Tetrahedron Lett.*, **1970**, 2297–2300.
31. S.H. Kang, S.B. Lee, *J. Chem. Soc., Chem. Commun.*, **1995**, 1017–1018.
32. D.H.R. Barton, J. Camara, X. Cheng, S.D. Géro, J.C. Jaszberenyi, B. Quiclet–Sire, *Tetrahedron*, **1992**, *48*, 9261–9276.
33. G. Just, M–I. Lim, *Can. J. Chem.*, **1977**, *55*, 2993–2997.
34. G. Just, T.J. Liak, M–I. Lim, P. Potvin, Y.S. Tsantrizos, *Can. J. Chem.*, **1980**, *58*, 2024–2033.
35. A.K. Saksena, A.K. Ganguly, *Tetrahedron Lett.*, **1981**, *22*, 5227–5230.
36. B.M. Trost, L.S. Kallander, *J. Org. Chem.*, **1999**, *64*, 5427–5435.
37. E. De Clercq, *Int. J. Antimicrob. Agents*, **1996**, *7*, 193–202.
38. J. Farkas, Z. Flegelova, F. Sorm, *Tetrahedron Lett.*, **1972**, 2279–2280.
39. S. De Bernardo, M. Weigele, *J. Org. Chem.*, **1976**, *41*, 287–290.
40. J.G. Buchanan, A. Stobie, R.H. Wightman, *J. Chem. Soc., Perkin Trans.1*, **1981**, 2374–2378.
41. N. Katagiri, K. Takashima, T. Haneda, T. Kato, *J. Chem. Soc., Perkin Trans.1*, **1984**, 553–560.
42. M. Regitz, *Angew. Chem., Int. Ed. Engl.*, **1967**, *6*, 733–749.
43. C.R. Petrie III, G.R. Revankar, N.K. Dalley, R.D. George, P.A. McKernan, R.L. Hamill, R.K. Robins, *J. Med. Chem.*, **1986**, *29*, 268–278.
44. X. Chen, S.W. Schneller, S. Ikeda, R. Snoeck, G. Andrei, J. Balzarini, E. De Clercq, *J. Med. Chem.*, **1993**, *36*, 3727–3730.
45. P.C. Srivastava, M.V. Pickering, L.B. Allen, D.G. Streeter, M.T. Campbell, J.T. Witkowski, R.W. Sidwell, R.K. Robins, *J. Med. Chem.*, **1977**, *20*, 256–262.
46. R.K. Robins, P.C. Srivastava, V.L. Naragyanan, J. Plowman, K.D. Paull, *J. Med. Chem.*, **1982**, *25*, 107–108.
47. H.N. Jayaram, R.L. Dion, R.I. Glazer, D.G. Johns, R.K. Robins, P.C. Srivastava, D.A. Cooney, *Biochem. Pharmacol.*, **1982**, *31*, 2371–2380.
48. B.M. Goldstein, F. Takusagawa, H.M. Berman, P.C. Srivastava, R.K. Robins, *J. Am. Chem. Soc.*, **1983**, *105*, 7416–7422.
49. W.J. Hennen, B.C. Hinshaw, T.A. Riley, S.G. Wood, R.K. Robins, *J. Org. Chem.*, **1985**, *50*, 1741–1746.
50. J. Matulic–Adamic, L. Beigelman, *Tetrahedron Lett.*, **1997**, *38*, 1669–1672.
51. I. Maeba, T. Iijima, Y. Matsuda, C. Ito, *J. Chem. Soc., Perkin Trans.1*, **1990**, 73–76.
52. S.Y.K. Tam, R.S. Klein, F.G. De Las Heras, J.J. Fox, *J. Org. Chem.*, **1979**, *44*, 4854–4862.
53. T. Sato, Y. Hayakawa, R. Noyori, *Bull. Chem. Soc. Jpn.*, **1984**, *57*, 2515–2525.
54. J. Ofengand, A. Bakin, *J. Mol. Biol.*, **1997**, *266*, 246–268.
55. B.G. Lane, J. Ofengand, M.W. Gray, *Biochimie*, **1995**, *77*, 7–15.
56. P.J. Grohar, C.S. Chow, *Tetrahedron Lett.*, **1999**, *40*, 2049–2052.
57. J.L. Frey, M. Orfanopoulos, M.G. Adlington, W.R. Dittman, *J. Org. Chem.*, **1978**, *43*, 374–375.
58. K. Hirota, K.A. Watanabe, J.J. Fox, *J. Org. Chem.*, **1978**, *43*, 1193–1197.
59. M. Weigele, S. De Bernardo, **1978**, U.S. Pat. 4,096, 321; *Chem. Abs.*, **1979**, *90*, 55251.
60. C.K. Chu, U. Reichman, K.A. Watanabe, J.J. Fox, *J. Org. Chem.*, **1977**, *42*, 711–714.
61. W. Liu, J.A. Walker II, J.J. Chen, D.S. Wise, L.B. Townsend, *Tetrahedron Lett.*, **1996**, *37*, 5325–5328.
62. P.C. Srivastava, R.K. Robins, *J. Med. Chem.*, **1981**, *24*, 1172–1177.
63. S. De Bernardo, M. Weigele, *J. Org. Chem.*, **1977**, *42*, 109–112.
64. M. Namikoshi, W.W. Carmichael, R. Sakai, E.A. Jareserijman, A.M. Kaup, K.L. Rinehart, *J. Am. Chem. Soc.*, **1993**, *115*, 2504–2505.
65. T. Huynh–Dinh, J. Igolen, J–P. Marquet, E. Bisagni, J.M. Lhoste, *J. Org. Chem.*, **1976**, *41*, 3124–3128.
66. P. Allard, T. Huynh–Dinh, C. Gouyette, J. Igolen, J–C. Chermann, F. Barre–Sinoussi, *J. Med. Chem.*, **1981**, *24*, 1291–1297.

67. J. Kobe, B. Brdar, J. Soric, *Nucleosides & Nucleotides*, **1986**, *5*, 135–151.
68. N.S. Girgis, M.A. Michael, D.F. Smee, H.A. Alghamandan, R.K. Robins, H.B. Cottam, *J. Med. Chem.*, **1990**, *33*, 2750–2755.
69. B.A. Otter, S.A. Patil, R.S. Klein, S.E. Ealick, *J. Am. Chem. Soc.*, **1992**, *114*, 668–671.
70. M.A.E. Sallam, S.M.E. Abdel Megid, *Carbohydr. Res.*, **1984**, *125*, 85–95.
71. S. Niitsuma, K. Kato, T. Takita, H. Umezawa, *Tetrahedron Lett.*, **1985**, *26*, 5785–5786.
72. S.A. Patil, B.A. Otter, R.S. Klein, *Tetrahedron Lett.*, **1994**, *35*, 5339–5342.
73. J.G. Buchanan, A.R. Edgar, R.J. Hutchison, A. Stobie, R.H. Wightmen, *J. Chem. Soc., Chem. Commun.*, **1980**, 237–238.
74. M.I. Lim, R.S. Klein, *Tetrahedron Lett.*, **1981**, *22*, 25–28.
75. J.M.S. Bartlett, J.J. Marr, S.F. Queener, R.S. Klein, J.W. Smith, *Antimicrob. Ag. Chemother.*, **1986**, *30*, 181–183.
76. M.I. Lim, W.Y. Ren, B.A. Otter, R.S. Klein, *J. Org. Chem.*, **1983**, *48*, 780–788.
77. B.K. Bhattacharya, B.A. Otter, R.L. Berens, R.S. Klein, *Nucleosides & Nucleotides*, **1990**, *9*, 1021–1043.
78. B.K. Bhattacharya, M.I. Lim, B.A. Otter, R.S. Klein, *Tetrahedron Lett.*, **1986**, *27*, 815–818.
79. S.P. Rao, K.V.B. Rao, B.A. Otter, R.S. Klein, W.Y. Ren, *Tetrahedron Lett.*, **1988**, *29*, 3537–3540.
80. M.H.D. Postema, *Tetrahedron*, **1992**, *48*, 8545–8599.
81. S. Hanessian, A.G. Pernet, *Adv. Carbohydr. Chem. Biochem.*, **1976**, *33*, 111–188.
82. G. Franz, M. Grun, *Planta Med.*, **1983**, *47*, 131–140.
83. K.M. Guckian, E.T. Kool, *Angew. Chem. Int. Ed. Engl.*, **1997**, *36*, 2825–2828.
84. N.C. Chaudhuri, X–F. Ren, E.T. Kool, *Synlett*, **1997**, 341–347.
85. B.A. Schweitzer, E.T. Kool, *J. Org. Chem.*, **1994**, *59*, 7283–7242.
86. N.C. Chaudhuri, E.T. Kool, *Tetrahedron Lett.*, **1995**, *36*, 1795–1798.
87. J.E. Nordlander, D.B. Catalane, K.D. Kotian, R.M. Stevens, J.E. Haky, *J. Org. Chem.*, **1981**, *46*, 778–782.

CHAPTER 7

CARBOCYCLIC NUCLEOSIDES

7.1 INTRODUCTION

Carbocyclic nucleosides, in which the D-ribose moiety of the nucleoside is replaced by a cyclopentane system, are stable towards hydrolysis by phosphorylases and often display enhanced biostability.[1] Isosteric replacement of the oxygen of furanose with a methylene group has been shown to result not only in improved enzymatic resistance, but also reduced toxicity of the carbocyclic nucleosides compared with conventional nucleosides.[2]

The bioactivity of the naturally occurring carbocyclic nucleosides (–)-aristeromycin (**7.1**)[3] and (–)-neplanocin A (**7.2**),[4] led to an interest in this class of compounds, resulting in the generation of a diverse range of antiviral carbocyclic nucleosides[5] such as Carbovir (**7.3**),[6] Abacavir (Ziagen™, **7.4**),[7] Lobucavir (**1.50**),[8] and more recently BMS-200475 (**7.5**) (Fig. 7.1).[9] Carbocyclic nucleosides are synthetically the most challenging class of nucleosides, requiring multistep and often elaborate syntheses to introduce the necessary stereochemistry. The numerous synthetic approaches have been reviewed,[10] a number of which will be described in relation to the chemotherapeutic carbocyclic nucleosides highlighted in this chapter.

7.2 (–)-ARISTEROMYCIN AND (–)-NEPLANOCIN A

The chemistry[11] and structure-activity relationships[5] of the potent inhibitors of AdoHcy hydrolase,[12,13] aristeromycin and neplanocin A, have been reviewed. Aristeromycin, a natural product isolated from *Streptomyces citricolor*,[14] is the carbocyclic analogue of adenosine and displays significant cytotoxicity in cell cultures.[3] Aristeromycin was first synthesised by Shealy and Clayton in 1966,[15] several other syntheses have been reported which allow enantioselective synthesis of the active (–)-aristeromycin enantiomer.[10,16-18] The method of Arita *et al*,[16] involves the synthesis of a cyclopentylamine (**7.13**) from the bicyclic [2,2,1] precursor (**7.6**), this method provides a fixed configuration at the C-1' and C-4' positions of the final carbocyclic nucleoside. The key step was the selective enzymatic deesterification of **7.6** using pig liver esterase to give the α-keto ester acid (**7.7**), which was then converted into the required carbocyclic skeleton (**7.8**) by decarboxylative ozonolysis. Further chemical transformation gave the versatile cyclopentylamine (**7.13**), which was converted into (–)-aristeromycin (**7.1**) (Scheme 7.1).

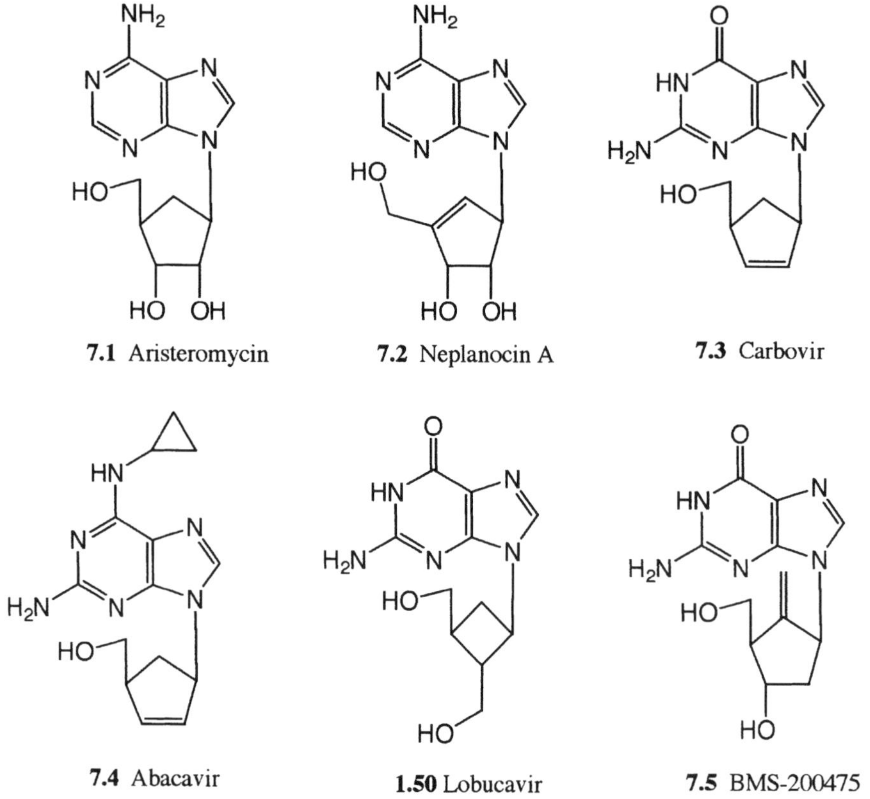

CARBOCYCLIC NUCLEOSIDES

7.1 Aristeromycin **7.2** Neplanocin A **7.3** Carbovir

7.4 Abacavir **1.50** Lobucavir **7.5** BMS-200475

FIGURE 7.1 CHEMOTHERAPEUTIC CARBOCYCLIC NUCLEOSIDES.

The (–)-Neplanocin A enantiomer can also be prepared by this method[16] starting from the δ-lactone (**7.10**), several other methods have also been described.[10,16,19,20] An efficient asymmetric synthesis starting from a readily accessible ribose sugar, involves a C-H insertion of alkylidenecarbene as the key step.[21] Reduction of the ribose sugar (**7.15**) and silyl protection of the primary hydroxy gave the acyclic compound (**7.16**), which was subjected to a Swern oxidation to give the ketone (**7.17**). Exposure of **7.17** to lithiotrimethylsilyldiazomethane[22] generated the carbene (**7.18**) which was inserted into the C-H bond adjacent to the protected hydroxyl group to give the cyclopentene (**7.19**). After removal of the silyl protecting group, the epimeric mixture was oxidized and then reduced stereoselectively to give (**7.20**) as a single stereoisomer. Treatment of **7.20** with adenine under Mitsunobu conditions followed by acidic hydrolysis of the protecting group gave (–)-Neplanocin A (**7.2**) (Scheme 7.2).

Neplanocin A (**7.2**), isolated from *Actinoplanacea ampullariella*,[4] exhibits significant anti-leukemia and antiviral activity,[23] however it is very toxic to host cells owing to phosphorylation by adenosine kinase[24] and is also rapidly deaminated by adenosine deaminase to give the inactive inosine derivative.[25]

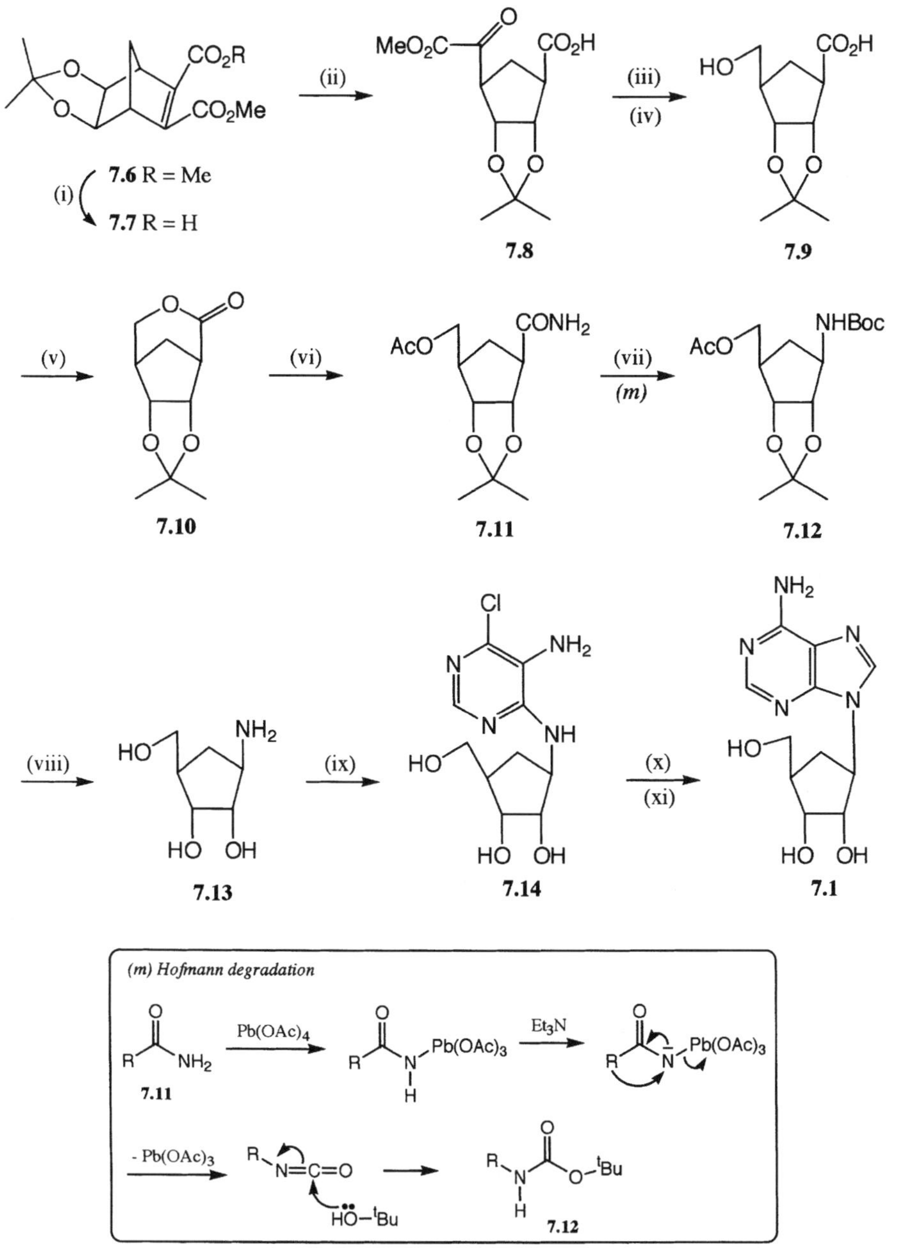

SCHEME 7.1 *REAGENTS AND CONDITIONS:* (I) PIG LIVER ESTERASE, ACETONE, H_2O, pH 8, 30–32°C (II) O_3, EtOAC, –78°C, 3 H (III) $NaBH_4$, MeOH, REFLUX, 3 H (IV) $NaIO_4$, MeOH, 30 MIN. THEN $NaBH_4$, 1 H (V) Ac_2O, PYRIDINE, O/N (VI) NH_3, MeOH, O/N (VII) $Pd(OAc)_4$, t-BuOH, Et_3N, REFLUX, 5 H (VIII) 2M HCl, MeOH, REFLUX, 1 H (IX) 5-AMINO-4,6-DICHLOROPYRIMIDINE (X) TRIETHYL ORTHOFORMATE (XI) NH_3.

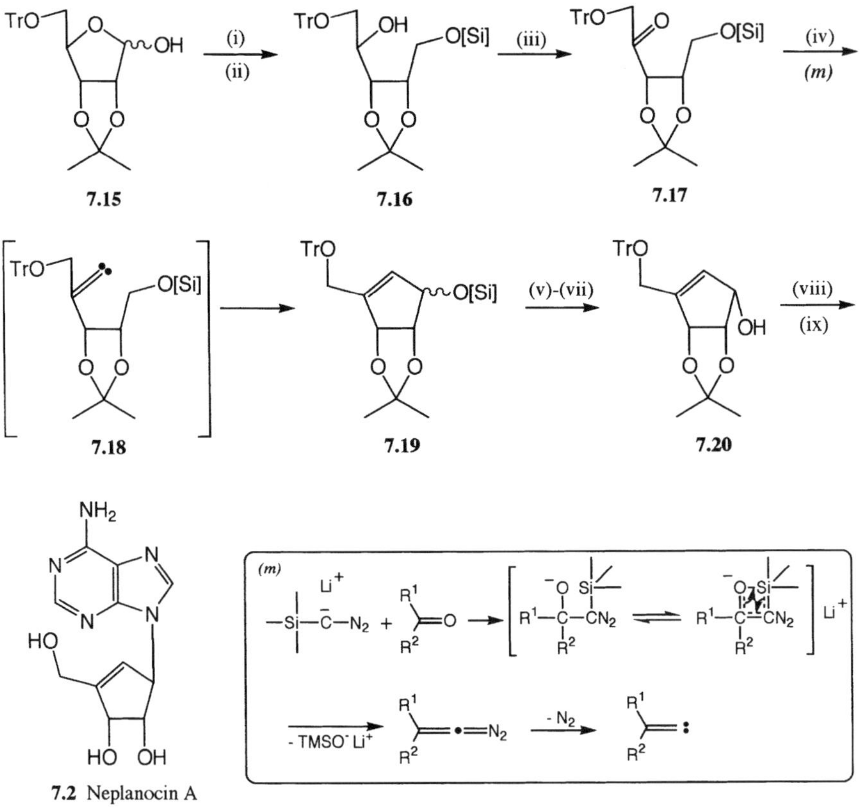

SCHEME 7.2 REAGENTS AND CONDITIONS: (I) LiAlH$_4$, Et$_2$O (II) TBDMSCl, IMIDAZOLE, DMF (III) OXALYL CHLORIDE, DMSO, Et$_3$N, CH$_2$Cl$_2$ (IV) TMSC(Li)N$_2$, THF, 0°C, 1 H (V) Bu$_4$NF, THF (VI) PYRIDINIUM DICHROMATE, CH$_2$Cl$_2$ (VII) LiAlH$_4$, THF (VIII) ADENINE (**1.3**), DEAD, Ph$_3$P, THF (IX) HCl, MeOH. [Si] = TBDMS.

Therefore effort has concentrated on the preparation of modified structures which are neither phosphorylated nor deaminated. The main advancements were the synthesis of (6'R)-6'-C-methylneplanocin A (RMNPA, **7.21**) which displayed greater selectivity and potency than the parent compound[25] and 6'-homoneplanocin (HNPA, **7.22**) which had an antiviral spectrum comparable with neplanocin A and was completely resistant to deamination by adenosine deaminase.[26]

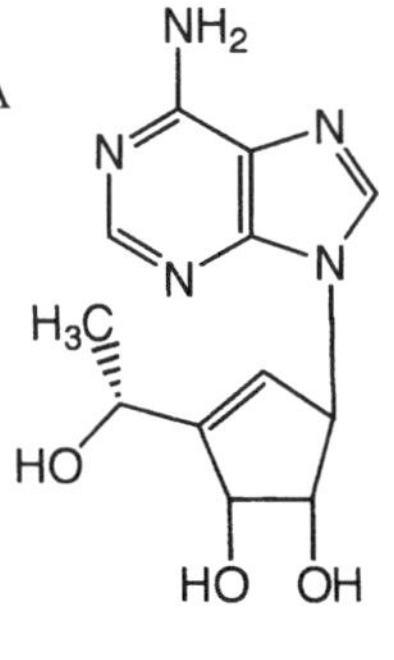

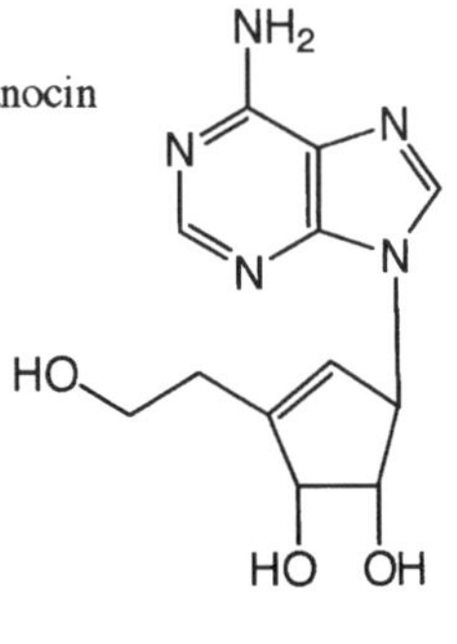

9-[(1*R*,2*S*,3*R*)-2,3-dihydroxy-4-(2-hydroxyethyl)-4-cyclopenten-1-yl]adenine (HNPA, **7.22**) was prepared from the cyclopentenone (**7.23**).[26] Stereoselective addition of the carbanion of ethyl acetate, prepared by treating ethyl acetate with LiN(TMS)$_2$,[27] gave **7.24** which was then reduced to the diol (**7.25**). Silyl protection of the primary hydroxy and acetyl protection of the secondary hydroxyl gave compound **7.26**, which underwent allylic rearrangement with Pd^{2+} catalyst.[28] The resulting cyclopentene (**7.27**) was converted to the mesylate (**7.28**), then reacted with the sodium salt of adenine to give the carbocyclic nucleoside (**7.29**), which was deprotected to give HNPA (**7.22**) (Scheme 7.3).

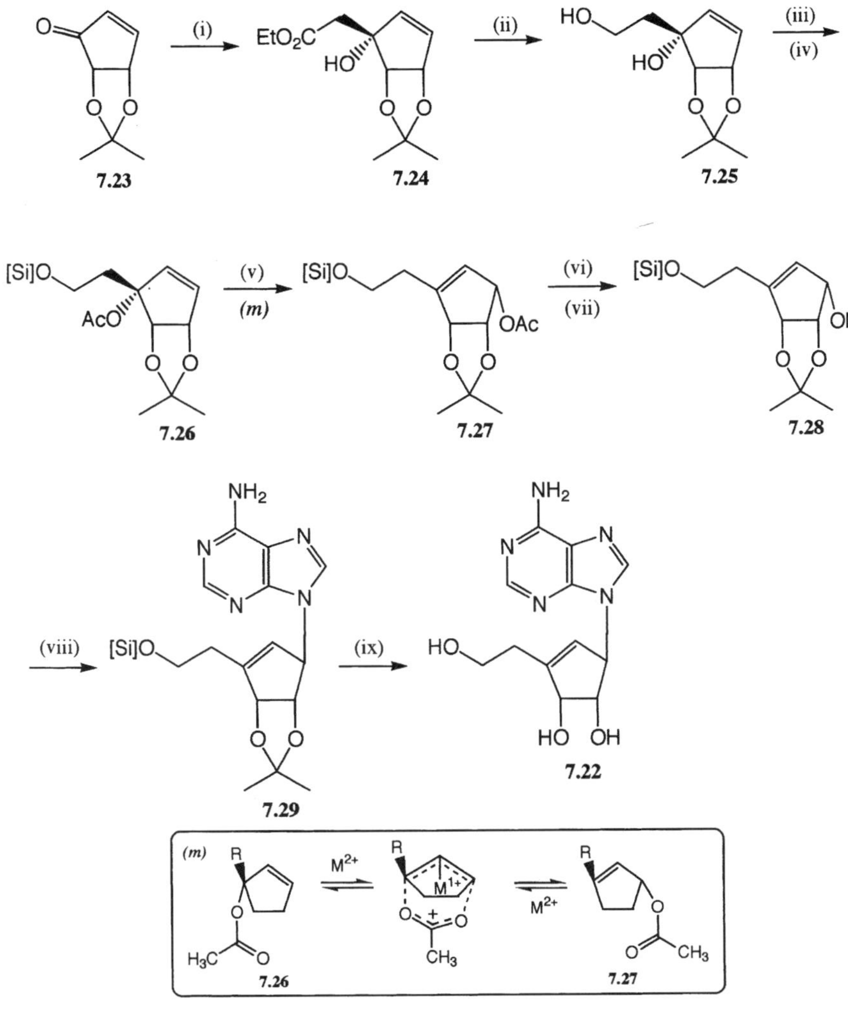

SCHEME 7.3 REAGENTS AND CONDITIONS: (I) (TMS)$_2$NH, BuLi, EtOAc, THF, −82°C, 1.5 H (II) LiBH$_4$, THF, R.T. 2 DAYS THEN REFLUX O/N (III) TBDMSCl, IMIDAZOLE, THF, O/N (IV) Ac$_2$O, DMAP, Et$_3$N, THF, O/N (V) PdCl$_2$(MeCN)$_2$, *P*-BENZOQUINONE, THF, REFLUX, O/N (VI) K$_2$CO$_3$, MeOH, 1.5 H (VII) MsCl, DMAP, CH$_2$Cl$_2$, 3.5 H (VIII) ADENINE, NaH, 15-CROWN-5, DMF, R.T., 15 H THEN 80°C, 2 H (IX) HCl, aq. MeOH, 2 H. [Si] = TBDMS.

Both Aristeromycin and Neplanocin A act as inhibitors of S-adenosyl-L-homocysteine hydrolase (AdoHcy), interfering with the 5'-capping and therefore maturation of viral mRNA (see section 1.7.3.1).

7.3 2'-DEOXYCARBOCYCLIC NUCLEOSIDES

The synthesis of the 2-deoxycarbocyclic component of 2'-deoxycarbocyclic nucleosides often requires multiple steps with very low overall yields.[29–31] Scheme 7.4 shows an efficient synthesis which allows the generation of either racemic or enantiomerically pure carbocyclic 2'-deoxyribonucleosides, involves 10–12 steps, with yields of 9–30%.[32] The key diol (±)-**7.33** was obtained from the cyclopentendiol (**7.30**) after a series of protection/deprotection steps, the hydroformylation (m, **7.31** →**7.32**) was performed under H_2/CO atmosphere in the presence of a rhodium catalyst.

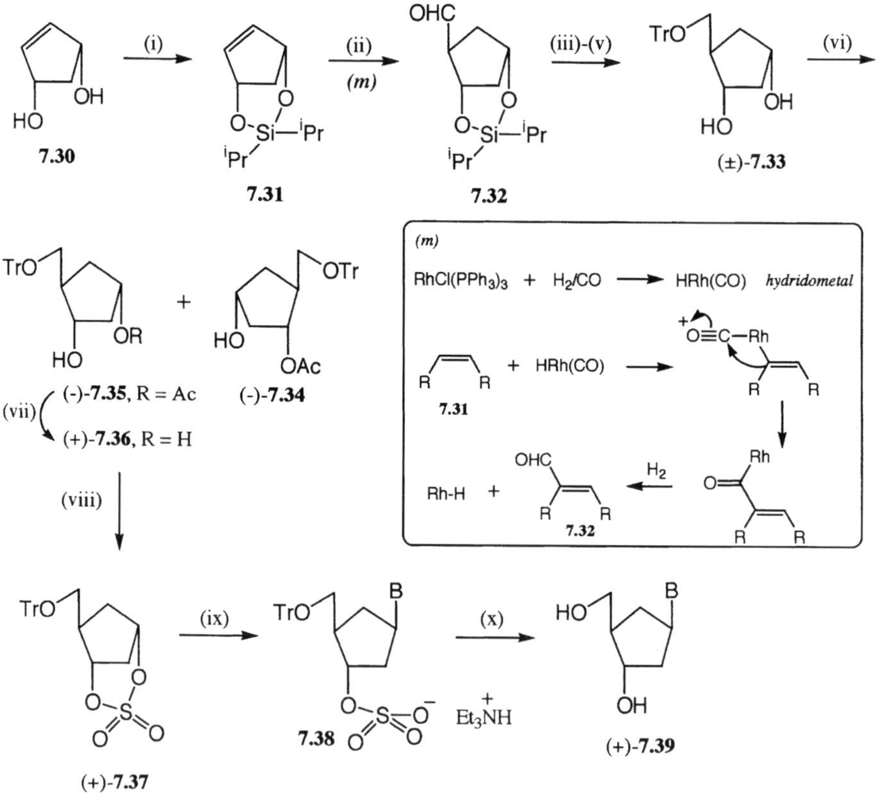

SCHEME 7.4 REAGENTS AND CONDITIONS: (I) $(t\text{-}Bu)_2Si(OSO_2CF_3)_2$, LUTIDINE, CH_2Cl_2, 0°C, 30 MIN. (II) $RhCl(PPh_3)_3$, THF, H_2/CO, 80 BAR, 80°C, 5 H (III) $NaBH_4$, THF, H_2O, 10 MIN. (IV) TrCl, DMAP, Et_3N, CH_2Cl_2, 18 H (V) $Bu_4NF\cdot 3H_2O$, THF, 5 H (VI) PFL, VINYL ACETATE, 50 H (VII) ETHYLENEDIAMINE, MeOH, 50°C, 15 H (VIII) $SOCl_2$, Et_3N, CH_2Cl_2, 0°C, 19 MIN. THEN CH_3CN/CCl_4, H_2O, $RuCl_3$, $NaIO_4$, 0°C, 1 H (IX) PURINE OR PYRIMIDINE BASE (B), DBU, CH_3CN (X) MeOH, aq. HCl.

Chiral resolution of (±)–**7.33** was achieved enzymatically using *Pseudomonas fluorescens* lipase (PFL) to give the regioisomeric monoacetates ((–)-**7.34** and (–)-**7.35**). After deprotection of the required monoacetate ((–)-**7.35**), the carbocyclic nucleosides ((+)–**7.39**) were prepared *via* the cyclic sulphate intermediate (**7.38**).

7.3.1 (+)-C-BVDU

The 2'-deoxycarbocyclic nucleoside (+)-1-[(1*R*,3*S*,4*R*-3-hydroxy-4-(hydroxy methylcyclopentyl]-5-(*E*)-(2-bromovinyl)-1*H*,3*H*-pyrimidin-2,4-dione, (+)-C-BVDU ((+)–**7.40**), has received considerable attention owing to its antiviral activity against VZV, HSV-1 and HSV-2.[33,34]

Preparation of (+)-C-BVDU from the aminosugar (**7.41**) can be achieved in a short high-yielding synthesis.[34] Reaction of **7.41** with the acyl carbamate (**7.42**) and subsequent acidic ring closure gave the 5-ethyluracil derivative (**7.44**, C-EDU[35]), the acetyl of which was converted to the 5-(*E*)-bromovinyl derivative ((+)–**7.40**) by bromination and final deacetylation (Scheme 7.5).

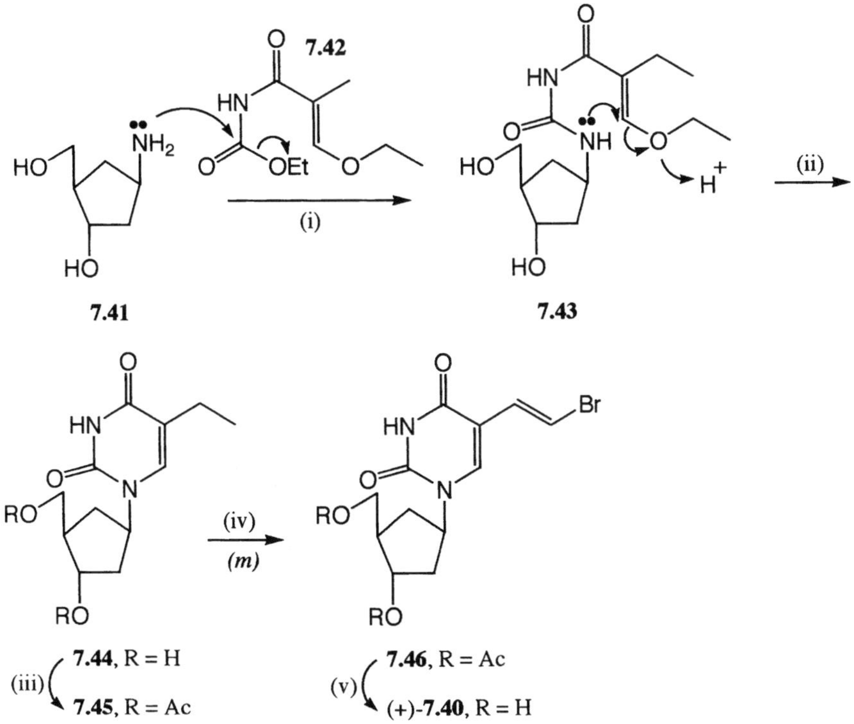

SCHEME 7.5 REAGENTS AND CONDITIONS: (I) Et₃N, DIOXANE, 100°C, 3 H (II) 2% 2N HCl, DIOXANE, 90°C, 13.5 H (III) Ac₂O, DMAP, DIOXANE, 16 H (IV) NBS, DIOXANE, Et₃N, 60°C, 1 H (V) aq. 2.5M NaOH, 70 MIN.

7.3.2 BMS-200475

Of more recent interest is the 2'-deoxyguanosine derivative BMS-200475 (**7.5**), which in its triphosphate form is a potent and selective antiviral agent active against Hepatitis B virus.[36] The chiral cyclopentyl epoxide (**7.48**) was reacted with 6-benzyloxy-2-aminopurine to give the N-9 adduct (**7.49**). The cyclopentyl alcohol (**7.50**) was then oxidized using Dess–Martin reagent,[37] with subsequent Nysted methylenation[38] yielding the methylene (**7.52**). Deprotection of the monomethoxytrityl (MMT) group and debenzylation gave the active methylenecyclopentyl nucleoside BMS-200475 (**7.5**) (Scheme 7.6).[9]

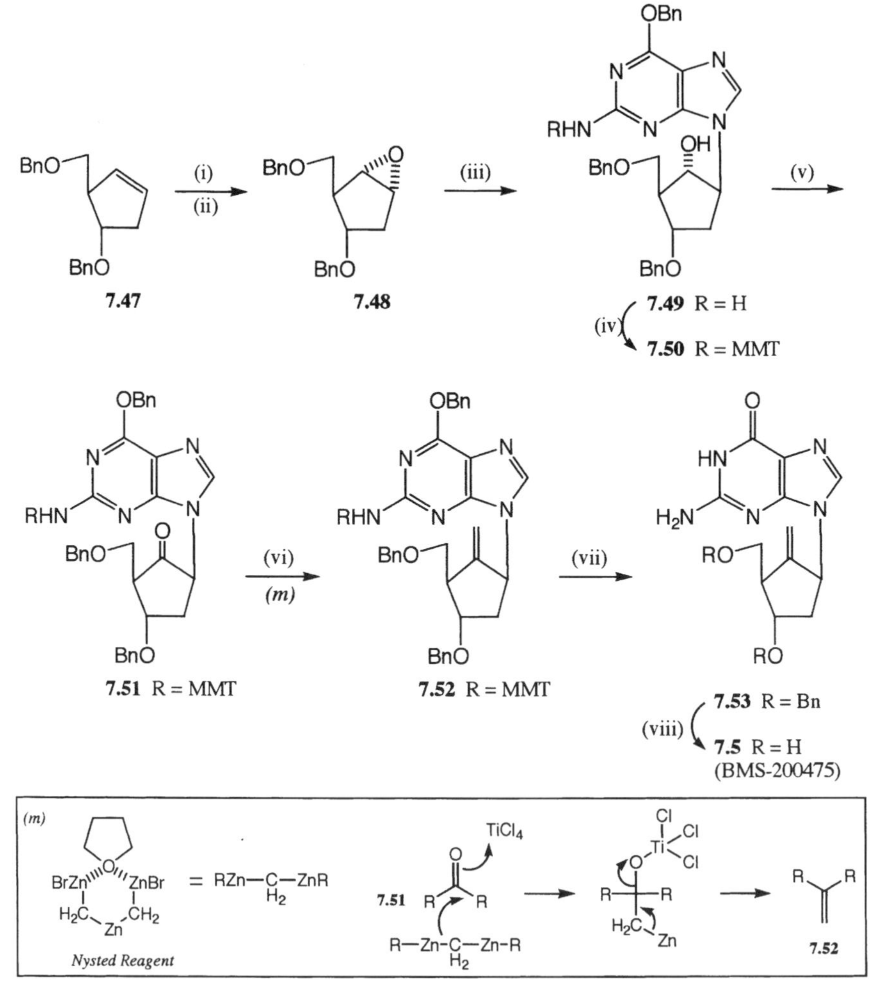

SCHEME 7.6 *REAGENTS AND CONDITIONS:* (I) VO(acac)$_2$, t-BuOOH, CH$_2$Cl$_2$ (II) BnBr, NaH, Bu$_4$NI, DMF (III) 6-BENZYLOXY-2-AMINOPURINE, LiH, DMF, 125°C (IV) MONOMETHOXYTRITYL CHLORIDE (MMT-CI), Et$_3$N, DMAP, CH$_2$Cl$_2$ (V) DESS–MARTIN REAGENT, t-BuOH, CH$_2$Cl$_2$ (VI) NYSTED REAGENT, TiCl$_4$, THF (VII) aq. HCl, THF, MeOH, 55°C (VIII) BCl$_3$, CH$_2$Cl$_2$, −78°C.

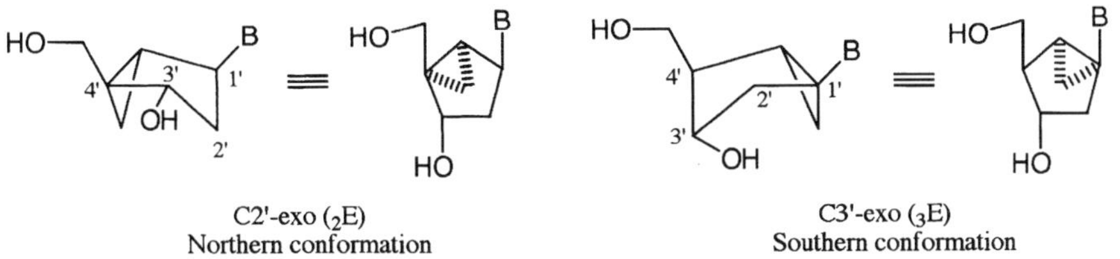

C2'-exo ($_2$E)
Northern conformation

C3'-exo ($_3$E)
Southern conformation

FIGURE 7.2   CONSTRAINED BICYCLO[3.2.1]HEXANE TEMPLATES.

7.3.3   Conformationally Constrained 2'-Deoxycarbocyclic Nucleosides

A bicyclo[3.1.0]hexane template has been developed which can adopt either a Northern or Southern conformation depending on the ring substituents (Fig. 7.2).[39–41] These constrained carbocyclic nucleosides more closely mimic the conformation of conventional nucleosides, that is C-2'exo ($_2$E, North) or C3'-exo ($_3$E, South) (see section 1.3.3).

Carbocyclic nucleosides constrained in the Northern conformation were the most active, with (N)-2'-deoxy-methanocarba-T (**7.54**) displaying potent antiherpetic activity whilst the Southern conformer was devoid of activity.[41] In the purine series, (N)-2'-deoxy-methanocarba-A (**7.55**), the methano-analogue of 2'-deoxyaristeromycin had significant activity against CMV and EBV.[42]

The (N)-2'-deoxy-methanocarba-nucleosides have been prepared from the chiral intermediate (**7.61**) which was obtained in six steps from the cyclopentenol (**7.56**) (Scheme 7.7).[42] Regioselective cleavage of the isopropylidene system of **7.56** gave the tert-butoxy-glycol (**7.57**) with subsequent silyl protection of the less hindered hydroxy group. Deoxygenation (see *(m)* Scheme 2.24) and silyl deprotection gave compound **7.60** with the allylic hydroxy group capable of directing the cyclopropanation to give the chiral bicyclo[3.2.1]hexane (**7.61**).

Coupling of the pyrimidine or purine bases with the chiral intermediate (**7.61**) was performed under Mitsunobu conditions, with subsequent conversion of **7.62**/**7.63** to the respective (N)-2'-deoxy-methanocarba-nucleosides (**7.54** and **7.55**) achieved by standard methods (Scheme 7.7).[41,42]

7.4   CARBOCYCLIC D$_4$-NUCLEOSIDES

The chemotherapeutic carbocyclic 2',3'-didehydro-2',3'-dideoxy (D$_4$)-nucleosides are active in their triphosphate forms and act as competitive inhibitors/chain terminators, in a manner similar to that described for conventional nucleosides (see section 1.7.1.1). Of the carbocyclic D$_4$-nucleosides, Carbovir (**7.3**)[6] and Abacavir (**7.4**)[7] are potent inhibitors of HIV reverse transcriptase, the L-D$_4$-adenosine derivative (**7.64**) exhibited potent anti-HBV activity and moderate anti-HIV activity.[43]

7.4.1   Carbovir and Abacavir

Although Carbovir, (±)-*cis*-2-amino-1,9-dihydro-9-[4-(hydroxymethyl)-2-cyclopenten-1-yl]-6*H*-purin-6-one,[6] showed much promise as an antiretroviral agent it was removed from clinical trials owing to toxicity problems. Abacavir,[7] the cyclopropylamine derivative of Carbovir, is activated through a unique pathway to

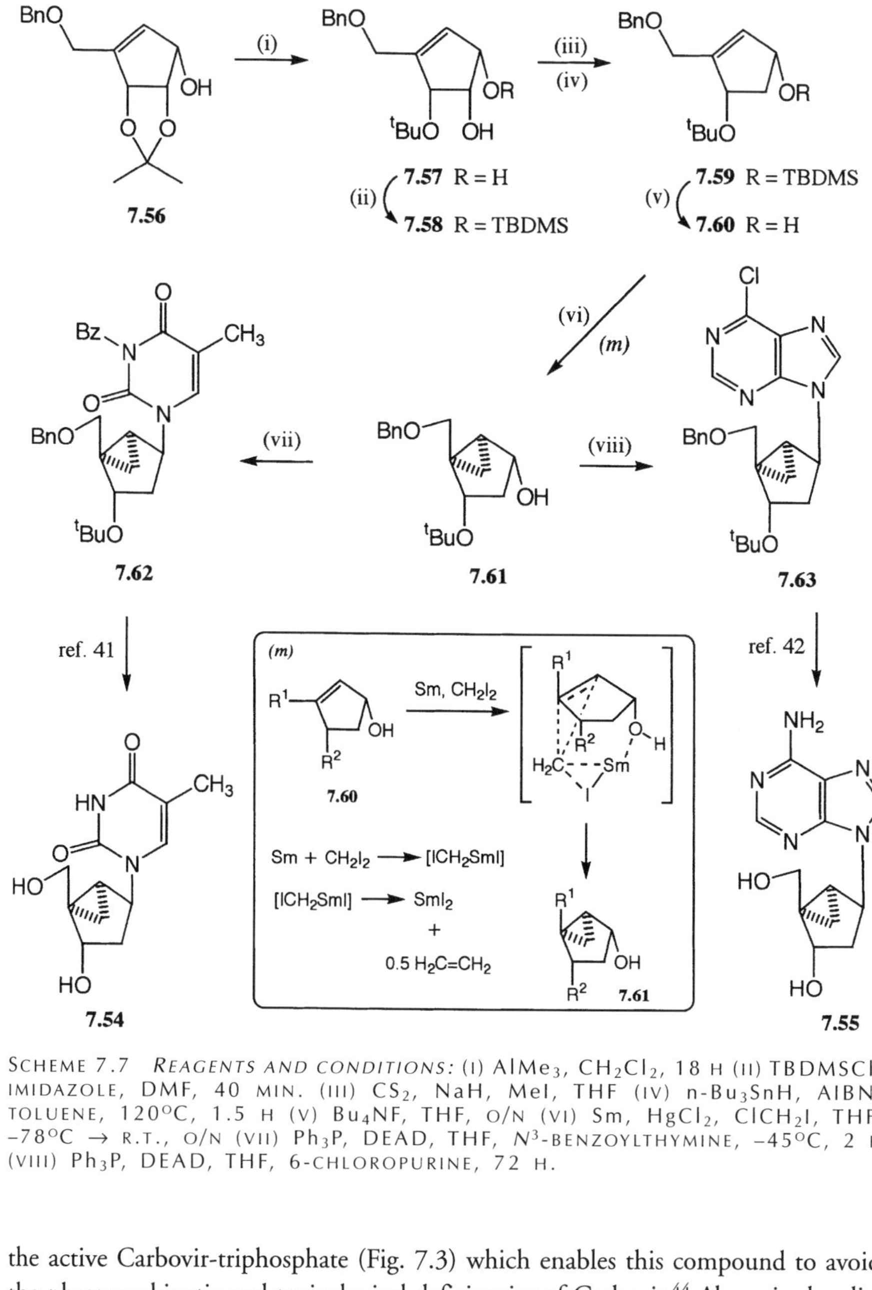

SCHEME 7.7 REAGENTS AND CONDITIONS: (I) AlMe$_3$, CH$_2$Cl$_2$, 18 H (II) TBDMSCl, IMIDAZOLE, DMF, 40 MIN. (III) CS$_2$, NaH, MeI, THF (IV) n-Bu$_3$SnH, AIBN, TOLUENE, 120°C, 1.5 H (V) Bu$_4$NF, THF, O/N (VI) Sm, HgCl$_2$, ClCH$_2$I, THF, −78°C → R.T., O/N (VII) Ph$_3$P, DEAD, THF, N^3-BENZOYLTHYMINE, −45°C, 2 H (VIII) Ph$_3$P, DEAD, THF, 6-CHLOROPURINE, 72 H.

the active Carbovir-triphosphate (Fig. 7.3) which enables this compound to avoid the pharmacokinetic and toxicological deficiencies of Carbovir.[44] Abacavir also displayed excellent oral bioavailability and CNS penetration and is undergoing phase III clinical trials for the treatment of HIV infections.[45]

Asymmetric syntheses of both (−)-Carbovir and (−)-Abacavir, the active enantiomers, have been described.[46–50] The efficient synthesis of Crimmins and King requires the preparation of enantiomerically pure 5-(hydroxymethyl)-2-cyclopenten-1-ol (**7.68**).[48] The asymmetric synthesis of **7.68** involves diastereoselective *syn* aldol condensation of the oxazolidinone (**7.65**) with acrolein using Evans' dialkylboron triflate protocol[51] to give the aldol product (**7.66**).

CARBOCYCLIC NUCLEOSIDES

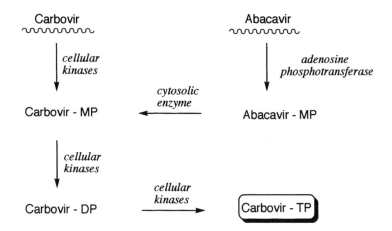

FIGURE 7.3 MECHANISM OF ACTIVATION OF CARBOVIR AND ABACAVIR.

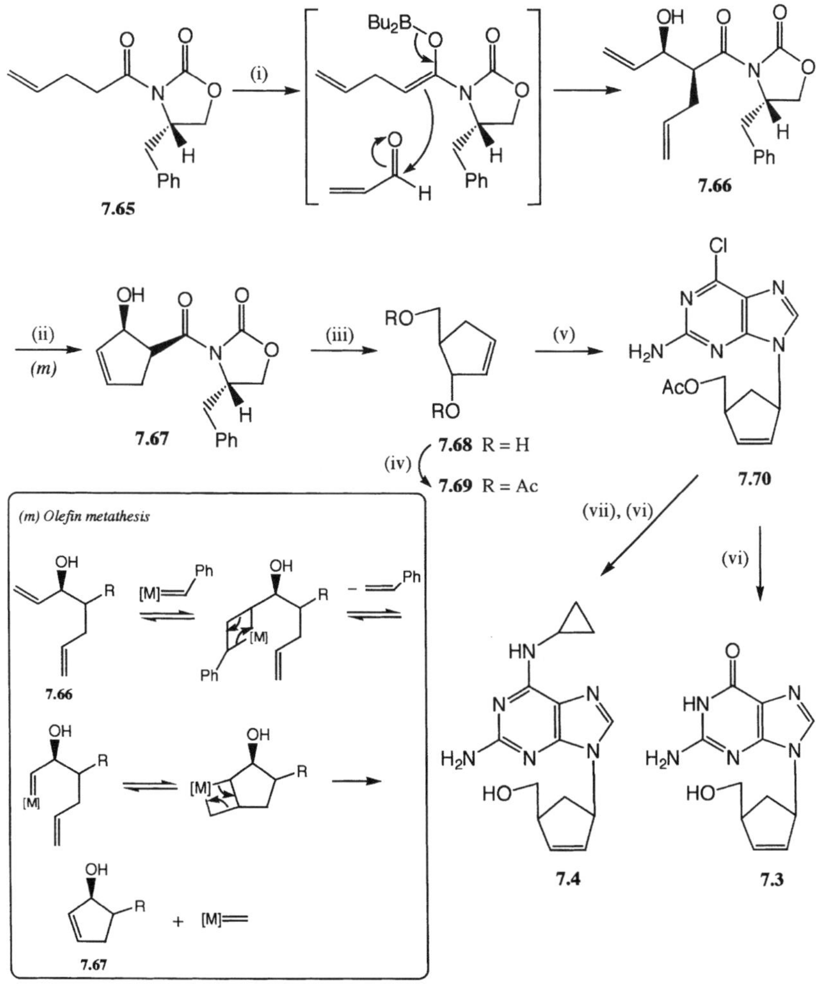

SCHEME 7.8 *REAGENTS AND CONDITIONS:* (I) Bu_2BOTf, Et_3N, CH_2Cl_2, CH_2=CHCHO, −78°C (II) GRUBBS CATALYST, CH_2Cl_2, 30 MIN. (III) $LiBH_4$, THF, MeOH (IV) Ac_2O, Et_3N, DMAP, CH_2Cl_2 (V) 2-AMINO-6-CHLOROPURINE, NaH, $Pd(PPh_3)_4$, THF, DMSO (VI) NaOH, H_2O (VII) CYCLOPROPYLAMINE, EtOH.

Ring closure was accomplished using Grubbs catalyst[52] with the chiral auxilary removed by treatment with lithium borohydride to yield the required cyclopentenol (**7.67**). The acetylated cyclopentenol (**7.69**) was coupled with 2-amino-6-chloropurine using Trost's palladium (0)-catalysed coupling method[53] to give the N^9-cyclopentene (**7.70**) as the major product. Hydrolysis of **7.70** with aqueous sodium hydroxide gave Carbovir (**7.3**). Treatment of the chloropurine (**7.70**) with cylopropylamine and hydrolysis of the acetyl protecting groups gave Abacavir (**7.4**) (Scheme 7.8).

7.4.2 Carbocyclic D₄-L-nucleosides

The asymmetric synthesis and anti-HIV and anti-HBV activities of carbocyclic β-L-2',3'-didehydro-2',3'-dideoxy and β-L-2',3'-dideoxy pyrimidine and purine nucleosides have been described.[54] Carbocyclic β-L-D₄A (**7.64**),[43] the most active in this class of carbocyclic nucleosides, was prepared from the protected cyclopentane (**7.71**) (Scheme 7.9).[43] Selective deprotection of the isopropylidene group gave the diol (**7.72**) which on treatment with trimethyl orthoformate gave the cyclic ortho ester (**7.73**). This 2-methoxy-1,3-dioxolane derivative (**7.73**) was subjected to an acetic acid catalyzed thermal elimination reaction,[55] resulting in the formation of the required cyclopentene (**7.74**) (see *m*, Scheme 7.9).

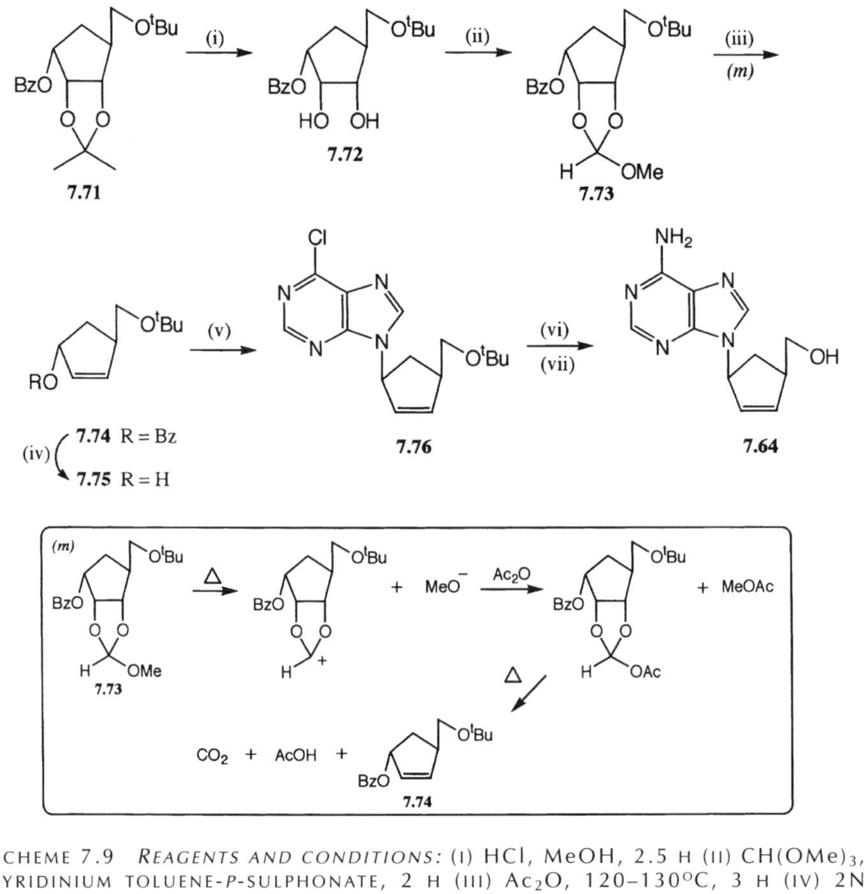

SCHEME 7.9 REAGENTS AND CONDITIONS: (I) HCl, MeOH, 2.5 H (II) CH(OMe)₃, PYRIDINIUM TOLUENE-*P*-SULPHONATE, 2 H (III) Ac₂O, 120–130°C, 3 H (IV) 2N NaOH, MeOH, 1.5 H (V) 6-CHLOROPURINE, DEAD, PPh₃, DIOXANE, 10 H (VI) NH₃, MeOH, 80–90°C, 20 H (VII) CF₃CO₂H, H₂O, 50°C, 3 H.

After removal of the benzoyl group, the alcohol (**7.75**) was reacted with 6-chloropurine under Mitsunobu conditions. The adenosine derivative, carbocyclic β-L-D$_4$A (**7.64**), was obtained by reaction of the 6-chloro-carbocyclic nucleoside (**7.76**) with methanolic ammonia followed by removal of the tert-butyl protecting group.

7.5 CYCLOPROPYL CARBOCYCLIC NUCLEOSIDES

Both D- and L-β-carbocyclic cyclopropyl nucleosides have been prepared, however no biological data has been provided, with the synthesis of the D- (or 1'S, 2'R) derivative commencing from 2,3-O-isopropylidene-D-glyceraldehyde,[56,57] and the L- (or 1'R, 2'S) derivative commencing from L-gulonic-γ-lactone.[58,59]

Scheme 7.10 describes the synthesis of the D-(1'S,2'R)-cyclopropyl pyrimidine nucleosides.[57] The α,β-unsaturated ester (**7.78**), prepared by Wittig reaction (see *m2*, Scheme 2.33) of glyceraldehyde (**7.77**) with Ph$_3$P=CHCO$_2$Me, was reduced to the alcohol (**7.79**) with DIBAL-H. Cyclopropanation was achieved by a modified Simmons–Smith reaction[60] to give the cyclopropyl methyl alcohol (**7.80**). Oxidation

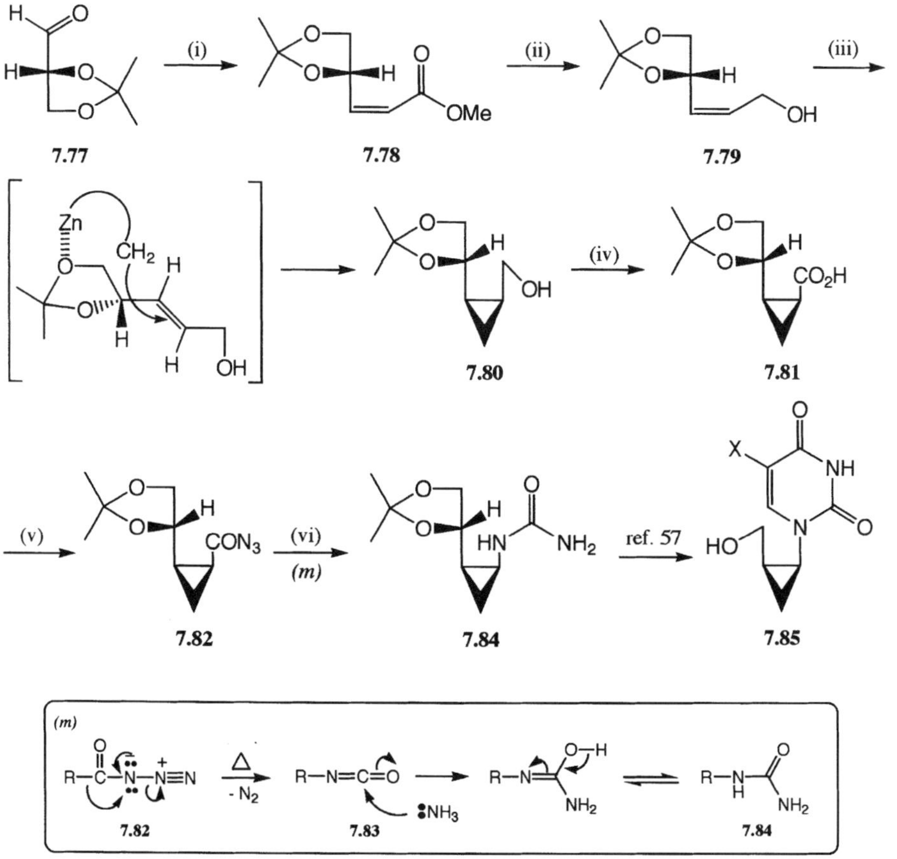

SCHEME 7.10 *REAGENTS AND CONDITIONS:* (I) Ph$_3$P=CHCO$_2$Me, MeOH, O/N (II) DIBAL-H, CH$_2$Cl$_2$, −78°C, 30 MIN. (III) Zn(Et)$_2$, CH$_2$I$_2$, DICHLOROETHANE, 0°C, 30 MIN. (IV) RuO$_2$, NaIO$_4$, CH$_3$CN, CHCl$_3$, H$_2$O, 16 H (V) ClCO$_2$Et, Et$_3$N, ACETONE, 1H THEN NaN$_3$, TOLUENE, 90–100°C, 1.5 H (VI) TOLUENE, 90–100°C, 1.5 H THEN NH$_3$ (G), Et$_2$O, 30 MIN.

of **7.80** with NaIO$_4$ in the presence of catalytic RuO$_2$ gave the acid (**7.81**), which was then converted to the azide (**7.82**). Curtius rearrangement (see *m*, Scheme 7.10) produced the isocyanate (**7.83**) which on treatment with ammonia gas gave the key cyclopropyl intermediate (**7.84**) for the preparation of the pyrimidine nucleosides (**7.85**).

7.6 CYCLOBUTYL CARBOCYCLIC NUCLEOSIDES (LOBUCAVIR)

Several carbocyclic cyclobutyl nucleosides have been prepared,[5,8,61–64] of these the most promising is [1R – 1α,2β,3α]-2-amino-9-[2,3-bis(hydroxymethyl)cyclobutyl] guanine (Lobucavir, (*R*)-BHCG, **1.50**).[65] In its active triphosphate form, Lobucavir is a potent inhibitor of the DNA polymerase of several strains of herpes virus (see section 1.7.2.1),[66] importantly Lobucavir is active against ganciclovir-resistant cytomegalovirus strains and is currently in Phase II clinical trials for the treatment of CMV infections.[67]

Asymmetric syntheses of Lobucavir have been described[65,68] as well as a method for regioselective coupling.[69] Asymmetric synthesis of the required cyclobutyl triflate (**7.91**) commenced from *trans*-3,3-diethoxy-1,2-cyclobutanedicarboxylic acid (**7.86**), which was coupled with *R*-(–)-2-phenylglycinol to give a mixture of diastereomeric bis-amides (**7.87(S)** and (**R**)) which were readily separated by crystallization. A three step reaction sequence transformed the bis-amide (**7.87(S)**) to the corresponding diol enantiomer (1*S*-*trans*)-3,3-diethoxy-1,2-cyclobutanedimethanol (**7.88**).[65] The benzoylated derivative (**7.89**) was converted to the cyclobutanone (**7.90**) which was then stereoselectively reduced and treated with triflic anhydride to give the required cyclobutyl triflate (**7.91**).[69] Regioselective coupling of **7.91** with the tetrabutylammonium salt of 6-iodo-9*H*-purin-2-amine (**7.92**) resulted in the formation of the N^9-product (**7.93**) from which Lobucavir (**1.50**) was obtained (Scheme 7.11).

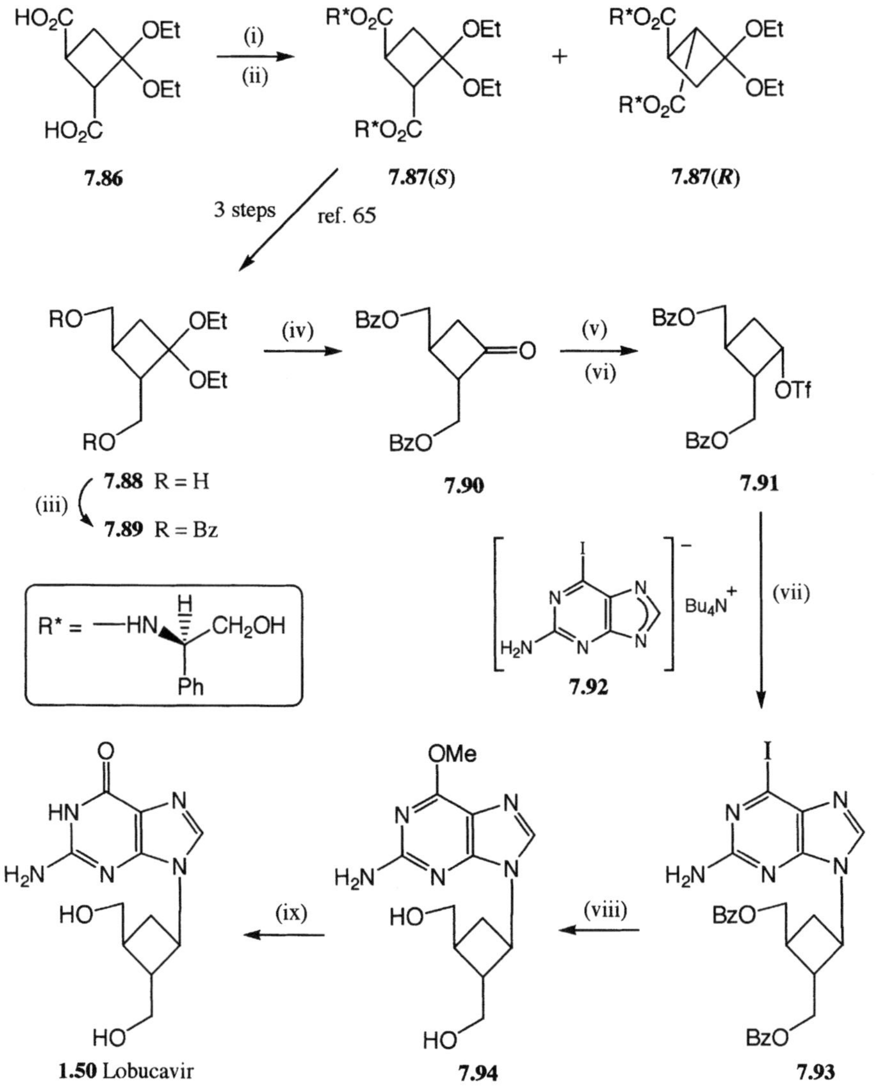

SCHEME 7.11 *REAGENTS AND CONDITIONS:* (I) R-(−)-2-PHENYLGLYCINOL, DCC, CH_2Cl_2, O/N (II) CRYSTALLIZE FROM CH_2Cl_2 TO OBTAIN **7.87(S)** THEN Et_2O/CH_2Cl_2 TO OBTAIN **7.87(R)** (III) BzCl, PYRIDINE, 2.5 H (IV) 0.5N H_2SO_4, CH_3CN, 17 H (V) LS-SELECTRIDE, THF, −78°C, 30 MIN. (VI) Tf_2O, CH_2Cl_2, PYRIDINE, 10 MIN. (VII) **7.92**, CH_2Cl_2, O/N (VIII) NaOMe, MeOH, REFLUX, 1.5 H (IX) aq. HCl, pH 0.5, 95°C, 3 H.

7.7 REFERENCES

1. H. Bricaud, P. Herdewijn, E. De Clercq, *Biochem. Pharmacol.*, **1983**, 3583–3590.
2. R. Cookson, P. Dudfield, R. Newton, P. Ravenscroft, D. Scopes, J. Cameron, *Eur. J. Med. Chem. Chim. Ther.*, **1985**, *20*, 375.
3. Y.F. Shealy, J.D. Clayton, *J. Pharm. Sci.*, **1973**, *62*, 858–859.
4. S. Yaginuma, N. Muto, M. Tsujino, Y. Sudate, M. Hayashi, M. Otani, *J. Antibiot.*, **1981**, *34*, 359–366.
5. V.E. Marquez, In *"Advances in antiviral drug design"*, E. DeClercq (ed.), JAI Press, Greenwich, **1996**, *2*, 89–146.

6. R. Vince, M. Hua, *J. Med. Chem.*, **1990**, *33*, 17–21.
7. D.W. Kimberlain, D.M. Coen, K.K. Biron, J.I. Cohen, R.A. Lamb, M. Mckinlay, E.A. Emini, R.J. Whitley, *Antiviral Res.*, **1995**, *26*, 369–401.
8. D.W. Norbeck, E. Kern, S. Hayashi, W. Rosenbrook, H. Sham, T. Herrin, J.J. Plattner, J. Erickson, J. Clement, R. Swanson, N. Shipkowitz, D. Hardy, K. Marsh, G. Arnett, W. Shannon, S. Broder, H. Mitsuya, *J. Med. Chem.*, **1990**, *33*, 1281–1285.
9. G.S. Bisacchi, S.T. Chao, C. Bachard, J.P. Daris, S. Innaimo, G.A. Jacobs, O. Kocy, P. Lapointe, A. Martel, Z. Merchant, W.A. Slusarchyk, J.E. Sundeen, M.G. Young, R. Colonno, R. Zahler, *Bioorg. Med. Chem. Lett.*, **1997**, *7*, 127–132.
10. L. Agrofoglio, E. Suhas, A. Farese, R. Condom, S.R. Challand, R.A. earl, R. Guedj, *Tetrahedron*, **1994**, *50*, 10611–10670.
11. A.D. Borthwick, K. Biggadike, *Tetrahedron*, **1992**, *48*, 571–623.
12. A. Guranowski, J.A. Montgomery, G.L. Cantoni, P.K. Chiang, *Biochemistry*, **1981**, *20*, 110–115.
13. R.T. Borchardt, B.T. Keller, U. Patel–Thombre, *J. Biol. Chem.*, **1984**, *259*, 4353–4358.
14. T. Kusaka, H. Yamamoto, M. Shibata, M. Muroi, T. Hicki, K. Mizuno, *J. Antiobiot.*, **1968**, *21*, 255–263.
15. Y.F. Shealy, J.D. Clayton, *J. Am. Chem. Soc.*, **1966**, *88*, 3885–3887.
16. M. Arita, K. Adachi, Y. Ito, H. Sawaki, M. Ohno, *J. Am. Chem. Soc.*, **1983**, *105*, 4049–4055.
17. S.J. Boyer, J.W. Leahy, *J. Org. Chem.*, **1997**, *62*, 3976–3980.
18. F. Burlina, A. Favre, J-L. Fourrey, M. Thomas, *Bioorg. Med. Chem. Lett.*, **1997**, *7*, 247–250.
19. N. Yoshida, T. Kamikubo, K. Ogasawara, *Tetrahedron Lett.*, **1998**, *39*, 4677–4678.
20. B.M. Trost, R. Madsen, S.D. Guile, *Tetrahedron Lett.*, **1997**, *38*, 1707–1710.
21. S. Ohira, T. Sawamoto, M. Yamamoto, *Tetrahedron Lett.*, **1995**, *36*, 1537–1538.
22. S. Ohira, K. Okai, T. Moritani, *J. Chem. Soc., Chem. Commun.*, **1992**, 721–722.
23. E. De Clercq, *Antimicrob. Ag. Chemother.*, **1985**, *28*, 84–89.
24. R.I. Glazer, M.C. Knode, *J. Biol. Chem.*, **1984**, *259*, 2964–2969.
25. S. Shuto, T. Obara, M. Toriya, M. Hosoya, R. Snoeck, G. Andrei, J. Balzarini, E. De Clercq, *J. Med. Chem.*, **1992**, *35*, 324–331.
26. S. Shuto, T. Obara, Y. Saito, G. Andrei, R. Snoeck, E. De Clercq, A. Matsuda, *J. Med. Chem.*, **1996**, *39*, 2392–2399.
27. M.W. Rathke, *J. Am. Chem. Soc.*, **1970**, *92*, 3222–3223.
28. A.C. Oehlschlager, P. Mishra, S. Dhami, *Can. J. Chem.*, **1984**, *62*, 791–797.
29. K. Biggadike, A.D. Bothwick, A.M. Exall, B.E. Kirk, S.M. Roberts, P. Youds, *J. Chem. Soc., Chem. Commun.*, **1987**, 1083–1084.
30. J. Béres, G. Sági, E. Baitz-Gács, I. Tömösközi, L. Ötvös, *Tetrahedron*, **1988**, *44*, 6207–6216.
31. J. Balzarini, H. Baumgartner, M. Bodenteich, E. De Clercq, H. Griengl, *J. Med. Chem.*, **1989**, *32*, 1861–1865.
32. H. Lang, H.E. Moser, *Helv. Chim. Acta*, **1994**, *77*, 1527–1540.
33. J. Balzarini, H. Baumgartner, M. Bodenteich, E. De Clercq, H. Griengl, *Nucleosides & Nucleotides*, **1989**, *8*, 855–858.
34. P.G. Wyatt, A.S. Anslow, B.A. Coomber, R.P.C. Cousins, D.N. Evans, V.S. Gilbert, D.C. Humber, I.L. Paternoster, S.L. Sollis, D.J. Tapolczay, G.G. Weingarten, *Nucleosides & Nucleotides*, **1995**, *14*, 2039–2049.
35. Y.F. Shealy, C.A. O'Dell, G. Arnett, W.M. Shannon, *J. Med. Chem.*, **1986**, *29*, 79–84.
36. G. Yamanaka, T. Wilson, S. Innaimo, G.S. Bisacchi, P. Egli, J.K. Rinehart, R. Zahler, R.J. Colonno, *Antimicrob. Ag. Chemother.*, **1999**, *43*, 190–193.
37. D.B. Dess, J.C. Martin, *J. Am. Chem. Soc.*, **1991**, *113*, 7277.
38. L.N. Nysted, *Aldrichimica Acta*, **1993**, *26*, 14.
39. K–H. Altmann, R. Kesselring, E. Francotte, G. Rihs, *Tetrahedron Lett.*, **1994**, *35*, 2331–2334.
38. A. Ezzitouni, J.J. Barchi Jr., V.E. Marquez, *J. Chem. Soc., Chem. Commun.*, **1995**, 1345–1346.
41. V.E. Marquez, M.A. Siddiqui, A. Ezzitouni, P. Russ, J. Wang, R.W. Wagner, M.D. Matteucci, *J. Med. Chem.*, **1996**, *39*, 3739–3747.
42. M.A. Siddiqui, H. Ford Jr., C. George, V.E. Marquez, *Nucleosides & Nucleotides*, **1996**, *15*, 235–250.
43. P. Wang, R.F. Schinazi, C.K. Chu, *Bioorg. Med. Chem. Lett.*, **1998**, *8*, 1585–1588.
44. M.B. Faletto, W.H. Miller, E.P. Garvey, M.H. St Clair, S.M. Daluge, *Antimicrob. Ag. Chemother.*, **1997**, *41*, 1099–1107.
45. R.H. Foster, D. Faulds, *Drugs*, **1998**, *55*, 729–736.
46. R. Vince, J. Brownell, Biochem. *Biophys. Res. Commun.*, **1990**, *168*, 912–916.
47. C.T. Evans, S.M. Evans, K.A. Shoberu, A.G. Sutherland, *J. Chem. Soc., Perkin Trans. 1*, **1992**, 589–592.
48. M.T. Crimmins, B.W. King, *J. Org. Chem.*, **1996**, *61*, 4192–4193.
49. N. Katagiri, M. Takebayashi, H. Kokufuda, C. Kaneko, K. Kanehira, M. Torihara, *J. Org. Chem.*, **1997**, *62*, 1580–1581.

50. H.F. Olivo, J. Yu, *J. Chem. Soc., Perkin Trans. 1*, **1998**, 391–392.
51. D.A. Evans, J. Bartroli, T.L. Shih, *J. Am. Chem. Soc.*, **1981**, *103*, 2127–2129.
52. P. Schwab, R.H. Grubbs, J. Ziller, *J. Am. Chem. Soc.*, **1996**, *118*, 100–110.
53. B.M. Trost, G-H. Kuo, T. Benneche, *J. Am. Chem. Soc.*, **1988**, *110*, 621–622.
54. P. Wang, B. Gullen, M.G. Newton, Y–C. Cheng, R.F. Schinazi, C.K. Chu, *J. Med. Chem.*, **1999**, *42*, 3390–3399.
55. M. Ando, H. Ohhara, K. Takase, *Chem. Lett.*, **1986**, 879–882.
56. Y. Zhao, T-F. Yang, M. Lee, B.K. Chun, J. Du, R.F. Schinazi, D. Lee, M.G. Newton, C.K. Chu, *Tetrahedron Lett.*, **1994**, *35*, 5405–5408.
57. Y. Zhao, T. Yang, M. Lee, D. Lee, M.G. Newton, C.K. Chu, *J. Org. Chem.*, **1995**, *60*, 5236–5242.
58. M. Lee, D. Lee, Y. Zhao, M.G. Newton, M.W. Chun, C.K. Chu, *Tetrahedron Lett.*, **1995**, *36*, 3499–3502.
59. M. Lee, J.F. Du, M.W. Chun, C.K. Chu, *J. Org. Chem.*, **1997**, *62*, 1991–1995.
60. S.E. Denmark, J.P. Edwards, *J. Org. Chem.*, **1991**, *56*, 6974–6981.
61. W.A. Slusarchyk, G.S. Bisacchi, A.K. Field, D.R. Hockstein, G.A. Jacobs, B.McGeeverrubin, J.A. Tino, A.V. Tuomari, G.A. Yamanaka, M.G. Young, R. Zahler, *J. Med. Chem.*, **1992**, *35*, 1799–1806.
62. B.L. Booth, P.R. Eastwood, *Tetrahedron Lett.*, **1993**, *34*, 5503–5506.
63. Y. Sato, T. Maruyama, *Chem. Pharm. Bull.*, **1995**, *43*, 91–95.
64. M. Frieden, M. Giraud, C.B. Reese, Q. Song, *J. Chem. Soc., Perkin Trans. 1*, **1998**, 2827–2832.
65. G.S. Bisacchi, A. Braitman, C.W. Cianci, J.M. Clark, A.K. Field, M.E. Hagen, D.R. Hockstein, M.F. Malley, T. Mitt, W.A. Slusarchyk, J.E. Sundeen, B.J. Terry, A.V. Tuomari, E.R. Weaver, M.G. Young, R. Zahler, *J. Med. Chem.*, **1991**, *34*, 1415–1421.
66. G. Yamanaka, A.V. Tuomari, M. Hagen, B.McGeeverrubin, B. Terry, M. Hafey, G.S. Bisacchi, A.K. Field, *Mol. Pharmacol.*, **1991**, *40*, 446–453.
67. D.J. Tenney, G. Yamanaka, S.M. Voss, C.W. Cianci, A.V. Tuomari, A.K. Sheaffer, M. Alam, R.J. Colonno, *Antimicrob. Agents Chemother.*, **1997**, *41*, 2680–2685.

CHAPTER 8

ACYCLIC NUCLEOSIDES

8.1 INTRODUCTION

Acyclic nucleosides differ from conventional nucleosides in that the sugar ring is replaced by an acyclic moiety. The first acyclic nucleoside to show selective inhibition of HSV replication was acyclovir (**1.47**, Zovirax®, ACV),[1,2] a guanosine based nucleoside whose clinical effectiveness[3] has stimulated considerable interest in acyclic nucleosides (Fig. 8.1).

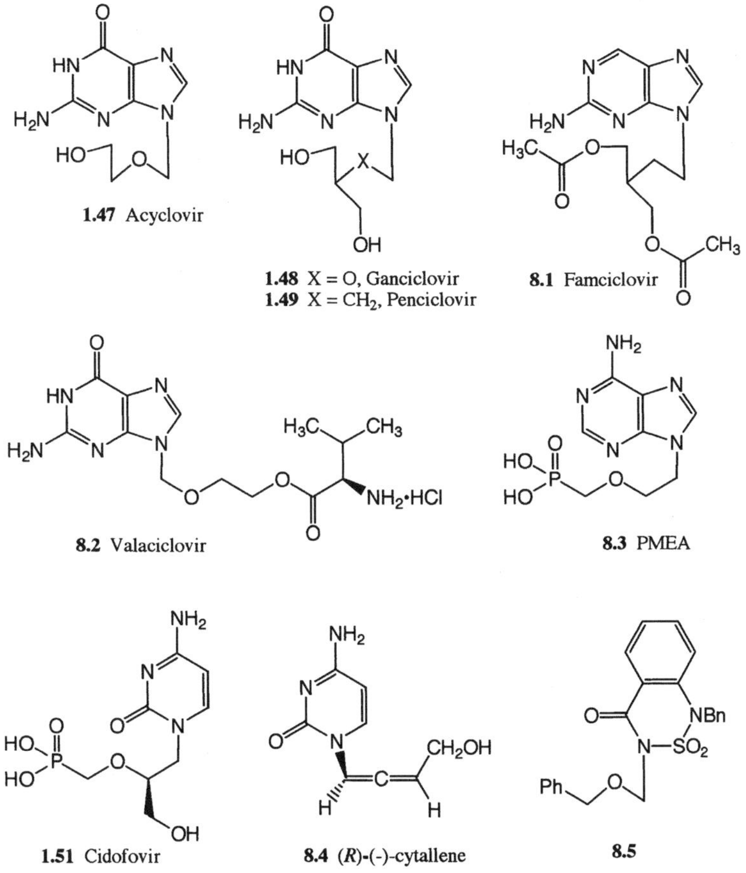

FIGURE 8.1 ACYCLIC NUCLEOSIDES.

The poor bioavailablity of ACV and the emergence of ACV-resistant strains of HSV and VZV[3] has led to the development of drugs with improved efficacy, such as ganciclovir (**1.48**, Cymevene®, GCV)[4] and penciclovir (**1.49**, Vectavir®, PCV),[5] the pro-drugs famciclovir (**8.1**, Famvir®, FCV)[6] and valaciclovir (**8.2**, Valtrex®)[7] which demonstrate improved oral bioavailability, and more recently the acyclic nucleoside phosphonates, such as PMEA (**8.3**, Adefovir®)[8] and Cidofovir (**1.51**, Vistide®, HPMPC)[9] which are active against ACV-resistant virus strains. Many other acyclic nucleosides have been prepared with modified acyclic chains, such as the conformationally restrained anti-HIV agent (R)-(–)-cytallene (**8.4**),[10] and modified heterocyclic bases, such as the benzothiadiazine derivative (**8.5**) which possesses activity against CMV.[11]

8.2 ACYCLOVIR

Acyclovir, 9-(2-hydroxyethoxymethyl)guanine, discovered in 1974,[1] is used in the treatment of mucosal, cutaneous and systemic HSV-1 and HSV-2 infections and in the first-line therapy of VZV.[12] Acylovir is active as its triphosphate, ACV-TP and the main target of ACV-TP is viral DNA polymerase[13,14] (see section 1.7.2.1).

8.2.1 Synthesis of Acyclovir

The initial synthesis of Schaeffer *et al.*[2] and a subsequent synthesis by Robins *et al.*[15] both involve the preparation of (2-acyloxyethoxy)methyl halides as the acyclic moiety, which can be achieved by acylation of the cyclic acetal 1,3-dioxolane[16] (**8.6**). Reaction of **8.6** with either benzoyl chloride or acetyl bromide produces the ethers **8.9** and **8.10** respectively *via* the oxonium intermediate **8.7** (Scheme 8.1).

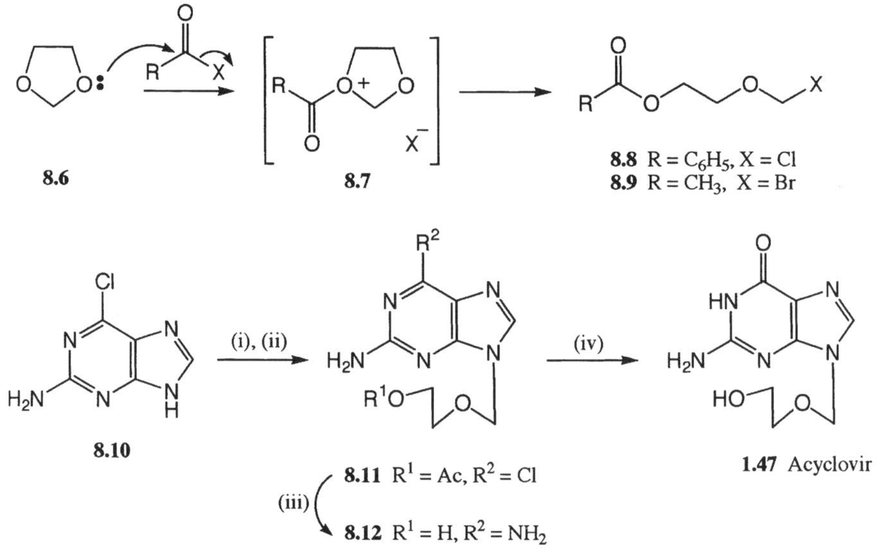

8.8 R = C$_6$H$_5$, X = Cl
8.9 R = CH$_3$, X = Br

8.11 R^1 = Ac, R^2 = Cl
8.12 R^1 = H, R^2 = NH$_2$

1.47 Acyclovir

SCHEME 8.1 *REAGENTS AND CONDITIONS*: (I) HEXAMETHYLDISILAZANE (II) **8.9**, Hg(CN)$_2$, BENZENE (III) MeOH/NH$_3$, 85°C (IV) ADENOSINE DEAMINASE, pH 7.5 PHOSPHATE BUFFER.

In the procedure of Robins *et al.*[15] condensation of **8.9** and 2-amino-6-chloropurine (**8.10**) in the presence of mercury (II) cyanide gave the *N*-9 product **8.11** in 89% yield. Ammonolysis at 85°C gave the 2,6-diaminopurine **8.12** with concurrent loss of the acetyl protecting group, with conversion of **8.12** to ACV (**1.47**) on treatment with adenosine deaminase (Scheme 8.1).

A similar preparation of the acylic moiety of ACV and the thioether analogue, described by Barrio *et al.*,[17] involves the rapid reaction of 1,3-dioxolane (**8.6**) or 1,3-oxathiolane (**8.13**) with trimethylsilyl iodide, to produce the iodomethylethers **8.15** and **8.16** respectively *via* the oxonium species **8.14** (Scheme 8.2).

Reaction of the iodomethylether **8.15** with the anion of 2-chloro-6-iodopurine (**8.17**), generated by reaction with sodium hydride, gave the *N*-9 product **8.18** in 75% yield and the corresponding 7-isomer in 10% yield. The greater susceptibility of the 6-iodopurines towards nucleophilic displacement allows the mild base hydrolysis of **8.18** to the 6-oxopurine **8.19** by treatment with aqueous potassium carbonate in dioxane. Reaction of **8.19** with liquid ammonia at 150°C in a steel bomb gave ACV (**1.47**) in 80% yield (Scheme 8.2).

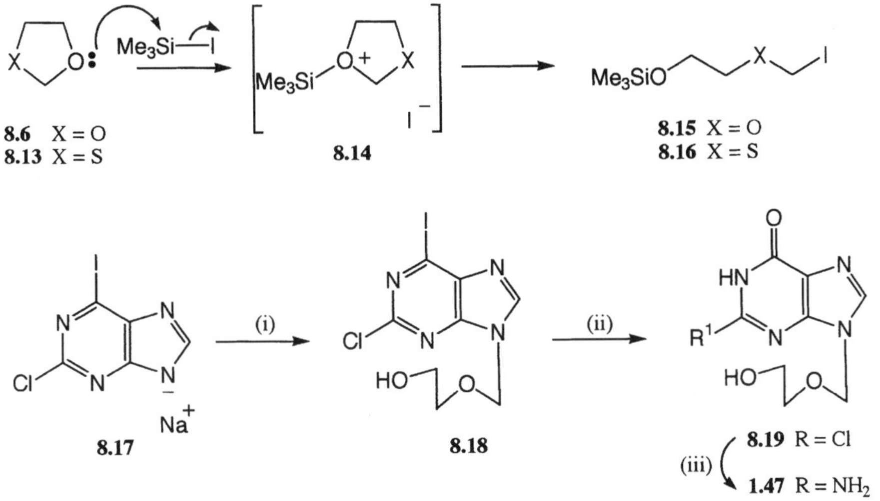

8.6 X = O
8.13 X = S
8.14
8.15 X = O
8.16 X = S
8.17
8.18
8.19 R = Cl
1.47 R = NH_2

SCHEME 8.2 *REAGENTS AND CONDITIONS:* (I) NaH, DMF, **8.15** (II) AQUEOUS K_2CO_3, DIOXANE (III) LIQUID NH_3, 150°C, STEEL BOMB.

Other methods for the synthesis of ACV have been described,[18–20] including routes *via* guanine and furazano[3,4-*d*]pyrimidines.

8.2.2 Acyclovir Prodrugs

The most promising prodrug of ACV is the L-valyl ester valaciclovir (**8.2**, Valtrex®),[7] an oral prodrug which undergoes rapid first-pass metabolism to produce the parent nucleoside ACV and L-valine. Valaciclovir is readily prepared[21] by the reaction of ACV (**1.47**) with the N-protected L-valine (**8.20**), using the coupling agent dicyclohexylcarbodiimide (DCC) to generate the reactive acyl isourea, and 4-dimethylaminopyridine (DMAP) as a catalyst to give the N-Cbz blocked ester

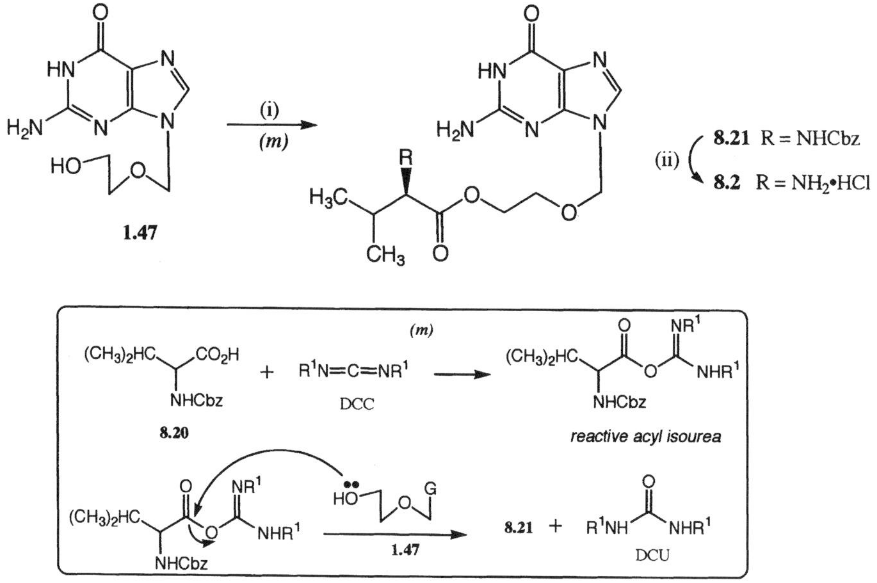

SCHEME 8.3 *REAGENTS AND CONDITIONS:* (I) **8.20**, DCC, DMAP, DMF (II) 0.5N AQUEOUS HCl, 5%Pd/C, H_2, MeOH, 50PSI. R^1 = CYCLOHEXYL.

derivative (**8.21**) and dicyclohexylurea (DCU). Deprotection by catalytic hydrogenation in the presence of HCl gave the target aminoacyl ester **8.2** (Scheme 8.3).

Valaciclovir administration is particularly beneficial where high concentrations of ACV are required, such as in the treatment of the less sensitive virus VZV. The phospholipid prodrug, acyclovir diphosphate dimyristoylglycerol (**8.23**), prepared by the reaction of ACV-MP with the morpholidate of phosphatidic acid,[22] (**8.22**) (Scheme 8.4), is active in TK-deficient (TK-) ACV-resistant strains of HSV owing to its ability to release ACV-MP intracellularly, by-passing the slow initial phosphorylation of ACV by viral-encoded TK.

Numerous compounds based on the acyclovir structure have been prepared and evaluated as antiviral agents.[23] Structure activity relationship studies have determined the portions of the molecule which are tolerant to chemical substitutions without (significant) loss of antiviral activity. A few of the key antiviral compounds are discussed in more detail in the following sections.

8.3 GANCICLOVIR

Ganciclovir (**1.48**, Cymevene®, GCV)[4] is active against all the herpes viruses and Epstein-Barr virus (EBV), targeting DNA polymerase as described for ACV (see section 1.7.2.1). However GCV can be distinguished from other guanosine acyclic analogues in that it demonstrates pronounced activity against cytomegalovirus (CMV).[24]

8.3.1 Mechanism of Action

CMV does not encode a viral specific TK, conversion of the acyclic nucleosides to their monophosphates is achieved by the CMV specified protein kinase UL97.[25] However of the guanosine acyclic analogues, only GCV is recognised as a substrate

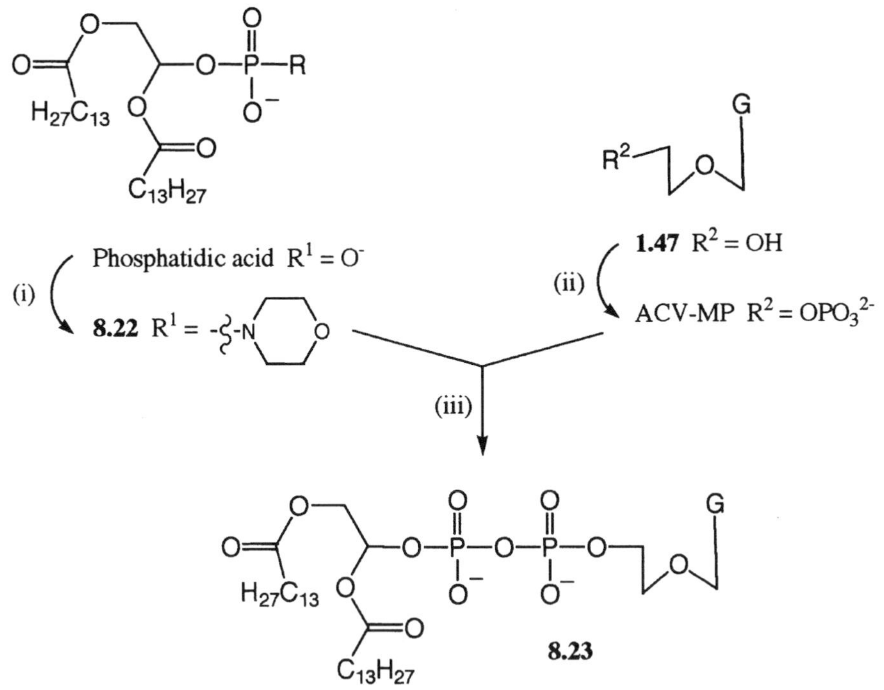

ACYCLIC NUCLEOSIDES

SCHEME 8.4 REAGENTS AND CONDITIONS: (I) TRIMETHYLPHOSPHATE, POCl$_3$ (II) MORPHOLINE, DCC, t-BuOH (III) PYRIDINE, 60°C.

by UL97 and can be successfully converted to GCV-MP. Conversion of GCV-MP to the active form GCV-TP is carried out by cellular kinases (Figure 8.2).

8.3.2 Synthesis of Ganciclovir

The syntheses of ganciclovir 9-(1,3-dihydroxy-2-propoxymethyl)guanine (**1.48**), involve the condensation of $N^{2,9}$-diacetylguanine (**8.24**) with either 1,3-dibenzyloxy-2-chloromethoxypropane[26] (**8.25a**), 2-acetoxymethoxy-1,3-diacetoxy propane[27] (**8.25b**) or 1,3-di-O-benzyl-2-acetoxymethoxyglycerol[28] (**8.25c**) (Scheme 8.5).

The synthesis of **8.25a** and **8.25c** involves base cleavage of the epoxide ring of epichlorohydrin (**8.26**) from the least hindered face followed by benzylation of the primary hydroxyl group to give the halohydrin (**8.27**). Base induced intramolecular substitution of **8.27** gives the epoxide (**8.28**) which is cleaved to give the diol, which undergoes selective benzylation of the primary alcohol group to give 1,3-di-O-

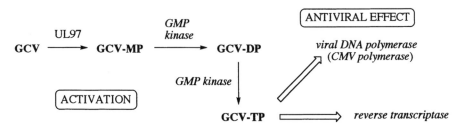

FIGURE 8.2 ACTIVATION AND ANTIVIRAL EFFECT OF GCV.

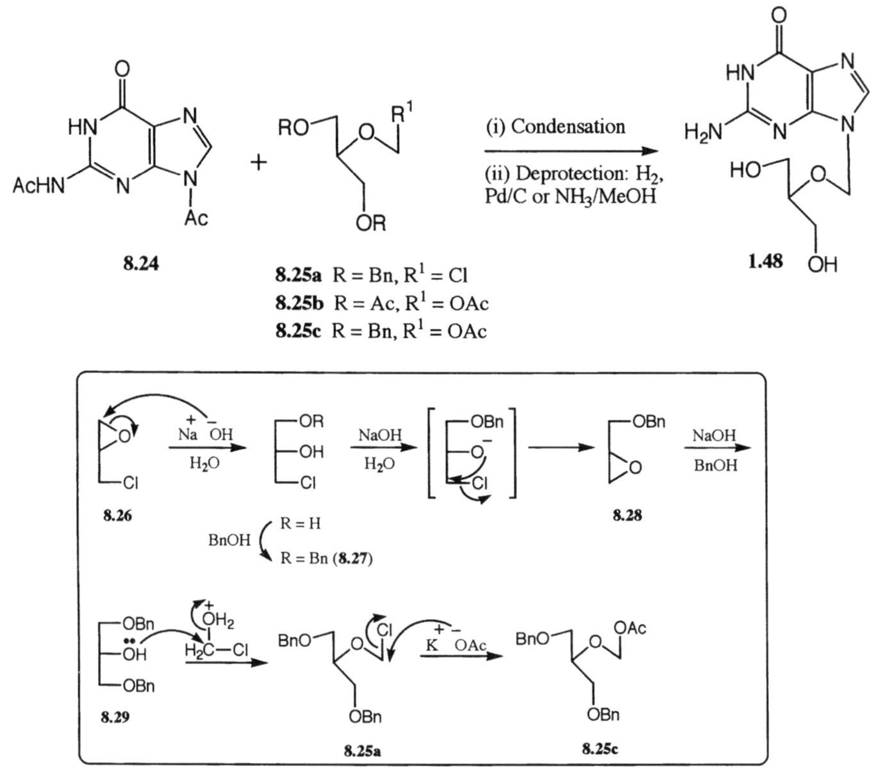

ACYCLIC NUCLEOSIDES

SCHEME 8.5 SYNTHESIS OF GANCICLOVIR.[26,27,28]

benzylglycerol (**8.29**). Reaction of paraformaldehyde with HCl produces the reactive electrophile, protonated chloromethyl alcohol, to carry out the chloromethylation of **8.29** producing the α-haloether **8.25a**, formation of **8.25c** occurs *via* displacement with acetate anion (Scheme 8.5).

8.4 PENCICLOVIR AND FAMCICLOVIR

Penciclovir (**1.49**, Vectavir®, PCV),[5] 9-(4-hydroxy-3-hydroxymethyl-but-1-yl)guanine, is the carboacyclic equivalent of GCV, with the ether oxygen replaced by a methylene unit. The antiviral activity of PCV is similar to ACV,[29] as is its mode of action.[30] The improved potency of PCV, compared with ACV, is attributed to the increased half-life of the major metabolite (S)-PCV-TP. The poor oral bioavailablity of PCV led to the design of the prodrug famciclovir (**8.4**, Famvir®, FCV),[6] diacetyl 6-deoxy-9-(4-hydroxy-3-hydroxymethyl-but-1-yl)guanine.

The mode of action of famciclovir, which has been reviewed,[31] involves initial hydrolysis of the 4'-acetyl group with subsequent removal of the remaining ester group in the liver to give 6-deoxy-PCV. The 6-deoxy-PCV is then oxidised in the liver by the molybdenum hydroxylase, aldehyde oxidase,[32] to give PCV (Fig. 8.3).

8.4.1 Synthesis of Penciclovir and Famciclovir
Several syntheses of PCV (**1.49**)[33–35] and FCV (**8.4**)[36,37] have been described. Two

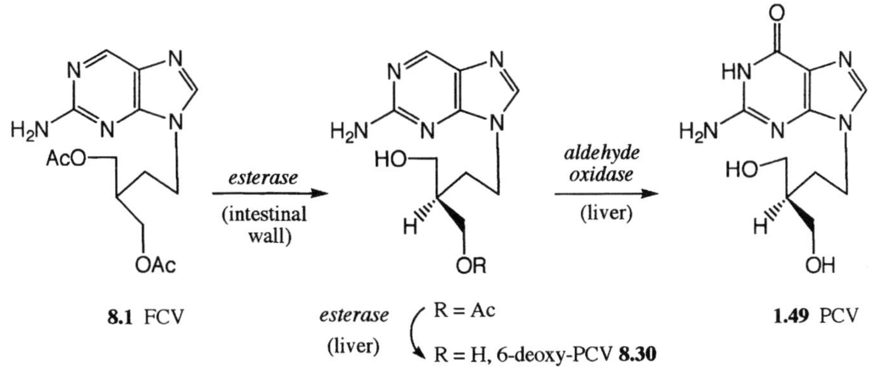

FIGURE 8.3 METABOLISM OF THE PRODRUG FCV.

convenient syntheses of both PCV and FCV from a common intermediate in good overall yields have been described.[38,39] The direct approach of Choudary et al.[38] involves the reaction of 2-amino-6-chloropurine (**8.10**) with triethyl 3-bromopropane-1,1,1-tricarboxylate (**8.31**), which acts as a surrogate for the unstable diethyl-2-(2-bromoethyl)malonate which has a tendency to undergo intramolecular cyclisation. Transesterification of **8.32** with sodium methoxide gave a 6.5:1 mixture of N^9:N^7 isomers, which were readily separated. Reduction of **8.33** and subsequent acetylation gave the common intermediate **8.34**. Conversion from **8.34** to PCV (**1.49**) involved acid hydrolysis, and from **8.34** to FCV (**8.4**) catalytic dehalogenation, with overall yields of 40–45% and 35–40% respectively (Scheme 8.6).

Several other prodrugs of PCV which show oral penciclovir bioavailability comparable with famciclovir have been described,[40,41] such as 2-amino-9-(3-acetoxy methyl-4-isopropoxycarbonyloxy-but-1-yl)purine (**8.35**).[40] Reaction of 6-deoxy-PCV (**8.30**) with trimethyl orthoacetate resulted in the formation of the cyclic orthoacetate (**8.36**) which gave selectively the 3-acetyl-derivative (**8.37**) on quenching with water, with subsequent introduction of the 4-isopropoxy-group on reaction with isopropyl 4-nitrophenyl carbonate (Scheme 8.7).

8.5 CONFORMATIONALLY RESTRICTED ACYCLIC NUCLEOSIDES

(R)-(–)-Adenallene (**8.38**, (R)-(–)-N^9-(4-hydroxy-1,2-butadien-1-yl)adenine)[42] and (R)-(–)-cytallene (**8.4**, (R)-(–)-N^9-(4-hydroxy-1,2-butadien-1-yl)cytosine)[43] are acyclic analogues of adenosine and cytidine with the chiral centres of the ribofuranose moiety replaced with the chiral axis of a 1,3-disubstituted allene.

The conformational rigidity, location of the heterocyclic base and hydroxy function, and orientation of the π-orbitals have all been shown to contribute to the anti-HIV activity of (R)-(–)-adenallene and (R)-(–)-cytallene. A-5021 (**8.39**, (1'S,2'R)-9-{[1',2'-bis(hydroxymethyl)cyclopropyl-1'-yl]methyl}guanine),[44] which displays potent and broad antiviral activity, is a conformationally restricted nucleoside, with two hydroxyl groups mimicking the 3'- and 5'-hydroxyl groups of the 2'-deoxyribose nucleosides and a cyclopropane ring in the 2,3-position of the side chain (Fig. 8.4).

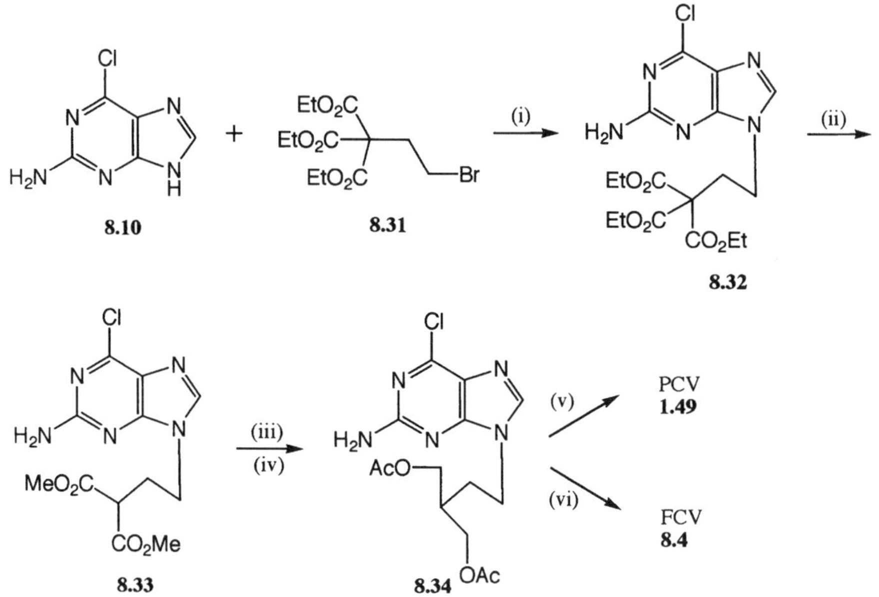

SCHEME 8.6 REAGENTS AND CONDITIONS: (I) K₂CO₃, DMF, 60°C, 22 H (II) NaOMe, MeOH, 1 H (III) NaBH₄, CH₂Cl₂, MeOH, 3.5 H (IV) Ac₂O, DMAP, CH₂Cl₂, REFLUX, 1.5 H (V) 2M aq. HCl, REFLUX, 3 H (VI) 5% Pd/C, H₂, Et₃N, EtOAc, AUTOCLAVE, 4 H.

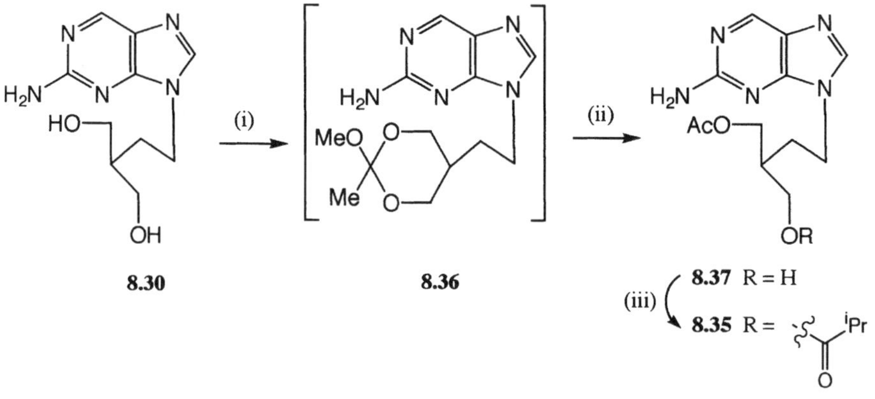

SCHEME 8.7 REAGENTS AND CONDITIONS: (I) TRIMETHYL ORTHOACETATE, P-TsOH•H₂O, DMF, 2 H (II) H₂O, 1 H (III) ISOPROPYL 4-NITROPHENYLCARBONATE, DMAP, PYRIDINE, 80°C, 24 H.

8.5.1 Adenallene and Cytallene

Preparation of (±)-adenallene (**8.40**) involved the alkylation of adenine (**1.3**) with 1,4-dichloro-2-butyne to give the corresponding N^9-alkyladenine (**8.41**) with >90% $N^9:N^7$ regioselectivity, subsequent acid hydrolysis gave the acetylene derivative (**8.42**). The base catalysed isomerisation of butyne (**8.42**) to the allene (**8.40**) with aqueous NaOH, resulted in a mixture of isomers containing 50% allene which was isolated by silica gel chromatography.[10] Resolution and separation of the enantiomers was achieved[42] by reaction with adenosine deaminase (ADA) to give (R)-(−)-adenallene

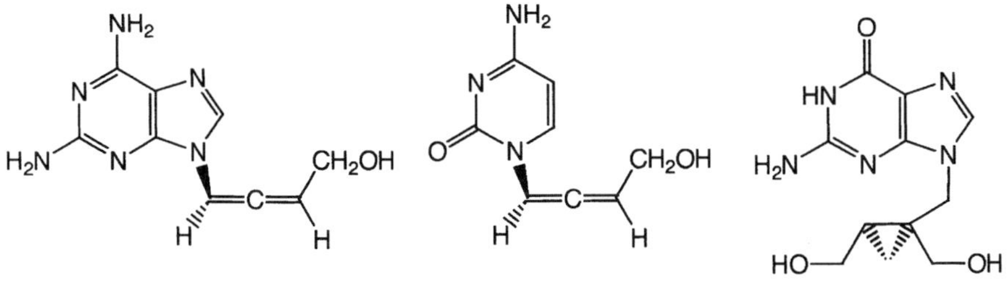

FIGURE 8.4 CONFORMATIONALLY RESTRAINED ACYCLIC NUCLEOSIDES.

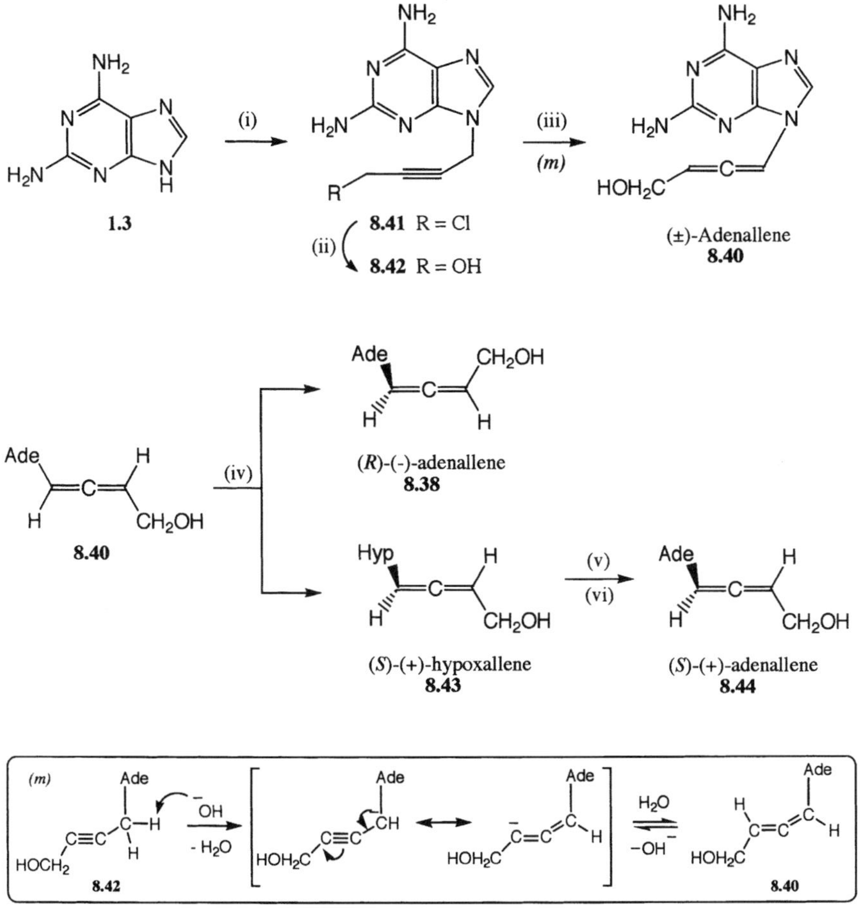

SCHEME 8.8 *REAGENTS AND CONDITIONS:* (I) 1,4-DICHLORO-2-BUTYNE, K_2CO_3, DMSO, 18 H (II) 0.1M HCl, REFLUX, 18 H (III) 0.1M NaOH, REFLUX, 30 MIN. (IV) ADA, 0.05M Na_2HPO_4 (PH 7.5), 26°C, 80 MIN. (V) Tf_2O, PYRIDINE, CH_2Cl_2, 7 H (VI) NH_3, DIOXANE, 1 H.

(**8.38**) and (*S*)-(+)-hypoxallene (**8.43**), with **8.43** then converted to (*S*)-(+)-adenallene (**8.44**) using standard procedures (Scheme 8.8).

(±)-Cytallene has been prepared in a similar manner to that described for **8.40**,[10,43] with resolution of enantiomers by enzymatic methods.[43,45] As with adenallene, the (*R*)-(–)-enantiomer of cytallene (**8.4**) was the most active, with a spectrum of antiviral activity (in particular HIV and HBV) and mechanism of action comparable with that observed for L-DDC (**5.22**).[46]

8.5.2 A-5021

A-5021 (**8.39**, (1'*S*,2'*R*)-9-{[1',2'-bis(hydroxymethyl)cyclopropyl-1'-yl]methyl} guanine), is a guanosine analogue with potent activity against HSV, VZV and CMV.[44,47] The strong antiviral activity of A-5021 was shown to result from a more rapid and stable accumulation of its triphosphate in infected cells compared with ACV, and on a stronger inhibition of viral DNA polymerase by its triphosphate compared with PCV.[48]

The asymmetric synthesis[44] of A-5021 commenced from the cyclopropane lactone (**8.46**), prepared by the reaction of (*R*)-(–)-epichlorohydrin (**8.45**) with diethylmalonate.[49] The lactone was then reduced to the diol (**8.47**) which was subsequently protected as the acetonide (**8.48**). The ester group of **8.48** was reduced to the alcohol (**8.49**) which, through a series of protection/deprotection reactions (**8.49** → **8.54**), gave the required tosylate (**8.54**). Coupling of **8.54** with 2-amino-6-(benzyloxy)purine gave the protected nucleoside (**8.55**) which was deprotected to give the 1'*R*,2'*S*-enantiomer, A-5021 (**8.39**) (Scheme 8.9).

8.6 ACYCLIC NUCLEOSIDE PHOSPHONATES

The acyclic nucleoside phosphonates (ANPs) (Fig. 8.5) are stable mimics of acyclic nucleosides which display both a broad spectrum of antiviral activity and prolonged antiviral action.[50] The unique feature of the ANPs is their ability to by-pass the first rate-limiting phosphorylation step, which is normally carried out by virus-encoded kinases.

With the emergence of virus strains in which the phosphorylating enzyme is altered or lacking, such as TK- strains of HSV, this 'by-pass' ability of ANPs is particularly valuable.

8.6.1 Mechanism of Action

ANPs are active as their diphosphates, the purine ANPs (*e.g.* PMEA, **8.3**) are either converted directly to the active diphosphate (PMEA-DP) by 5-phosphoribosyl-1-pyrophosphate (PRPP) synthetase[51] or in a stepwise fashion by the action of adenosine monophosphate (AMP) kinase.[52] Conversion of the pyrimidine ANPs (*e.g.* HPMPC, **1.51**) to the active diphosphate (HPMPC-DP) is catalysed by pyrimidine nucleoside monophosphate (PNMP) kinase (Fig. 8.6, see also section 1.7.2.1).[53]

The diphosphorylated ANP (a triphosphate mimic) has a long intracellular half-life and selectively targets the viral DNA polymerase.[54] However, other enzymes may be inhibited by the ANP compounds, such as HSV-1 encoded ribonucleotide reductase, and certain ANP compounds also target reverse transcriptase and act as chain terminators.[55]

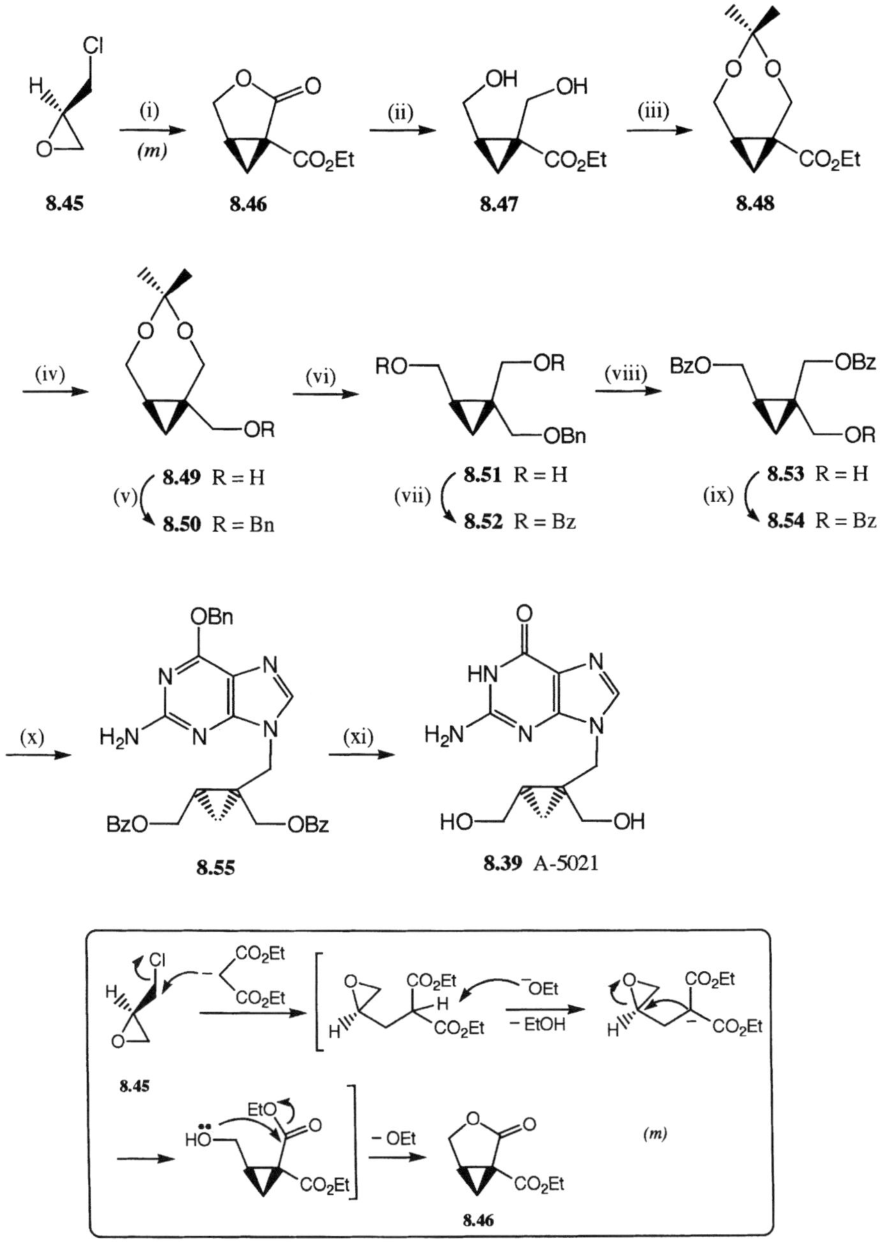

SCHEME 8.9 *REAGENTS AND CONDITIONS:* (I) DIETHYLMALONATE, NaOEt, 75°C, 20 H (II) NaBH$_4$, EtOH, 2 H (III) 2,2-DIMETHOXYPROPANE, *P*-TsOH•H$_2$O, DMF, 12 H (IV) LiBH$_4$, THF, 72°C, 12 H (V) BnBr, NaH, DMF, 14 H (VI) 1N HCl, THF, 0°C, 30 MIN. (VII) BzCl, CHCl$_3$, PYRIDINE, 0°C, 12 H (VIII) Pd/C, EtOH, ACOH, 3 DAYS (IX) *P*-TsCl, DMAP, CH$_2$Cl$_2$, 0°C, 1 H (X) 2-AMINO-6-(BENZYLOXY)PURINE, 18-CROWN-6, DMF, 60°C, 2 H (XI) NaH, MeOH, 30 MIN. THEN 1N HCl, 50°C, 30 MIN.

ACYCLIC NUCLEOSIDES

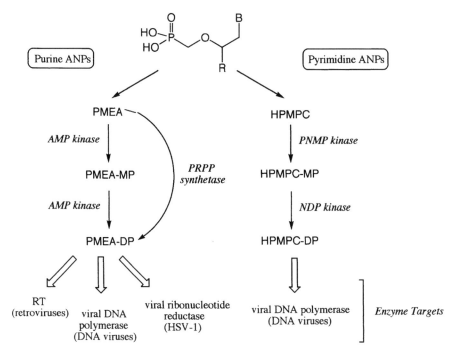

8.3 B = A, X = H, PMEA
8.56 B = G, X = H, PMEG
8.57 B = DAP, X = H, PMEDAP
8.58, B = A, X = F, F-PMEA

DAP = 2,6-diaminopurin-9-yl

8.59 B = A, X = OH,(S)-HPMPA
8.60 B = G, X = OH, (S)-HPMPG
8.61 B = DAP, X = OH, (S)-HPMPDAP
1.51 B = C, X = OH, (S)-HPMPC
8.62 B = A, X = H, (R)-PMPA
8.63 B = DAP, X = H, (R)-PMPDAP
8.64 B = A, X = F, (S)-FPMPA

FIGURE 8.5 ACYCLIC NUCLEOSIDE PHOSPHONATES.

FIGURE 8.6 MECHANISM OF ACTIVATION AND ENZYME TARGETS OF THE ANPs.

8.6.2 Synthesis of PMEA

The earlier syntheses of 9-(2-phosphonomethoxyethyl)adenine (**8.3**, PMEA) suffered from poor N^9:N^7 regioselectivity on coupling adenine with an appropriate electrophile.[56,57] This problem has been alleviated in the regio-defined synthesis of Dang et al.,[58] which involves substitution of 5-amino-4,6-dichloropyrimidine (**8.65**) with the phosphonate amine (**8.66**) to give compound **8.67** which was cyclised on treatment with diethoxymethyl acetate.[59] The 6-amino group was then introduced

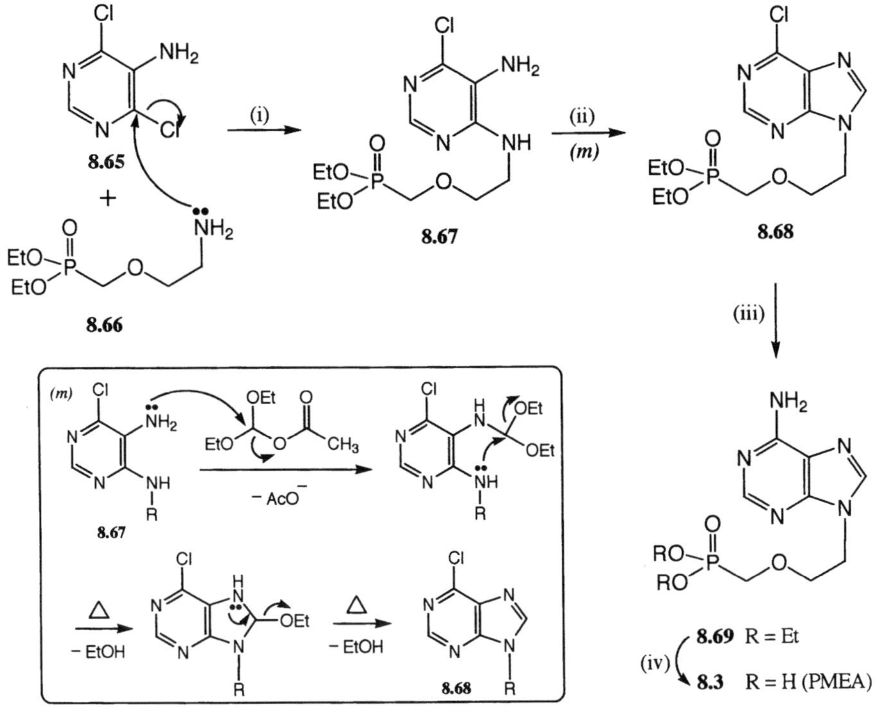

SCHEME 8.10 REAGENTS AND CONDITIONS: (I) Et₃N, EtOH, REFLUX, 20 H (II) DIETHOXYMETHYL ACETATE, 120°C, 3 H (III) NH₃ (lq.), THF-DMSO, STEEL BOMB, 24 H (IV) TMSBr, MeCN, 16 H.

by careful amination, with deprotection of the PMEA diethyl ether (**8.69**) achieved with TMSBr (Scheme 8.10).

Of the PME analogues prepared, PMEG (**8.56**) showed excellent activity against papilloma viruses,[60] PMEDAP (**8.57**) displayed antiviral activity against HIV, CMV and murine sarcoma virus (MSV).[61] One of the more recently prepared PME derivatives, the monoammonium salt of F-PMEA (**8.58**) was active against CMV, Epstein–Barr virus and measles.[62]

8.6.3 Synthesis of HPMPC and PMPA

Several syntheses of HPMPC (**1.51**, Cidofovir)[63–65] and PMPA (**8.62**) have been reported.[66–69] The inexpensive (R)- and (S)-glycidol (**8.70R** and **8.70S**) have proved valuable starting material for efficient, multi-gramme synthesis of both (S)-HPMPC (39% overall yield) and (R)-PMPA (30–35% overall yield),[65,69] using methodology which is applicable to the synthesis of other PMP/HPMP-pyrimidine and purine ANPs (Scheme 8.11).

The acyclic nucleosides (**8.73/8.76**) were obtained by reaction of the acyclic precursors (**8.71** and **8.75**) with the appropriate nucleobase (Scheme 8.11). The (S)-trityl ether (**8.71**) was readily prepared by reaction of (R)-glycidol with trityl chloride; subsequent coupling with benzoylated cytosine (**8.72**) in the presence of catalytic potassium carbonate gave **8.73** selectively owing to regiospecific opening

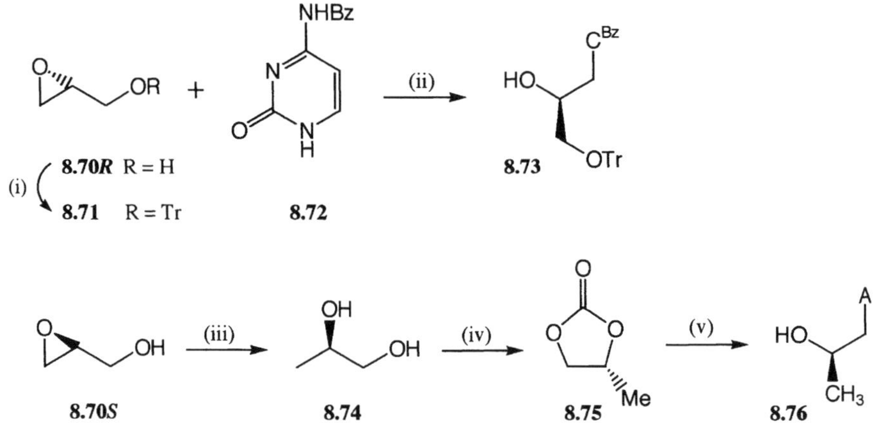

SCHEME 8.11 REAGENTS AND CONDITIONS: (I) TrCl, Et₃N, CH₂Cl₂, 3 H (II) K₂CO₃, DMF, 72°C, 5 H (III) H₂, 5% Pd/C, NaOH, EtOH (IV) (EtO)₂CO, NaOEt (V) ADENINE (1.3), NaOH, DMF, 140°C, 20 H.

of the epoxide. (*R*)-Propylene carbonate (**8.75**) was prepared in two steps from (*S*)-glycidol, involving initial hydrogenation to give (*R*)-1,2-propanediol (**8.74**). Condensation of **8.74** with diethyl carbonate afforded **8.75** which was coupled with adenine to give the acyclic nucleoside (**8.76**).

The resulting acyclic nucleosides (**8.73/8.76**) were reacted with tosyloxymethylphosphonate (**8.77**) to give the nucleotide esters (**8.78/8.79**). Deprotection of **8.78** was achieved in three steps to give (*S*)-HPMPC, while (*R*)-PMPA was obtained in one step by TMSBr cleavage of the ester groups of **8.79** (Scheme 8.12).

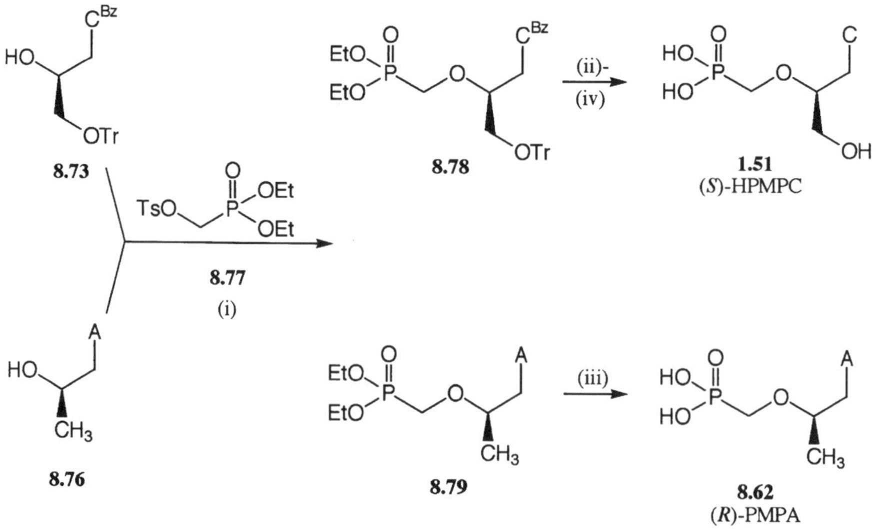

SCHEME 8.12 REAGENTS AND CONDITIONS: (I) TsOCH₂P(O)(OEt)₂ **8.77**, NaH OR ᵗBuOLi (II) HCl (G), CH₂Cl₂, 0–5°C, 10 MIN. (III) TMSBr, CH₂Cl₂, 18 H OR CH₃CN, REFLUX (IV) NH₄OH, 4 H.

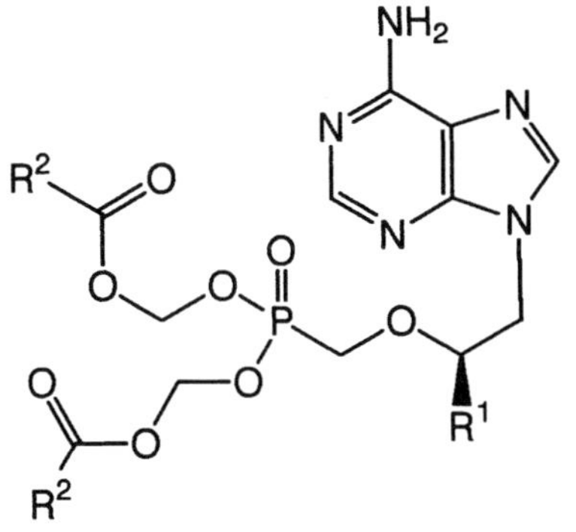

8.80 bis(POM)-PMEA : R^1 = H, R^2 = *t*-Bu
8.82 bis(POC)-PMPA : R^1 = Me, R^2 = OCH(Me)$_2$

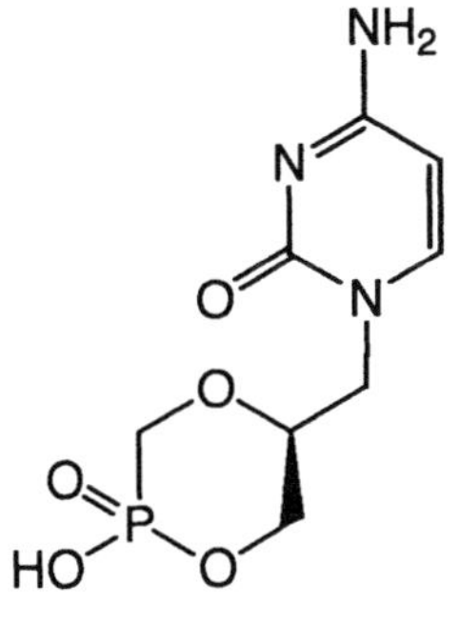

8.81 cHPMPC

FIGURE 8.7 ANP PRODRUGS IN CLINICAL TRIALS.

8.6.4 ANP Prodrugs

The ANPs suffer from poor oral bioavailability and also have a number of toxicity implications (*e.g.* hepatic toxicity, neutropenia and nephrotoxicity). To improve both the bioavailability and safety profile of the ANPs, a number of prodrugs have been developed, most notably, bis(POM)-PMEA (Adefovir dipivoxil®, **8.80**),[70] cyclic HPMPC (cHPMPC, **8.81**)[71] and bis(POC)-PMPA (**8.82**) (Fig. 8.7).[72] These prodrugs (**8.80–8.82**) are currently in clinical trials for the treatment of hepatitis B infections, CMV retinitis and HIV/AIDS respectively.

8.7 BASE-MODIFIED ACYLIC NUCLEOSIDES

A number of base-modified acyclic nucleosides have been prepared,[23,73] of these 1-[(2-hydroxyethoxy)methyl]-6-(phenylthio)thymine (**8.83**, HEPT)[74] and more recently the benzothiadiazine dioxide acyclic nucleoside (**8.5**)[11] and the imidazo[1,2-*a*]pyridine acyclo-*C*-nucleosides (*e.g.* **8.84**)[75] have shown the most promise as therapeutic agents (Fig. 8.8).

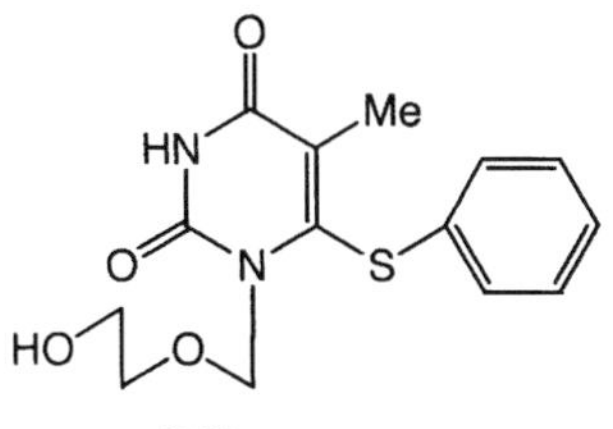

8.83 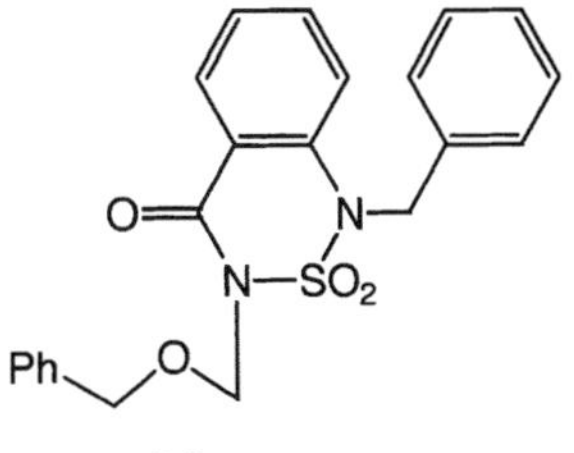   **8.5** 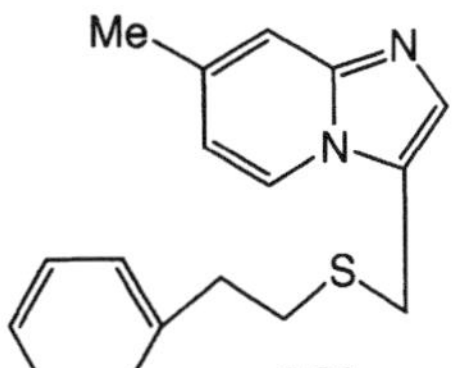   **8.84**

FIGURE 8.8 BASE-MODIFIED ACYLIC NUCLEOSIDES.

8.7.1 HEPT Acyclic Nucleosides

HEPT is a selective HIV-1 RT inhibitor which, as seen previously for TSAO-T (**1.46**), acts as a NNRTI.[76] Comprehensive structure-activity relationship studies have been undertaken with HEPT,[77-79] with potent activity observed for a number of compounds including 5-ethyl-1-ethoxymethyl-6-(phenylthio)uracil (**8.93**, E-EPU) and 5-isopropyl-1-ethoxymethyl-6-(benzyl)uracil (**8.96**, I-EBU).[80]

HEPT and its derivatives E-EPU and I-EBU are prepared by condensation of an appropriate acyclic moiety with a nucleobase (**1.7**, **8.85** and **8.86** respectively) to give the acyclic nucleosides (**8.87**, **8.90** and **8.91**), in the case of E-EPU and I-EBU

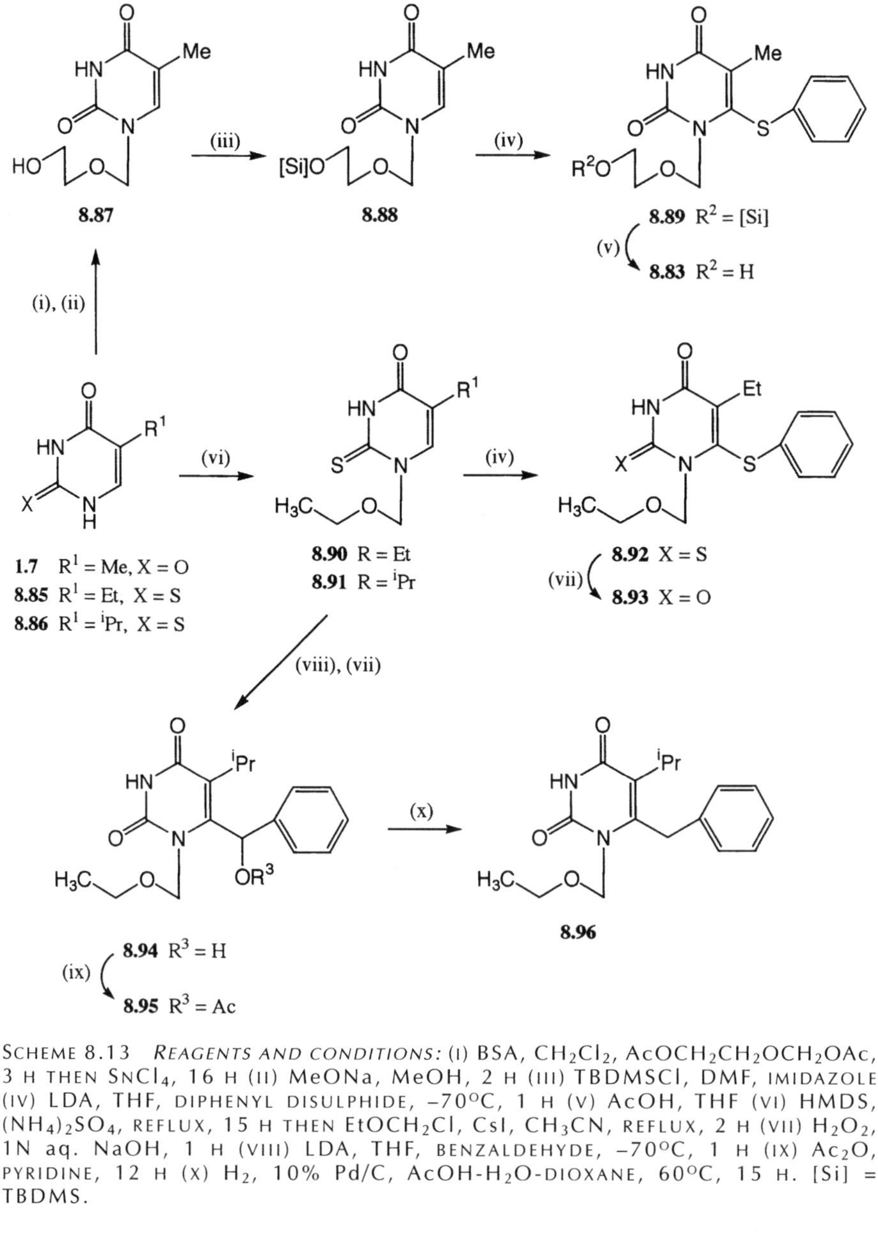

SCHEME 8.13 *REAGENTS AND CONDITIONS:* (I) BSA, CH_2Cl_2, $AcOCH_2CH_2OCH_2OAc$, 3 H THEN $SnCl_4$, 16 H (II) MeONa, MeOH, 2 H (III) TBDMSCl, DMF, IMIDAZOLE (IV) LDA, THF, DIPHENYL DISULPHIDE, −70°C, 1 H (V) AcOH, THF (VI) HMDS, $(NH_4)_2SO_4$, REFLUX, 15 H THEN $EtOCH_2Cl$, CsI, CH_3CN, REFLUX, 2 H (VII) H_2O_2, 1N aq. NaOH, 1 H (VIII) LDA, THF, BENZALDEHYDE, −70°C, 1 H (IX) Ac_2O, PYRIDINE, 12 H (X) H_2, 10% Pd/C, $AcOH-H_2O$-DIOXANE, 60°C, 15 H. [Si] = TBDMS.

the 2-thiouracil nucleobases were used to improve the yield of the condensation step.[81,82] Introduction of the C-6 moiety, phenylthio- or benzyl-, was achieved by initial formation of the C-6 lithiated species with subsequent reaction with diphenyl disulphide or benzaldehyde to give the 6-substituted derivatives (**8.89**, **8.92** and **8.94**). Acidic deprotection of **8.89** gave HEPT (**8.83**), while oxidative hydrolysis of the thione function of **8.92** gave E-EPU (**8.93**). The acyclic nucleoside (**8.94**) was first converted to its acetate (**8.95**) which was then subjected to hydrogenolysis in the presence of 10% Pd/C to give the benzyl derivative I-EBU (**8.96**) (Scheme 8.13).

However, resistance to HEPT and HEPT derivatives, resulting from mutations at multiple sites in the HIV-1 reverse transcriptase, has been found.[83] The rapid emergence of drug-resistant isolates, both in cell cultures and patients treated with this class of compound, represents a major obstacle to the further development of HEPT acyclic nucleosides.

8.7.2 Benzothiadiazine Dioxide Acyclic Nucleosides

Of more recent interest are the benzothiadiazine dioxide acyclonucleosides, such as **8.5** which displays activity against both CMV and HIV (types 1 and 2). The target site of the benzothiadiazine dioxide acyclonucleosides is as yet undetermined, however its structure would suggest this class of compounds might act as NNRTIs.

The benzothiadiazine dioxide ring (**8.99**) was prepared by the reaction of methyl anthranylate (**8.7**) with sulphamoyl chloride (**8.98**). Coupling of silylated **8.99** with acetoxymethylbenzyl ether under Lewis acid conditions resulted in the regioselective synthesis of the *N*-3 acyclic nucleoside (**8.100**), which was benzylated to yield the product (**8.5**) (Scheme 8.14).[11]

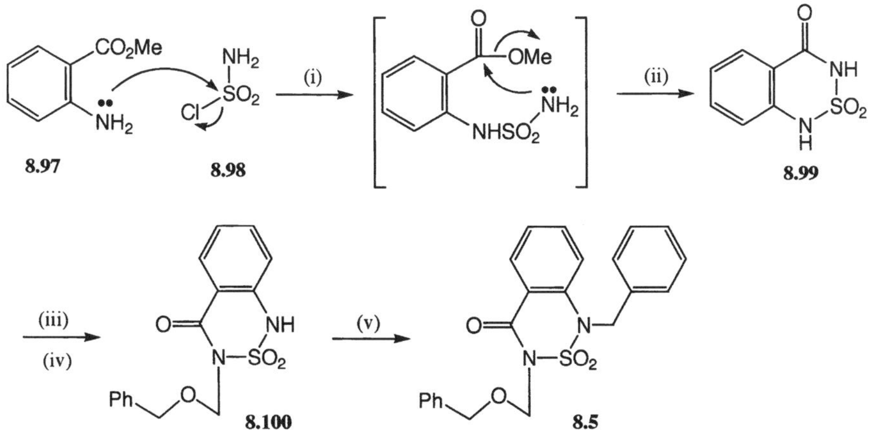

SCHEME 8.14 *REAGENTS AND CONDITIONS:* (I) C_6H_6 (II) 6N NaOH (III) HMDS, $(NH_4)_2SO_4$, CH_3CN, REFLUX (IV) ACETOXYMETHYLBENZYL ETHER, CH_2Cl_2, $BF_3 \bullet OEt_2$ (V) BnBr, aq. $NaHCO_3$, REFLUX, 4 H.

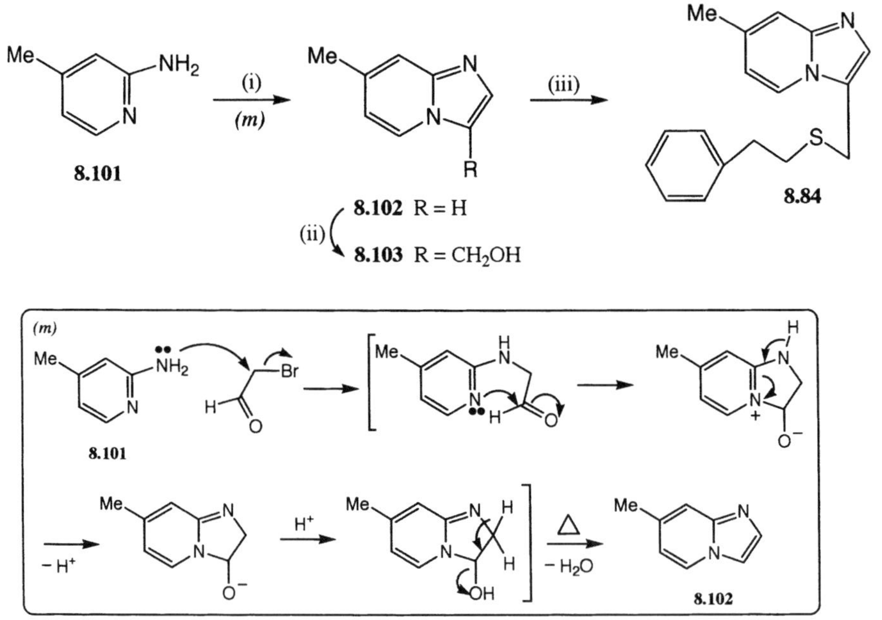

SCHEME 8.15 REAGENTS AND CONDITIONS: (I) BrCH$_2$CHO, EtOH, REFLUX (II) HCHO, AcOH, AcONa, REFLUX, 4 H (III) BnSH, AcOH, 100°C, 2 H.

8.7.3 Imidazo[1,2-a]pyridine Acyclic C-Nucleosides

The heterocyclic moiety, 7-methyl-imidazo[1,2-*a*]pyridine (**8.102**) was prepared by condensation of 2-amino-4-methylpyridine (**8.101**) with bromoacetaldehyde.[84] Prolonged refluxing of **8.102** with formaldehyde, sodium acetate and acetic acid resulted in direct hydroxymethylation[85] to give the acyclic *C*-nucleoside (**8.103**), which was converted to the thioether (**8.84**) on reaction with benzylmercaptan (Scheme 8.15).[75] The imidazo[1,2-*a*]pyridine acyclic *C*-nucleosides exhibited pronounced activity against CMV and VZV, although the initial compounds in this series displayed considerable toxicity.[84]

Structure-activity relationship studies have allowed identification of those functional moieties required for activity and those responsible for toxicity resulting in the development of compounds, such as **8.84**, with improved toxicity profiles.[75] A large number of acyclic nucleosides have been prepared with considerable variation in both the heterocyclic and acyclic components, the syntheses of which have been described in the comprehensive reviews of E.S.H. El Ashry and Y. El Kilany.[73, 86–87]

The antiviral activity of acyclic nucleosides has also been reviewed by a number of authors.[88,89]

8.8 REFERENCES

1. G.B. Elion, P.A. Furman, J.A. Fyfe, P. de Miranda, L. Beauchamp, H.J. Schaeffer, *Proc. Natl. Acad. Sci. USA*, **1977**, *74*, 5716–5720.
2. H.J. Schaeffer, L. Beauchamp, P. de Miranda, G.B. Elion, D.J. Bauer, P. Collins, *Nature*, **1978**, *272*, 583–585.
3. A.J. Wagstaff, D. Faulds, K.L. Goa, *Drugs*, **1994**, *47*, 153–205.

4. K.O. Smith, K.S. Galloway, W.L. Kennell, K.K. Ogilvie, B.K. Radatus, *Antimicrob. Ag. Chemother.*, **1982**, *22*, 55–61.
5. M.R. Boyd, T.H. Bacon, D. Sutton, M. Cole, *Antimicrob. Ag. Chemother.*, **1987**, *31*,1238–1242.
6. C.M. Perry, A.J. Wagstaff, *Drugs*, **1995**, *50*, 396–415.
7. C.M. Perry, D. Faulds, *Drugs*, **1996**, *52*, 754–772.
8. E. De Clercq, A. Holy, I. Rosenberg, T. Sakuma, J. Balzarini, P.C. Maudgal, *Nature*, **1986**, *323*, 464–467.
9. E. De Clercq, T. Sakuma, M. Baba, R. Pauwells, J. Balzarini, I. Rosenberg, A. Holy, *Antiviral Res.*, **1987**, *8*, 261–272.
10. S. Phadtare, J. Zemlicka, *J. Am. Chem. Soc.*, **1989**, *111*, 5925–5931.
11. A. Martinez, A.I. Esteban, A. castro, C. Gil, S. Conde, G. Andrei, R. Snoeck, J. Balzarini, E. De Clercq, *J. Med. Chem.*, **1999**, *42*, 1145–1150.
12. J. Balzarini, R.F. Schinazi, D. Kinchington, *Int. Antiviral News*,**1997**, *5*, 74–81.
13. K. Stenberg, A. Larsson, R. Datema, *J. Biol. Chem.*, **1986**, *261*, 2134–2139.
14. J.E. Reardon, T. Spector, *J. Biol. Chem.*, **1989**, *264*, 7405–7411.
15. M.J. Robins, P.W. Hatfield, *Can. J. Chem.*, **1982**, *60*, 547–553.
16. M. Senkus, *J. Am. Chem. Soc.*, **1946**, *68*, 734–736.
17. J.R. Barrio, J.D. Bryant, G.E. Keyser, *J. Med. Chem.*, **1980**, *23*, 572–574.
18. H. Shiragami, Y. Koguchi, Y. Tanaka, S. Takamatsu, Y. Uchida, T. Ineyama, K. Izawa, *Nucleosides & Nucleotides*, **1995**, *14*, 337–340.
19. J.L. Kelly, H.J. Schaeffer, *J. Heterocyclic Chem.*, **1986**, *23*, 271–273.
20. H. Matsumoto, C. Kaneko, K. Yamada, T. Takeuchi, T. Mori, Y. Mizuno, *Chem. Pharm. Bull.*, **1988**, *36*, 1153–1157.
21. L.M. Beauchamp, G.F. Orr, P. de Miranda, T. Burnette, T.A. Krenitsky, *Antiviral Chem. Chemother.*, **1992**, *3*, 157–164.
22. (a) G.M.T. van Wijk, K.Y. Hostetler, H. van den Bosch, *J. Lipid Res.*, **1992**, *33*, 1211–1219 (b) K.Y. Hostetler, S. Parker, C.N. Sridhar, M.J. Martin, J–L. Li, L.M. Stuhmiller, G.M.T. van Wijk, H. van den Bosch, M.F. Gardner, K.A. Aldern, D.D. Richman, *Proc. Natl. Acad. Sci. USA*, **1993**, *90*, 11835–11839.
23. E. De Clercq, R.T. Walker, In *Progress in Medicinal Chemistry*, G.P. Ellis, G.B. West (Eds.); Elsevier Science: Amsterdam, **1986**; *23*, pp 187–218 and references cited therein.
24. A. Markham, D. Faulds, *Drugs*, **1994**, *48*, 455–484.
25. E. Littler, A.D. Stuart, M.S. Chee, *Nature*, **1992**, *358*, 160–162.
26. K.K. Ogilvie, U.O. Cheriyan, B.K. Radatus, K.O. Smith, K.S. Galloway, W.L. Kennel, *Can. J. Chem.*, **1982**, *60*, 3005–3010.
27. A.K. Field, M.E. Davies, C. DeWitt, H.C. Perry, R. Liou, J. Germershausen, J.D. Karkas, W.T. Ashton, D.B.R. Johnston, R.L. Tolman, *Proc. Natl. Acad. Sci. USA*, **1983**, *80*, 4139–4143.
28. J.C. Martin, C.A. Dvorak, D.F. Smee, T.R. Matthews, J.P.H. Verheyden, *J. Med. Chem.*, **1983**, *26*, 759–761.
29. P. Ertl, W. Snowden, D. Lowe, W. Miller, P. Collins, E. Littler, *Antiviral Chem. Chemother.*, **1995**, *6*, 89–97.
30. R.A. Vere Hodges, Y-C. Cheng, *Antiviral Chem. Chemother.*, **1993**, *4 (Suppl. 1)*, 13–24.
31. R.A. Vere Hodge, *Antiviral Chem. Chemother.*, **1993**, *4*, 67–84.
32. M.R. Rashidi, J.A. Smith, S.E. Clarke, C. Beedham, *Drug Metabolism and Disposition*, **1997**, *25*, 805–813.
33. J. Hannah, R.L. Tolman, J.D. karkas, R. Liou, H.C. Perry, A.K. Field, *J. Het. Chem.*, **1989**, *26*, 1261–1271.
34. M.R. Harnden, R.L. Jarvest, T.H. Bacon, M.R. Boyd, *J. Med. Chem.*, **1987**, *30*, 1636–1642.
35. B.M. Choudary, G.R. Geen, T.J. Grinter, F.S. MacBeath, M.J. Parratt, *Nucleosides & Nucleotides*, **1994**, *13*, 979–996.
36. G.R. Geen, P.M. Kincey, *Tetrahedron Lett.*, **1992**, *33*, 4609–4612.
37. G.R. Geen, T.J. Grinter, P.M. Kincey, R.L. Jarvest, *Tetrahedron*, **1990**, *46*, 6903–6914.
38. B.M. Choudary, G.R. Geen, P.M. Kincey, M.J. Parratt, *Nucleosides & Nucleotides*, **1996**, *15*, 981–994.
39. B. Brand, C.B. Reese, Q. Song, C. Visintin, *Tetrahedron*, **1999**, *55*, 5239–5252.
40. D-K. Kim, M. Lee, D.H. Ryu, Y–W. Kim, J–S. Kim, K. Chang, G–J. Im, W–S. Choi, Y–B. Cho, K.H. Kim, D. Colledge, S. Locarnini, *Bioorg. Med. Chem.*, **1999**, *7*, 1715–1725.
41. D-K. Kim, N. Lee, Y-W. Kim, K. Chang, J-S. Kim, G-J. Im, W-S. Choi, W.S. Jung, T-S. Kim, Y-Y. Hwang, D-S. Min, K.A. Um, Y-B. Cho, K.H. Kim, *J. Med. Chem.*, **1998**, *41*, 3435–3441.
42. S. Megati, Z. Goren, J.V. Silverton, J. Orlina, H. Nishimura, T. Shirasaki, H. Mitsuya, J. Zemlicka, *J. Med. Chem.*, **1992**, *35*, 4098–4104.
43. B.C.N.M. Jones, J.V. Silverton, C. Simons, S. Megati, H. Nishimura, Y. Maeda, H. Mitsuya, J. Zemlicka, *J.Med.Chem.*, **1995**, *38*, 1397–1405.
44. T. Sekiyama, S. Hatsuya, Y. Tanaka, M. Uchiyama, N. Ono, S. Iwayama, M. Oikawa, K. Suzuki, M. Okunishi, T. Tsuji, *J. Med. Chem.*, **1998**, *41*, 1284–1298.
45. B.C.N.M. Jones, C. Simons, H. Nishimura, J. Zemlicka, *Nucleosides & Nucleotides*, **1995**, *14*, 431–434.

46. Y.L. Zhu, S.B. Pai, S.H. Liu, K.L. Grove, B.C.N.M. Jones, C. Simons, J. Zemlicka, *Antimicrobial Agents Chemother.*, **1997**, *41*, 1755–1760.
47. S. Iwayama, N. Ono, Y. Ohmura, K. Suzuki, M. Aoki, H. Nakazawa, M. Oikawa, T. Kato, M. Okunishi, Y. Nishiyama, K. Yamanishi, *Antimicrobial Agents Chemother.*, **1998**, *42*, 1666–1670.
48. N. Ono, S. Iwayama, K. Suzuki, T. Sekiyama, H. Nakazawa, T. Tsuji, M. Okunishi, T. Daikoku, Y. Nishiyama, *Antimicrobial Agents Chemother.*, **1998**, *42*, 2095–2102.
49. M.C. Pirrung, S.E. Dunlap, U.P. Trinks, *Helv. Chim. Acta*, **1989**, *72*, 1301–1310.
50. L. Naesens, E. De Clercq, *Nucleosides & Nucleotides*, **1997**, *16*, 983–992.
51. J. Balzarini, E. De Clercq, *J. Biol. Chem.*, **1991**, *266*, 8686–8689.
52. A. Merta, J. Votruba, J. Jindrich, A. Holy, T. Cihlár, I. Rosenberg, M. Otmar, T.Y. Herve, *Biochem. Pharmacol.*, **1992**, *44*, 2067–2077.
53. H–T. Ho, K.L. Woods, J.J. Bronson, H. De Boeck, J.C. Martin, M.J.M. Hitchcock, *Molec. Pharmacol.*, **1992**, *41*, 197–202.
54. J.M. Cherrington, S.J.W. Allen, N. Bischofberger, R.E. Benveniste, R. Black, *Antiviral Chem. Chemother.*, **1995**, *6*, 217–221.
55. J. Cerny, I. Votruba, V. Vonka, I. Rosenberg, M. Otmar, A. Holy, *Antiviral Res.*, **1990**, *13*, 253–264.
56. A. Holy, I. Rosenberg, *Collect. Czech. Chem. Commun.*, **1987**, *52*, 2801–2809.
57. A. Holy, I. Rosenberg, H. Dvorakova, *Collect. Czech. Chem. Commun.*, **1989**, *54*, 2190–2210.
58. Q. Dang, Y. Liu, M.D. Erion, *Nucleosides & Nucleotides*, **1998**, *17*, 1445–1451.
59. J.A. Montgomery, C. Temple, *J. Am. Chem. Soc.*, **1957**, *79*, 5238–5242.
60. J.W. Kreider, K. Balogh, R.O. Olson, J.C. Martin, *Antiviral Res.*, **1990**, *14*, 51–58.
61. L. Naesens, J. Neyts, J. Balzarini, A. Holy, I. Rosenberg, E. De Clercq, *J. Med. Virol.*, **1993**, *39*, 167–172.
62. W. Chen, M.T. Flavin, R. Filler, Z–Q. Xu, *J. Chem. Soc., Perkin Trans. 1*, **1998**, 3979–3988.
63. A. Holy, I. Rosenberg, *Collect. Czech. Chem. Commun.*, **1982**, *47*, 3447–3463.
64. A. Holy, M. Masojudkova, *Collect. Czech. Chem. Commun.*, **1995**, *60*, 1196–1212.
65. L.M. Schultze, H.H. Chapman, N.J.P. Dubree, R.J. Jones, K.M. Kent, T.T. Lee, M.S. Louie, M.J. Postich, E.J. Prisbe, J.C. Rohloff, R.H. Yu, *Tetrahedron Lett.*, **1998**, *39*, 1853–1856.
66. J.J. Bronson, I. Guazzouli, M.J.M. Hitchcock, R.R. Webb, J.C. Martin, *J. Med. Chem.*, **1989**, *32*, 1457–1463.
67. J.J. Bronson, L.M. Ferrara, H.G. Howell, P.R. Brodfeuhrer, J.C. Martin, *Nucleosides & Nucleotides*, **1990**, *9*, 745–769.
68. A. Holy, I. Rosenberg, H. Dvorakova, *Collect. Czech. Chem. Commun.*, **1989**, *54*, 2470–2501.
69. P.R. Brodfeuhrer, H.G. Howell, C.Sapino, P. Vemishetti, *Tetrahedron Lett.*, **1994**, *35*, 3243–3246.
70. L. Naesens, J. Balzarini, N. Bischofberger, E. De Clercq, *Antimicrobial Agents Chemother.*, **1996**, *40*, 22–28.
71. N. Bischofberger, M.J.M. Hitchcock, M.S. Chen, D.B. Barkhimer, K.C. Cundy, K.M. Kent, S.A. Lacy, W.A. Lee, Z.H. Li, D.B. Mendel, D.F. Smee, J.L. Smith, *Antimicrobial Agents Chemother.*, **1994**, *38*, 2387–2391.
72. R.V. Srinivas, A. Fridland, *Antimicrob. Agents Chemother.*, **1998**, *42*, 1484–1487.
73. E.S.H. El Ashry, Y. El Kilany, *Adv. Het. Chem.*, **1998**, *69*, 129–215.
74. T. Miyasaka, H. Tanaka, M. Baba, H. Hayakawa, R.T. Walker, J. Balzarini, E. De Clercq, *J. Med. Chem.*, **1989**, *32*, 2507–2509.
75. A. Gueiffier, S. Mavel, M. Lhassani, A. Elhakmaoui, R. Snoeck, G. Andrei, O. Chavignon, J–C. Teulade, M. Witvrouw, J. Balzarini, E. De Clercq, J-P. Chapat, *J. Med. Chem.*, **1998**, *41*, 5108–5112.
76. M. Baba, H. Tanaka, E. De Clercq, R. Pauwels, J. Balzarini, D. Schols, H. Nakashima, C–F. Perno, R.T. Walker, T. Miyasaka, *Biochem. Biophys. Res. Commun.*, **1989**, *165*, 1375–1381.
77. H. Tanaka, M. Baba, H. Hayakawa, T. Sakamaki, T. Miyasaka, M. Ubasawa, H. Takashima, K. Sekiya, I. Nitta, S. Shigeta, R.T. Walker, J. Balzarini, E. De Clercq, *J. Med. Chem.*, **1991**, *34*, 349–357.
78. H. Tanaka, M. Baba, M. Ubasawa, H. Takashima, K. Sekiya, I. Nitta, S. Shigeta, R.T. Walker, E. De Clercq, T. Miyasaka, *J. Med. Chem.*, **1991**, *34*, 1394–1399.
79. H. Tanaka, H. Takashima, M. Ubasawa, K. Sekiya, I. Nitta, M. Baba, S. Shigeta, R.T. Walker, E. De Clercq, T. Miyasaka, *J. Med. Chem.*, **1992**, *35*, 337–345.
80. J. Balzarini, M. Baba, E. De Clercq, *Antimicrob. Agents Chemother.*, **1995**, *39*, 998–1002.
81. H. Tanaka, H. Takashima, M. Ubasawa, K. Sekiya, I. Nitta, M. Baba, S. Shigeta, R.T. Walker, E. De Clercq, T. Miyasaka, *J. Med. Chem.*, **1992**, *35*, 4713–4719.
82. H. Tanaka, H. Takashima, M. Ubasawa, K. Sekiya, N. Inouye, M. Baba, S. Shigeta, R.T. Walker, E. De Clercq, T. Miyasaka, *J. Med. Chem.*, **1995**, *38*, 2860–2865.
83. R.W. Buckheit, V. Fliakas-Boltz, S. Yeagy-Bargo, O. Weislow, D.L. Mayers, P.L. Boyer, S.H. Hughes, B-C. Pan, S-H. Chu, J.P. Bader, *Virology*, **1995**, *210*, 186–193.
84. A. Gueiffier, M. Lhassani, A. Elhakmaoui, R. Snoeck, G. Andrei, O. Chavignon, J-C. Teulade, A. Kerbal, E. Mokhtar Essassi, J-C. Debouzy, M. Witvrouw, Y. Blache, J. Balzarini, E. De Clercq, J-P. Chapat, *J. Med. Chem.*, **1996**, *39*, 2865–2859.

85. J-C. Teulade, P.A. Bonnet, J.N. Rieu, H. Viols, J-P. Chapat, G. Grassy, A. Carpy, *J. Chem. Res. (S)*, **1986**, 202–203.
86. E.S.H. El Ashry, Y. El Kilany, *Adv. Het. Chem.*, **1997**, *67*, 391–438.
87. E.S.H. El Ashry, Y. El Kilany, *Adv. Het. Chem.*, **1997**, *68*, 1–88.
88. N.G. Johansson, *Adv. Antiviral Drug Res.*, **1993**, *1*, 87.
89. S. Freeman, J.M. Gardiner, *Mol. Biotechnology*, **1996**, *5*, 125–137.

INDEX

Abacavir (Ziagen™), 137, 145, 146, 148
Acadenosine® (see AICAR)
Acetobromofuranosyl nucleosides, 48
Acetoxymethoxybenzyl ether, 171
2-Acetoxymethoxy-1,3-diacetoxypropane, 159
Acetylation, 114, 161
N-Acetyl cytosine, 39, 105
Acetylene, 85, 87
Acid-catalyzed fusion procedure, 85, 110
ACV (see Acyclovir)
Acyclic nucleosides, 22, 155, 156, 164, 167, 168, 169, 170, 171, 172
   conformationally constrained, 156, 161
Acyclic C-nucleosides, 172
Acyclovir (ACV, Zovirax®), 155, 156, 157, 158
   diphosphate dimyristoylglycerol, 158
   mechanism of action, 22
   monophosphate, 158
   prodrugs, 157
   triphosphate, 156, 160, 164
Acyclic nucleoside phosphonates, 156, 164, 167, 169
   mechanism of action, 22, 164
   prodrugs, 169
Acylated glycosyl halides, 30
   acylated arabinosyl halide, 31
   acylated ribosyl bromide, 29
   acylated ribosyl chloride, 29

Acylation, 77, 156
Acyl glycosides, 30
   acetylated ribose, 93
   acylated ribose, 85, 129
Adefovir® (see PMEA)
Adefovir dipivoxil® (see Bis(POM)-PMEA)
Adenallene, 161, 162, 164
Adenine (6-aminopurine), 3, 5, 19, 65, 105, 138, 141, 162, 166, 168
Adenosine, 13, 25, 46, 48, 91, 127, 129, 133, 137, 149, 161
   5'-diphosphate (ADP), 14
   5'-monophosphate (AMP), 24
   synthesis, 29
   5'-triphosphate (ATP), 13, 15
Adenosine deaminase, 88, 89, 94, 129, 138, 140, 157, 162
Adenosine kinase, 138
Adenosine monophosphate (AMP) kinase, 164
Adenosine regulation, 83
S-Adenosyl-L-homocysteine (see AdoHcy hydolase)
S-Adenosylmethionine (SAM), 25, 26, 92
AdoHcy hydrolase, 24, 25, 137, 142
   inhibition of, 26
AICAR (Acadenosine®), 15, 83, 87
AICAR transformylase, 120
AIDS (Acquired Immunodeficiency Syndrome), 1, 20, 47, 169
Aldehyde oxidase, 160

INDEX

Aldol condensation, 146
Alkene, 38, 58
5-Alkenyl-nucleosides, 58
N^9-Alkyladenine, 162
Alkylation, 69, 74, 162
Alkylidenecarbene, 138
5-Alkyne-2'-deoxyuridine nucleosides, 97
5-Alkynyl-nucleosides, 58
Allene, 161, 162
Allylic rearrangement, 141
Amination, 167
2-Amino-6-(benzyloxy)purine, 164
4-Amino-6-bromo-5-cyanopyrrolo[2,3-d]pyrimidine, 89
2-Amino-6-chloropurine, 41, 105, 148, 157, 161
2-Amino-2-cyanoacetate, 122
2-Amino-4,6-dichloropyrimidine, 41
5-Amino-4,6-dichloropyrimidine, 166
2-Amino-4-methylpyrimidine, 172
2-Amino-6-oxypurine (see Guanine)
6-Aminopurine (see Adenine)
4-Aminopyrrolo[2,1-f]1,2,4-triazin-7-yl-C-nucleosides, 127
Aminosugar, 40, 41, 49, 143
Ammonolysis, 157
Anchimeric assistance, 39
1,5-Anhydro-4,6-O-benzylidene-3-deoxy-D-glucitol, 77
1,6-Anhydro-gulopyranose, 105
1,6-D-Anhydrohexitol nucleosides, 75, 77
L-Anhydrohexitol nucleosides, 112
  3-azido-L-hexapyranosyl nucleoside, 114
1,6-Anhydro-D-mannopyranose, 71
Anhydronucleosides, 44, 46, 47, 51
  2,2'-anhydronucleosides, 44, 49
    2,2'-anhydro-β-L-uridine, 107
  2,3'-anhydronucleosides, 44, 49
Anisyl tellurium sugar, 118
Annulation, 129
Anomeric carbon/centre, 3, 33

Anomeric configuration, 30
Antibacterial, 91, 94, 103, 118, 126, 129
Antibiotic, 1, 4, 75, 96
Anticancer, 1, 2, 4, 24, 25, 46, 51, 55, 63, 83, 93, 94, 97, 126, 127, 129
Antifungal, 120
Antiherpetic, 3, 51, 145
Antileukaemia, 97, 124, 138
Antiprotozoan, 129
Antiretroviral, 145
Antitumour, 53, 54, 88, 89, 118, 120, 122, 124, 129
Antiviral, 1, 2, 7, 14, 19, 20, 24, 25, 60, 68, 70, 72, 75, 78, 83, 85, 91, 94, 96, 97, 103, 105, 110, 111, 114, 118, 120, 122, 127, 129, 137, 138, 140, 143, 144, 158, 160, 161, 164, 172
D-Arabinofuranose, 3
L-Arabinofuranosyl nucleosides, 109, 111
Arabinonucleosides, 53
  Ara-A (Vidarabine®, arabinosyladenosine), 1, 3, 46
  Ara-C (arabinosylcytidine), 1, 55
  Ara-FU (arabinosyl-5-fluorouridine), 44
  Arabinosylguanine, 24
  Ara-U (arabinosyluridine), 55
L-Arabinose, 106, 107, 110
Aristeromycin, 137, 142
  2'-deoxyaristeromycin, 145
Aromatisation, 121
Aspartate, 15
Asymmetric synthesis, 70, 71, 104, 105, 138, 146, 148, 150, 164
8-Azaadenine, 94
2-Azaadenosine (2-AzaAR), 94
5-Azacytidine, 83, 96
  2'-deoxy-5-azacytidine, 96
2-Aza-3-deazapurine nucleosides, 94
8-Aza-1-deazapurine nucleosides, 94

INDEX

8-Azaguanine, 93
4'-Azanucleosides, 66, 68, 72
    4'-aza-ribose nucleosides, 68
    4'-aza-xylose nucleosides, 68
8-Azapurine nucleosides, 93, 94
    8-azaadenosine, 93, 94
    8-azainosine, 1, 93
    8-azapurine 2'-deoxyribonucleosides, 94
    8-azapurine 2',3'-dideoxyribonucleosides, 94
Azapyrimidine nucleosides, 94
6-Azapyrimidine nucleosides, 96
    6-azacytidine, 97
    6-azathymidine, 97
    6-azauridine, 96, 97
Azetidine nucleosides, 77
Azidonucleosides, 49
    5'-Azido-5'-deoxy-thymidine, 50
    AZT (Zidovudine®), 1, 14, 49
    mechanism of action, 20
Azidosugars, 49, 87
Azole C-nucleosides, 118
AZT (Zidovudine®, see Azidonucleosides)

Barton-McCombie procedure, 47
BCH-189 (see 2',3'-Dideoxy-3'-thiacytidine)
Benzimidazole nucleosides, 83, 87
    1263W94, 110
    BDCRB, 87, 88
    TCRB, 83, 87, 88
Benzothiadiazine dioxide acyclic nucleosides, 156, 169, 171
Benzoyloxyacetaldehyde, 71
Benzoyl peroxide, 71
Benzylation, 159, 171
6-Benzyloxy-2-aminopurine, 144
BHCG (Lobucavir), 150
    mechanism of action, 22
Bicyclo-isoxazole nucleosides, 73, 74
Biosynthesis, 19
    de novo synthesis, 15
    of purine nucleotides, 24
    of pyrimidine nucleotides, 26
    of UMP, 26
Bis(benzonitrile)palladium dichloride, 85
Bis(*para*-nitrophenyl)phosphate, 85
Bis(POM)-PMEA (Adefovir dipivoxil®), 169
Bis(POC)-PMPA, 169
N,O-Bis(trimethylsilyl)acetamide (BSA), 35
Blood-platelet aggregation, 89
Borane-methyl sulphide complex, 107
Boron trifluoride, 123
Bredinin (Mizoribine®), 83
Bromination, 143
Bromoacetaldehyde, 71
8-Bromoadenosine, 60
5-Bromocytosine, 105
Bromosugar, 48, 114, 121
γ-Butyrolactone, 107
BVDU, (*E*)-5-(2-bromovinyl)-2'-deoxyuridine, 68
    mechanism of action, 22
    4'-S-BVDU, 68

4(5)-Carbamoylimidazolium-5(4)-olate, 85
Carba-oxetanocin G (see BHCG)
Carbene, 138
Carboacyclic nucleosides, 22, 160
Carbocyclic nucleosides, 40, 137, 141, 143, 145
    BMS-200475, 137, 144
    conformationally constrained, 145
    cyclobutyl, 150
    cyclopropyl, 149
    DHCeA, 26
    D_4–D-nucleosides, 145
    D_4–L-nucleosides, 148, 149
Carbon tetrahalides, 51
N,N'-Carbonyldiimidazole (CDI), 126
Carbovir, 137, 145, 146, 148
    triphosphate, 146

Carboxyamidation, 60
C-arylglycosides, 132, 133
(+)-C-BVDU, 143
C-EDU, 143
Cell division/cycle, 23
Cellular kinases, 22
Cesium carbonate, 74
Chain termination, 19, 20, 22, 164
C-H insertion, 138
Chiral auxillary, 148
Chiral resolution, 143, 162, 164
Chiral template, 71
Chlorination, 88
4-Chlorobenzene diazonium chloride, 41
5-Chlorocytosine, 105
7-Chloro-1*H*-imidazo[4,5-*b*]pyridin-5-acetylamine, 91
2-Chloro-6-iodopurine, 157
Chloromercuri procedure (see Nucleoside synthesis)
Chloromethylation, 160
6-Chloropurine, 35, 105, 149
4-Chloropyrrolo[2,3-*d*]pyrimidine, 88
Chlorosugar, 35, 40
Cidofovir (see HPMPC)
Cine-substitution, 129
Clinical trials, 51, 110, 120, 145, 146, 150, 169
C-Nucleosides, 117, 118, 127
2'/3'-*C*-nucleosides, 53
  CNDAC, 53, 55
  DMDC, 53, 55
  ECyd, 53, 54
  EUrd, 53, 54
  mechanism of action, 54
Combination therapy, 103
Competitive inhibitor, 2, 19, 20, 22, 85, 92, 145
Conformation, 5, 6, 7, 9, 10, 13, 66, 77, 109, 111, 145
  anti/syn, 6
  C4'–C5' bond geometry, 6
  envelope and twist, 6, 145

North (N)/South (S) type, 7, 10, 13, 145
  ring puckering, 6, 7
Copper iodide, 97
Cordycepin, 4
Corey-Winter procedure, 48
Coupling methods (see Nucleoside synthesis)
Coxsackie B1 virus, 94
Cross-coupling reaction, 85
Curtius rearrangement, 150
Cyanogen bromide, 88
5-Cyanomethyl-4-carbomethoxyimidazole, 93
Cyclisation, 41, 64, 66, 74, 88, 89, 91, 92, 109, 111, 129, 133, 161, 166
Cycloadditions, 74, 75
Cyclopentylcytosine (C-Cyd), 27
Cyclopropanation, 145, 149
Cyclopropane, 161
Cymevene® (see Ganciclovir)
Cytallene, 156, 161, 162, 164
Cytidine, 10, 14, 51, 55, 98
  5'-diphosphate (CDP), 27
  synthesis, 29
  5'-triphosphate (CTP), 17, 26, 27
Cytidine deaminase, 29, 106
Cytidine 5'-triphosphate (CTP) synthetase, 17, 26, 27, 97
Cytomegalovirus (CMV), 22, 83, 87, 94, 97, 124, 145, 150, 156, 158, 164, 169, 171, 172
Cytosine (2-oxy-4-aminopyrimidine), 3, 5, 19, 55, 71, 72, 97, 105, 161

DAST (diethylaminosulphur trioxide), 51, 52
DDC (Zalcitabine®, see 2',3'-Dideoxy-D-nucleosides)
DDI (Didanosine®, see 2',3'-Dideoxy-D-nucleosides)
DDQ (2,3-dichloro-5,6-dicyano-1,4-benzoquinone), 74

Deacetylation, 29, 143
Deamination, 106, 138, 140
3-Deazaadenosylhomocysteine, 92
9-Deazapurine C-nucleosides, 127, 129
    9-deazaadenosine, 127, 129
    9-deazainosine, 129
1-Deazapurine nucleosides, 89
    1-deazaadenosine, 89, 91
    1-deazaguanosine, 91
3-Deazapurine nucleosides, 91
    3-deazaadenosine, 91, 92
    3-deazaguanosine, 93
7-Deazapurine nucleosides, 88
3-Deazapyrimidine nucleosides, 94
    3-deazacytidine, 97
    3-deazathymidine, 97
    3-deazauridine, 97
Debenzylation, 124, 144
Debromination, 89
Decarbomethoxylation, 118
Decarboxylation, 15, 71
Decarboxylative ozonolysis, 137
Deformylation, 129
Dehalogenation, 29, 88, 114, 161
Dehydration, 86
De novo pathway, 54, 97
    cytidine nucleotide biosynthesis, 15, 17
    purine ribonucleotide biosynthesis, 4, 15, 24, 83, 91, 120
    pyrimidine synthesis, 15, 96, 120
    thymidine nucleotide biosynthesis, 15
2'-Deoxy-carbocyclic nucleosides, 142
2'-Deoxycytidine 5'-diphosphate (dCDP), 27
Deoxycytidine kinase, 106
Deoxycytidylate kinase, 106
2'-Deoxycytidine 5'-triphosphate (dCTP), 27
3-Deoxy-diacetone-D-glucose, 77
Deoxygenation, 46, 47, 53, 65, 74, 107, 145
2'-Deoxy-2',2'-difluoro-L-erythro-pentofuranosyl nucleosides, 111

3-Deoxy-L-glucose, 114
2-Deoxy-D-glycosyl halide, 32
2'-Deoxyguanosine 5'-triphosphate (dGTP), 24
(N)-2'-Deoxy-methanocarba-nucleosides, 145
    (N)-2'-deoxy-methanocarba-A, 145
    (N)-2'-deoxy-methanocarba-T, 145
2'-Deoxy-3'-oxa-4'-thiocytidine (dOTC), 72
5'-Deoxypyrazofurin, 122
2-Deoxy-β-D-ribofuranose (2-Deoxyribose), 2, 3
3-Deoxy-β-D-ribofuranose, 4
2'-Deoxy-L-ribofuranosyl nucleosides, 110
3'-Deoxy-L-ribofuranosyl nucleosides, 111
1-Deoxyribose, 65
2-Deoxyribose nucleosides, 161
    ^{13}C NMR, 13
    conformation, 7
    2'-deoxyadenosine, 13, 35, 46
    2'-deoxyguanosine, 13
    2'-deoxyuridine, 55
    2'-deoxy-6-thioguanosine, 27
    ^{1}H NMR, 10
    synthesis, 35, 37, 38, 43
2'-Deoxy-D-4'-thionucleosides, 68
2'-Deoxy-L-4'-thionucleosides, 111
2-Deoxy-4-thioribose, 66, 67
2'-Deoxyuridine 5'-monophosphate (dUMP) 15, 17, 19
Dess-Martin reagent, 144
Dethioacetalization, 66
Dethiophenylation, 118
Detritylation, 44
$N^{2,9}$-Diacetylguanine, 159
3,4-Diamino-2,6-dichloropyrimidine, 92
2,6-Diaminopurine, 71, 157
3,4-Diaminopyrimidine, 89
Diarylcadmium reagents, 132
1,5-Diazabicyclo[4.3.0]nonene-5 (DBN), 129

INDEX

2,8-Diaza-3-deazapurine nucleosides, 94
1,4-Diazine C-nucleosides (see Pyrazine C-nucleosides)
Diazotization, 29, 85, 88
Diazo transfer, 121
1,3-Dibenzyloxy-2-chloromethoxypropane, 159
1,3-Di-O-benzyl-2-acetoxymethoxyglycerol, 159
2,8-Dichloroadenine, 29
1,4-Dichloro-2-butyne, 162
4,5-Dichloro-o-phenylenediamine, 88
2,6-Dichloropyrimidine, 29
Dicyclohexylcarbodiimide (DCC), 157
Didanosine® (DDI, see 2',3'-Dideoxy-D-nucleosides)
2',3'-Dideoxy-4'-azanucleosides, 68
2,3-Dideoxy-4-azasugars, 69
2',3'-Dideoxy-2',3'-didehydro-D-nucleosides (D_4-D-nucleosides), 47
  D_4A, 48
  D_4C, 48
  D_4T (Stavudine®), 1, 20
    mechanism of action, 20, 52
  D_4U, 48
2',3'-Dideoxy-2',3'-didehydro-L-nucleosides (D_4-L-nucleosides), 107, 108
  β-L-D_4C, 108
  β-L-2'-FD_4A, 108
  β-L-FD_4C, 108
  β-L-D_4U, 107
2',3'-Dideoxy-D-nucleosides, 20, 47, 69, 70
  2',3'-dideoxycytidine (DDC, Zalcitabine®), 1, 14, 48 106
    mechanism of action, 20
  2',3'-dideoxyinosine (DDI, Didanosine®), 1, 24
    mechanism of action, 20
  2',3'-dideoxythymidine (DDT), 52
    enzymatic synthesis, 43
    sulphur-mediated synthesis, 39

2',3'-Dideoxy-L-nucleosides, 70, 106
  β-L-DDC, 106, 164
  β-L-DDU, 107
  β-L-FDDC, 106
2,3-Dideoxy-L-ribose, 106
  2',3'-Dideoxy-3'-thiacytidine (BCH-189), 72
  2,3-Dideoxy-4-thiosugars, 69
  2,3-dideoxy-3-C-(hydroxymethyl)-4-thioribose, 66
  2,3-dideoxy-4-thioribose, 66
Diethoxymethyl acetate, 166
2,6-Diethyoxypyrimidine, 29
Diethyl 1,3-acetonedicarboxylate, 121
Diethyl carbonate, 168
Difluorinated nucleosides, 51
  2',2'-difluoro-2'-deoxycytidine (dFdC), 51
5,6-Dihydro-5-azathymidine (DHAdT), 96
2,3-Dihydrofuran[2,3-d]pyrimidin-2-one nucleosides, 97
Dihydrogen sulphide, 89
Dihydro-isoxazole nucleosides, 73, 74
9-(1,3-Dihydroxy-2-propoxymethyl) guanosine (see Ganciclovir)
Diiodomethane, 85
2,3:5,6-Di-O-isopropylidene-L-gulofuranose, 105
2,4-Dimethoxypyrimidin-5-yl C-riboside, 124
4-Dimethyaminopyridine (DMAP), 47, 157
4,6-Dimethylindole, 133
Dimethyl(methylthio)sulphonium tetrafluoroborate ($Me_2S(SMe)BF_4$), 65
L-Dioxolane nucleosides, 104, 105
1,3-Dioxolane, 71, 156, 157
1,3-Dioxolanyl-D-nucleosides, 70
  DAPD, 71
  dioxolane-T, 71
2,4-Dioxy-5-methylpyrimidine (see Thymine)

2,6-Dioxypurine (see Xanthine)
2,4-Dioxypyrimidine (see Uracil)
Diphenylphosphoryl azide (DPPA), 50
Diphenyl disulphide, 171
1,3-Dipolar cycloaddition, 74, 87
Directing group, 32, 37, 39, 67, 145
Dithiolane sugar, 71
DMDC (see 2'/3'-C-nucleosides)
DNA, 1, 14, 15, 19, 20, 23, 24, 87, 89, 106, 132
   composition of, 2, 3, 4
   mitochondrial, 106
   strand breakage, 22
   viruses, 22, 24, 94, 96, 103, 112
DNA polymerase, 22
DNA synthesis, inhibition of, 51, 54
D_4T (Stavudine®, see 2',3'-Dideoxy-2',3'-didehydro-D-nucleosides)
Duck hepatitis B virus, 74

ECyd (see 2'/3'-C-nucleosides)
EICAR, 24, 83, 85
Elimination, 39, 48, 65, 67, 77, 107, 120, 122, 132, 133
Enantioselective synthesis, 137
2',3'-Ene-nucleosides (see 2',3'-Dideoxy-2',3'-didehydro-D-nucleosides)
Enzymatic deesterification, 137
Enzymatic synthesis (see Nucleoside synthesis)
Epichlorohydrin, 159, 164
Epimerisation, 121
Epoxide, 44, 46, 75, 159, 168
Epstein Barr virus (EBV), 22, 105, 112, 158, 167
Ethane-1,2-dithiol, 71
Ethylaluminium dichloride, 107
Ethylbromodifluoroacetate, 111
Ethyl cyanoformate, 124
Ethyl 3,5-dichloro-6-(β-D-ribofuranosyl)pyrazine-2-carboxylate, 117
5-Ethyl-1-ethoxymethyl-6-(phenylthio)uracil (E-EPU, see HEPT)
O-Ethylxanthic acid, 104

EUrd (see 2'/3'-C-nucleosides)
Evans' dialkylboron triflate protocol, 146

Famciclovir (FCV, Famvir®), 156, 160, 161
F-C-Ado, 26
FCV (see Famciclovir)
FDA (Federal Drug Administration), 103
Fischer-Helferich procedure, 29, 35
Flavazol-3-yl C-nucleosides, 127
Fluorinated nucleosides, 51, 52
   Fiacitabine (FIAC), 51
   Fialuridine, (FIAU), 51
2-Fluoro-L-*arabino*-furanoside, 112
N-Fluorobenzenesulphonimide $(FN(SO_2Ph)_2)$, 108
5-Fluorocytosine, 105, 108
5-Fluoro-2'-deoxyuridine (FUDR), 1
β-L-2'-Fluoro-5-methyl-arabinofuranosyluracil (see L-FMAU)
Fluorosugar, 112
2'-Fluoro-2',3'-unsaturated L-nucleosides, 108
5-Fluorouracil, 108
L-FMAU, 103, 111, 112
Formaldehyde, 172
Formamidine acetate, 129
Formycin, 117, 127, 129
Formycin B (Laurusin), 127, 129
Formylation, 124, 127
(+)-FTC, 72
β-L-FTC, 106
Furazano[3,4-*d*]pyrimidines, 157
Furo[3,2-*d*]inosine C-nucleosides, 129
Furo[3,2-*d*]pyrimidine C-nucleosides, 127
Fused pyrimidine nucleosides, 94, 97

Ganciclovir (GCV, Cymevene®), 156, 158, 159, 160
   mechanism of action, 22, 158
   monophosphate, 159
   triphosphate, 159

Gauche effect, 7
GCV (see Ganciclovir)
Glucose, 65, 75
D-Glutamic acid, 106
Glutamine, 15
Glycal procedure (see Nucleoside synthesis)
D-Glyceraldehyde, 106, 108
Glycidol, 72, 167, 168
Glycosidic bond/link, 2, 9, 43, 64, 117, 118
Glycosidic cleavage, 9
Glycosidic conformation, 10
Glycosylamine procedure (see Nucleoside synthesis)
Glycosylation, 29, 30, 36, 37, 39, 65, 93, 94, 110
Glycosyl halide, 30, 132
Grignard reaction, 132
Grubbs catalyst, 148
Guanine (2-amino-6-oxypurine), 3, 14, 19, 105, 122, 157
Guanosine, 13, 24, 60, 155, 158, 164
   5'-monophosphate (GMP), 15, 24
   synthesis of, 29, 36, 37
   5'-triphosphate (GTP), 14, 15, 24
L-Gulonic-γ-lactone, 149
L-Gulose, 104

Halogenation, 38, 51, 112
5'-Halonucleosides, 51
8-Halo-purine nucleosides, 60
1-Halo-4-thiosugars, 66, 67
Hepatitis B virus (HBV), 22, 83, 85, 98, 103, 104, 106, 108, 112, 144, 145, 148, 164, 169
HEPT, 2, 21, 169, 170, 171
   derivatives, 170, 171
      E-EPU, 170, 171
      I-EBU, 170, 171
   mechanism of action, 21
Herpes simplex virus (HSV), 22, 46, 96, 97, 110, 111, 143, 155, 156, 158, 164

Hexamethyldisilazane (HMDS), 35
   lithium (LiHMDS), 108
   sodium (NaHMDS), 109
L-Hexopyranosyl-nucleosides, 109, 114
5-Hexynyl-2'-deoxyuridine, 97
Hilbert and Johnson procedure, 29
HIV (Human Immunodeficiency Virus), 1, 14, 19, 20, 21, 48, 49, 64, 72, 73, 74, 75, 77, 103, 104, 105, 106, 108, 110, 112, 114, 145, 146, 148, 156, 161, 164, 167, 169, 170, 171
   specific RT inhibitors of, 21, 78
L-Homocysteine (Hcy), 25, 26
Horner-Emmons reaction, 109
HPMPC (Cidofovir, Vistide®), 156, 164, 167, 168
   cyclic (cHPMPC), 169
   diphosphate, 164
   mechanism of action, 22
Hydrazine, 99
Hydrazoic acid, 50
Hydroformylation, 142
Hydrogenation, 29, 40, 41, 48, 70, 107, 126, 158, 168, 171
Hydrogen bonding, 5, 10, 14, 89, 132, 133
Hydrogen peroxide (H_2O_2), 85, 89
Hydrolysis, 44, 50, 65, 69, 112, 118, 126, 137, 138, 148, 157, 160, 161, 162, 171
9-(2-Hydroxyethoxymethyl)guanine (see Acyclovir)
1-[(2-Hydroxyethoxy)methyl]-6-(phenylthio)thymine (see HEPT)
9-(4-Hydroxy-3-hydroxymethyl-but-1-yl)guanine (see Penciclovir)
Hydroxymethylation, 172
5-(Hydroxymethyl)-2-cyclopenten-1-ol, 146
Hypoxallene, 164
Hypoxanthine (6-oxypurine), 4, 19

Imidazole, 15, 85

Imidazo[4,5-d]isothiazole nucleosides, 94
Imidazole C-nucleosides, 118
Imidazole nucleosides, 83, 93
Imidazo[1,2-a]pyridine acyclo-C-nucleosides, 169, 172
Imidazo[1,2-c]pyrimidine nucleosides, 98
   dioxolane, 98
   oxathiolane, 98
Imidazo[4,5-b]quinolin-2-one nucleosides, 94
Immunostimulant, 127
Immunosuppressant, 83, 85
Indole, 132, 133
Influenza, 27, 83, 87, 122
Inosine, 24, 60, 91, 127, 129, 138
   5'-monophosphate (IMP), 4, 15, 24
Inosine monophosphate dehydrogenase (IMPDH), 24, 83, 85, 122
Iodocyclisation, 75
5-Iodocytosine, 105
5-Iodo-2'-deoxyuridine, 58, 97
6-Iodo-9H-purin-2-amine, 150
N-Iodosuccinimide (NIS), 38, 39
5-Iodouracil, 77
Isoamyl nitrite, 85
Isocyanate method, 96
1-Isocyanate riboside, 96
Isomerisation, 43, 162
2'-Isonucleosides, 64, 65
   2'-iso-DDA, 64
   2'-iso-DDG, 64
   synthesis, 64
3'-Isonucleosides, 65
   3'-iso-DDA, 65
   synthesis, 65
5-Isopropyl-1-ethoxymethyl-6-(benzyl)uracil (I-EBU, see HEPT)
2,3-O-Isopropylidene-D-glyceraldehyde, 149
2,3-O-Isopropylidene-L-glyceraldehyde, 111
2,3-O-Isopropylidene-L-lyxopyranoside, 68
Isostere, 129, 132, 133
Isosteric mimic, 118, 122, 127, 137
2-Isosugar, 64
3-Isosugar, 65
Isoxazolidine nucleosides, 72, 73, 78

Ketonucleosides, 47, 52, 53
Ketosugar, 54

L1210 cancer cell line, 94, 96, 122
Lamivudine® (3TC, see 1,3-Oxathiolane-L-nucleosides)
Lassa fever, 85
Laurusin (see Formycin B)
Lawesson's reagent, 55
Lead tetraacetate (Pb(OAc)$_4$), 71
Leukaemia, 24, 25, 53, 96
Lewis acid, 35, 36, 39, 70, 107, 112, 123, 171
Lithiotrimethylsilyldiazomethane, 138
Lithium bis(trimethylsilyl)amide (LiN(TMS)$_2$), 141
5-Lithium-2,4-dimethoxypyrimidine, 123
Lithium 2,2,6,6-tetramethylpiperidine (LTMP), 124
Lobucavir (see BHCG)
L-Lyxo-furanosyl-nucleosides, 109
L-Lyxose, 66

Magnesium methoxide, 47
Maleimide, 118, 120
D-Mannitol, 71
Markiewicz reagent, 46
Mass spectroscopy, 8
   chemical ionisation (CI), 9
   fast atom bombardment (FAB), 9
Maximum torsional angle (τ_{max}), 6
Measles, 85, 167
Mercaptoacetaldehyde dimethylacetal, 71
Mercaptoacetic acid, 74

Mercury (Hg), 30, 35
Mercury cyanide, 157
Mercury sulphate, 97
N^6-Methyladenine, 105
Methyl 2,5-anhydroallonothioate, 122
Methyl anthranylate, 171
N^7-Methylated inosine, 43
N^7-Methylated guanosine, 43
Methylation, 25, 26
N-Methylbiuret, 96
5-Methylcytosine, 105
Methyl donor, 25
Methylenation, 144
Methyl propiolate, 87
Methyl transferases, 92
Minimycin (see Oxazinomycin)
Mitsunobu reaction, 49, 50, 66, 77, 111, 138, 145, 149
Mizoribine® (see Bredinin)
Molybdenum hydroxylase, 160
Mumps, 85
Murine sarcoma virus (MSV), 167

Naphtho[2,3-d]imidazole nucleosides, 94
Neighbouring group participation, 31
(−)-Neplanocin A, 137, 138, 140, 142
  6'-homoneplanocin (HNPA), 140, 141
  (6'R)-6'-C-methylneplanocin A (RMNPA), 140
Nicotinamide, 122
Nicotinamide adenine dinucleotide (NAD), 14, 122
Nitrous acid deamination, 106
Nomenclature, 4, 7
Non-nucleoside reverse transcriptase inhibitors (NNRTI's), 2, 21, 78, 170, 171
Nuclear magnetic resonance (NMR), 7–9
  ^{13}C NMR, 13
  ^{1}H NMR, 9, 10

Nuclear Overhauser Effect (N.O.E.), 9, 10, 13
Nucleic acid, 1
Nuclein, 1
Nucleocidin, 1
L-Nucleosides, 2, 103, 107, 109, 110, 112
  mechanism of action, 106
Nucleoside diphosphate kinase, 54, 106
Nucleoside phosphorylase, 43, 68
Nucleoside synthesis
  chloromercuri procedure, 33, 35
  enzymatic synthesis, 43, 68, 85
  glycal procedure, 38
    NIS-mediated reaction, 39
    selenium-mediated reaction, 39
    sulphur-mediated reaction, 39
  glycosylamine procedure, 40
    purine synthesis from 1-aminosugars, 41
    pyrimidine synthesis from 1-aminosugars, 40
  sodium-salt-glycosylation procedure, 35, 88, 133, 141
  stereoselectivity, 37, 38
  stereospecificity, 30
  Vorbrüggen procedure, 35, 54, 68, 70, 71, 77, 85, 88, 91, 97
    mechanism 2'-deoxynucleoside synthesis, 37
    mechanism purine nucleoside synthesis, 36
    mechanism pyrimidine nucleoside synthesis, 35
Nucleotide, 14, 19, 51
  de novo biosynthesis, 15, 122
  diphosphokinase, 17
  metabolising enzymes, 23, 24, 26
Numbering, 3
Nysted reagent, 144

OMP decarboxylase, 26, 27, 96, 120
Orotate, 15, 19

Orotate phosphoribosyl transferase, 15
Orotidine 5'-monophosphate (OMP), 15, 96
1,2,4-Oxadiazole-C-nucleosides, 118
Oxanosine, 4
1,3,4-Oxathiazole-C-nucleosides, 118
1,3-Oxathiolane, 104, 157
1,3-Oxathiolane-D-nucleosides, 71
1,3-Oxathiolane-L-nucleosides, 103, 104, 105
   1,3-oxathiolane-L-cytosine (3TC, Lamivudine®), 1, 2, 103, 104, 106
1,3-Oxazine-C-nucleosides, 122, 126
Oxazinomycin (Minimycin), 117, 122, 126
Oxazolidine nucleosides, 72
Oxetane, 75, 77
Oxetanocin-A, 75, 77
Oxidation, 39, 53, 71, 74, 118, 120, 129, 138, 144, 149, 160, 171
Oxidative cleavage, 70, 71, 104, 111
Oxidative decarboxylation, 71
Oxidative elimination, 108
2-Oxy-4-aminopyrimidine (see Cytosine)
6-Oxypurine (see Hypoxanthine)
Ozonolysis, 75, 137

P388 cancer cell line, 122
Papilloma virus, 167
Paraformaldehyde, 160
PCP (see Pneumocystis carnii pneumonia)
PCV (see Penciclovir)
Penciclovir (PCV, Vectavir®), 156, 160, 161, 164
   6-deoxy-PCV, 160, 161
   mechanism of action, 22
   triphosphate, 160
Pentofuranosyl donor, 43
Peracetylated sugar, 85
Peracylated sugar, 35
Phenylchlorothionocarbonate, 47

Phenylselenyl chloride. 39
Phenylsulphenyl chloride, 39
Phosgene, 86
Phosphatidic acid, 158
9-(2-Phosphonomethoxyethyl)adenine (see PMEA)
5-Phosphoribosyl-pyrophosphate (PRPP), 15, 19, 164
Phosphorus pentasulphide, 55
Phosphoryl transferases (PRTase), 19
Photoadduct, 75
[2 + 2] Photocyloaddition, 75
Photolysis, 118
Pig liver esterase (PLE), 137
Platelet aggregation, 14
PMEA (Adefovir®), 156, 164, 166, 167
   diphosphate, 164
   F-PMEA, ammonium salt, 167
PMEDAP, 167
PMEG, 167
PMPA, 167, 168
Pneumocystis carnii pneumonia (PCP), 129
PNP (see Purine nucleoside phosphorylase)
Polio (Sb-1) virus, 94
Potassium nonafluoro-1-butane sulphonate, 107
Potassium tert-butoxide, 48, 108
Propylene carbonate, 168
Protein synthesis, 93
Proton abstraction, 48, 78
PRPP (see 5-Phosphoribosyl-pyrophosphate)
PRTase (see Phosphoryl transferase)
Pseudomonas fluorescens lipase (PFL), 143
Pseudorotational phase angle (P), 6
Pummerer rearrangement, 71, 118
Purine mimetic, 127, 132
Purine nucleoside phosphorylase (PNP), 24, 25, 43, 85, 88, 129
Pyranose nucleosides, 77
Pyrazine-C-nucleosides, 122

Pyrazofurin, 27, 117, 118, 120, 122
   5'-deoxypyrazofurin, 122
Pyrazole-3-carboxamide nucleosides, 87
Pyrazole C-nucleosides, 129
Pyrazolo[3,4-*b*]pyridine nucleosides, 91
Pyrazolo[4,5-*b*]pyridine nucleosides (see 1-Deazapurine nucleosides)
Pyridazine-C-nucleosides, 122
Pyridine-C-nucleosides, 122
Pyrimidine mimetic, 132
Pyrimidine-C-nucleosides, 123
   pseudoisocytidine, 123, 124
   pseudouridine, 117, 122, 124
Pyrimidine nucleoside monophosphate (PNMP) kinase, 164
Pyrrole C-nucleosides, 129
Pyrrolo[3,2-*d*]oxazin-3-yl C-nucleosides, 127
Pyrrolo[2,3-*d*]pyrimidine nucleosides (see 7-deazapurine nucleosides)
Pyrrolo[2,3-*d*]pyrimidin-6-yl C-nucleosides, 127, 129
Pyrrolosine, 127, 129

Radical, 65, 118
Radical exchange, 118
RDPR (see Ribonucleoside diphosphate reductase)
Reduction, 41, 47, 48, 68, 71, 91, 107, 123, 138, 139, 141, 149, 150, 161, 164
Reductive debromination, 89
Reductive dediazotization, 122
Reductive deoxygenation, 74
Reductive elimination, 48
Reformatsky reaction, 111
Regioselective synthesis, 36, 37
Resistance, 2, 156, 158, 171
Respiratory syncytial virus (RSV), 3, 83, 85
Retroviruses, 121
Reverse transcriptase (RT), 1, 2, 15, 19, 20, 21, 77, 145, 164, 170, 171

Rheumatoid arthritis, 25
Rhino 1-A virus, 91
Ribavirin (Virazole®), 3, 24, 83, 85, 87
β-D-Ribofuranose (D-Ribose), 2, 3, 5, 31, 66, 75, 110, 118, 120, 137, 138, 161
L-Ribofuranosyl donor, 110
2-(β-D-Ribofuranosyl)maleimide (see Showdomycin)
L-Ribofuranosyl nucleosides, 109, 110
2-(β-D-Ribofuranosyl)selenazole-4-carboxamide (see Selenazofurinin)
2-(β-D-Ribofuranosyl)thiazole-4-carboxamide (see Tiazofurinin)
1-β-D-Ribofuranosyl-1*H*-1,2,4-triazole-3-carboxamidine (see TCNR)
Ribonucleoside diphosphate reductase (RDPR), 54, 55
Ribonucleotide reductase, 51, 164
L-Ribose, 110, 112
Ribose-1-phosphate, 43
Ribosyl donor, 43
Ring contractions, 75
RNA, 1, 3, 4, 14, 15, 19, 23, 24, 26, 93, 106, 123, 132
   composition, 2
   hydrogen bonding, 5, 132
   messenger (mRNA), 4, 24, 26, 142
   polymerase, 54
   ribosomal (rRNA), 123
   synthesis inhibitors, 54, 129
   transfer (tRNA), 123
   viruses, 24, 103, 112
Ruthenium oxide (RuO_4), 71
Rydon's reagent, 129

SAH hydrolase (see AdoHcy hydrolase)
Salvage pathways, 19, 24
SAM (see *S*-Adenosylmethionine)
Sandmeyer reaction, 85
Sangivamycin, 88, 89

Selenazofurin, 117, 122
Selenium, 38, 39, 64, 66, 122
4'-Selenium nucleosides, 66
Semliki Forest virus, 127
Showdomycin, 117, 118, 120
　derivatives, 120
Silver salt, 29, 30, 33, 35
Silver triflate (AgOTf), 39
Simmons-Smith reaction, 149
Sindbis virus, 127
Sodium azide, 41
Sodium borohydride, 47
Sodium cyanoborohydride, 74
Sodium periodate, 70, 150
Sodium-salt-glycosylation procedure
　(see Nucleoside synthesis)
Sonogashiri method, 58, 97
Spiro-nucleosides, 78
Stavudine® (D_4T, see 2',3'-Dideoxy-
　2',3'-didehydro-D-nucleosides)
Stereoselectivity, 38, 39, 54
Stille coupling, 58
Stroke, 83
Structure-activity relationship (SAR)
　studies, 70, 78, 97, 137, 158, 170, 172
2'-Substituted L-nucleosides, 109
3'-Substituted L-nucleosides, 109
2-Sulphamoyl chloride, 171
2-Sulphanylethanol, 71
Sulphonyl nucleoside, 46, 47
Sulphur, 38, 51, 64, 66, 122
Swern oxidation, 53, 138

Tautomerism, 5, 10
3TC (see 1,3-Oxathiolane-
　L-nucleosides)
T-cell lymphocytes, 87
TCNR, 83, 85
　5-amino-TCNR, 85, 86
Tetra-O-acetyl-D-ribose, 36
Thermal elimination, 148
Thermal rearrangement, 129
1,3,4-Thiadiazole C-nucleosides, 118

Thiation, 55
1,3-Thiazine C-nucleosides, 122
1,3-Thiazole C-nucleosides, 118
Thiazolidine nucleosides, 74
1",3"-Thiazolidine-2"-spiro-3'-3'-
　deoxyuridine nucleosides, 78
Thiazolidinone nucleosides, 74
Thieno[3,4-d]pyrimidine C-nucleosides, 117, 127, 129
　Thieno[3,4-d]inosine C-nucleoside, 129
Thietane nucleoside, 177
1,6-Thioanhydro-L-gulopyranose, 104
1,6-Thioanhydro-D-mannopyranose, 71
Thioglycoside, 65, 118
Thionocarbamate, 37
2',3'-O-Thionocarbonates, 48
4'-Thionucleosides, 66, 68
Thiophene C-nucleoside, 129
Thiophile, 48
2-Thiopyridone ethanoate, 118
4-Thiopyrimidine nucleosides, 55
4-Thioribose, 66
Thiosangivamycin, 89
Thiosugar, 68
2-Thiouracil, 171
Thymine (2,4-dioxy-5-methylpyrimidine), 3, 19, 40, 72, 112
D-Thymidine, 13, 14, 39, 43, 49, 52, 55, 132, 133
　5'-monophosphate (TMP), 15, 17
L-Thymidine, 110
Thymidine kinase (TK), 22, 77, 110, 158, 164
Thymidine phosphorylase, 43
Thymidylate synthase, 17
Tiazofurin, 117, 122
　adenine-dinucleotide, 122
Tin (IV) chloride ($SnCl_4$), 35, 36, 39, 94, 112
T-4 Lymphoma, 96
TMSOTf, 35, 36, 37, 112
Torsional angle, 5, 6

Tosyl azide, 121
Tosyloxymethylphosphonate, 168
Toxicity, 2, 33, 44, 51, 54, 66, 68, 97, 103, 104, 106, 120, 137, 138, 145, 169, 172
Toyocamycin, 1, 88, 89
Trans-*N*-deoxyribosylase, 68
Transesterification, 72, 161
Transglycosylation, 108
Transmethylation, 92
Trans rule, 30, 31
Triazine, 96
1,2,4-Triazine *C*-nucleosides, 122
Triazole nucleosides, 85, 87
1,2,4-Triazole *C*-nucleosides, 118
1,2,4-Triazolo[1,5-*a*]pyridin-2-yl *C*-nucleosides, 127
1,2,4-Triazolo[4,3-*a*]pyridin-3-yl *C*-nucleosides, 127
1,2,3-Triazolo[4,5-*d*]pyrimidine nucleosides (see 8-Azapurine nucleosides)
1,2,4-Triazolo[4,3-*c*]pyrimidone nucleosides, 99
1,3,5-Tri-*O*-benzoyl-α-L-ribofuranose, 112
Tributyltin hydride (Bu_3SnH), 39, 47, 114
Triethyl 3-bromopropane-1,1,1-tricarboxylate, 161
Triethyl-α-fluorophosphonacetate, 109
Triethylorthoformate, 41, 89
Triethylsilane, 123
Trimethyl orthoacetate, 161
Trimethyl orthoformate, 148
Trimethylphosphite, 48
Trimethylsilyl bromide, 167, 168
Trimethylsilyl iodide, 157
Triphenylphosphine, 50, 51
Trost's palladium (0)-catalysed coupling, 148
TSAO-T, 2, 21, 78, 170
 mechanism of action, 21

1,2,3-triazole-TSAO, 87
Tubercidin, 83, 85
 2'-deoxy-2'-*ara*-fluoro-tubercidin, 88

UL89 gene product, 87
UL97 protein kinase, 158, 159
Uracil (2,4-dioxypyrimidine), 3, 14, 19, 53
Uridine, 10, 13, 55, 108
 5'-monophosphate (UMP), 15, 17, 26, 96
 synthesis, 29
 5'-triphosphate (UTP), 15, 17, 26

Vaccinia virus, 96
Valaciclovir (Valtrex®), 156, 157, 158
L-Valine, 157
Varicella Zoster virus (VZV), 22, 83, 96, 97, 143, 156, 158, 164, 172
Vectavir® (see Penciclovir)
Vidarabine® (Ara-A, see Arabinonucleosides)
Vinyltributyltin, 58
Viral packaging, 87
Viral processing, 87
Virazole® (see Ribavirin)
Vistide® (see HPMPC)
Vorbrüggen procedure (see Nucleoside synthesis)

Waldon inversion, 30, 33, 44
Watson-Crick base pairing, 14, 89
Wittig reaction, 53, 55, 118, 149
Wolfe oxidation, 71

Xanthine, 5
Xanthosine, 4, 60, 129
 monophosphate, 24
X-ray crystallography, 7
Xylo-bromosugars, 48
L-*Xylo*-furanosyl nucleosides, 109, 112
D-Xylose (D-*xylo*-furanose), 46
L-Xylose (L-*xylo*-furanose), 110, 111, 112

Zalcitabine® (DDC, see 2',3'-Dideoxy-D-nucleosides)
Ziagen® (see Abacavir)
Zidovudine® (AZT, see Azidonucleosides)

Zinc chloride (ZnCl$_2$), 71
Zn/Cu couple, 48
Zovirax® (see Acyclovir)